AF580806

THE SOLAR TACHOCLINE

Helioseismology has enabled us to probe the internal structure and dynamics of the Sun, including how its rotation varies in the solar interior. The unexpected discovery of an abrupt transition – the tachocline – between the differentially rotating convection zone and the uniformly rotating radiative interior has generated considerable interest and raised many fundamental issues. This volume contains invited reviews from distinguished speakers at the first meeting devoted to the tachocline, held at the Isaac Newton Institute. It provides the only comprehensive account of the current understanding of the properties and dynamics of the tachocline, including both observational results and major theoretical issues, involving both hydrodynamic and magnetohydrodynamic behaviour.

The Solar Tachocline is a valuable reference for researchers and graduate students in astrophysics, heliospheric physics and geophysics, and the dynamics of fluids and plasmas.

DAVID HUGHES is Professor of Applied Mathematics at the University of Leeds.

ROBERT ROSNER is William E. Wrather Distinguished Service Professor at the University of Chicago, and Director of Argonne National Laboratory.

NIGEL WEISS is Emeritus Professor of Mathematical Astrophysics at the University of Cambridge.

THE SOLAR TACHOCLINE

Edited by

DAVID W. HUGHES
University of Leeds

ROBERT ROSNER
University of Chicago and Argonne National Laboratory

NIGEL O. WEISS
University of Cambridge

CAMBRIDGE
UNIVERSITY PRESS

CAMBRIDGE UNIVERSITY PRESS
Cambridge, New York, Melbourne, Madrid, Cape Town, Singapore, São Paulo

Cambridge University Press
The Edinburgh Building, Cambridge CB2 8RU, UK

Published in the United States of America by Cambridge University Press, New York

www.cambridge.org
Information on this title: www.cambridge.org/9780521861014

First published 2007

Printed in the United Kingdom at the University Press, Cambridge

A catalogue record for this publication is available from the British Library

ISBN-13 978-0-521-86101-4 hardback

Contents

Contributors

P. S. Cally
Centre for Stellar and
Planetary Astrophysics
School of Mathematical Sciences
Monash University
Melbourne, Victoria 3800
Australia

J. Christensen-Dalsgaard
Institute of Physics and Astronomy
Aarhus University
Aarhus, DK-8000
Denmark

P. H. Diamond
Department of Physics
University of California, San Diego
La Jolla, CA 92093
USA

P. Garaud
Department of Applied Mathematics and
Statistics
Baskin School of Engineering
University of California
Santa Cruz, CA 95064
USA

P. A. Gilman
High Altitude Observatory
National Center for
Atmospheric Research
Boulder, CO 80301
USA

D. O. Gough
Institute of Astronomy
University of Cambridge
Cambridge CB3 0HA
UK

D. W. Hughes
Department of Applied Mathematics
University of Leeds
Leeds LS2 9JT
UK

K. Itoh
National Institute for Fusion Science
Toki, Gifu 509-5292
Japan

S.-I. Itoh
Research Institute for Applied Mechanics
Kyushu University
Fukuoka 816-8580
Japan

L. L. Kitchatinov
Institute for Solar-Terrestrial Physics
P. O. Box 4026
Irkutsk 664033
Russia

M. E. McIntyre
Department of Applied Mathematics and
Theoretical Physics
University of Cambridge
Cambridge CB3 0WA
UK

M. S. Miesch
High Altitude Observatory
National Center for Atmospheric Research
Boulder, CO 80301
USA

G. I. Ogilvie
Department of Applied Mathematics and Theoretical Physics
University of Cambridge
Cambridge CB3 0WA
UK

R. Rosner
Department of Astronomy and Astrophysics
University of Chicago
Chicago, IL 60637
USA

G. Rüdiger
Astrophysikalisches Institut Potsdam
An der Sternwarte 16
Potsdam, D-14482
Germany

L. J. Silvers
Department of Physics
University of California, San Diego
La Jolla, CA 92093
USA

E. A. Spiegel
Department of Astronomy
Columbia University
New York, NY 10027
USA

M. J. Thompson
Department of Applied Mathematics
University of Sheffield
Sheffield S3 7RH
UK

S. M. Tobias
Department of Applied Mathematics
University of Leeds
Leeds LS2 9JT
UK

N. O. Weiss
Department of Applied Mathematics and Theoretical Physics
University of Cambridge
Cambridge CB3 0WA
UK

J.-P. Zahn
LUTH
Observatoire de Paris
Meudon, F-92195
France

Preface

Over the past 25 years helioseismology has at last enabled us to probe the internal structure and dynamics of our local star, the Sun. Perhaps its greatest triumph has been to determine how the rotation varies in the solar interior. Although the bulk of the radiative zone, occupying the innermost 70% by radius, rotates more or less uniformly, the known variation with latitude of angular velocity at the surface persists down to the base of the outer convective envelope. Since it had previously been supposed that the Sun rotates sufficiently rapidly for the angular velocity to be constant on cylindrical surfaces in the convection zone it was a surprise to find that it is actually constant on conical surfaces. It came as an even greater surprise to discover that the transition between the differentially rotating exterior and the uniformly rotating interior is effected through an extremely thin layer – *the tachocline* – whose thickness is less than 4% of the solar radius.

This unexpectedly abrupt transition has forced us all to refine our ideas on the interactions between turbulent convection, rotation and magnetic fields, for it seems that these last play a key role in preventing the tachocline from spreading downwards into the radiative zone. To describe the internal structure of the tachocline requires an understanding of convective penetration, turbulent diffusion, mixing and angular momentum transport. This shear layer also hosts a great range of potential instabilities and, furthermore, is the most likely seat for the dynamo responsible for the cyclic magnetic activity that is observed at the surface of the Sun. These physical processes are all of interest in themselves and not only raise major issues in astrophysical fluid dynamics, but also relate to significant problems in the physics of the Earth's oceans and atmosphere, in planetary physics, and in plasma confinement. More importantly, an understanding of the dynamics of the tachocline is essential not just in order to match the interior to the exterior of the Sun – and hence to establish its global properties – but also to explain the structure, evolution and magnetism of all similar stars with deep convection zones.

The programme on *Magnetohydrodynamics of Stellar Interiors* held at the Issac Newton Institute for Mathematical Sciences in Cambridge, which we organized in

2004, provided a timely opportunity for holding a workshop on *Tachocline Dynamics*. This one week meeting (8–12 November 2004) was the first ever to be entirely devoted to the subject and it brought almost all the key players together to discuss it. The workshop was informally structured, our aim being to maximize opportunities for argument and discussion. The number of invited lectures was therefore restricted, so as to leave plenty of time for structured discussions, led and organized by appropriate experts. In all, there were a dozen talks that introduced the principal topics, and eight hours of scheduled discussion, which continued informally among the 48 participants outside the lecture room. This format worked extremely well and the workshop was felt to be a great success. It certainly identified all the main issues, although no consensus on detailed models was expected or achieved.

There was a general feeling among the participants that, while tachocline dynamics remained a young and rapidly developing subject, it had now reached a stage where it was ready for description in a book. The workshop provided the ideal basis for such a volume and this is the result. Rather than produce a volume of proceedings, it was agreed that we should edit a book containing invited chapters from selected participants, each of which would be refereed by another participant. We are pleased that there is a wide range of age and experience among the authors; most of them are known for their work in Astrophysical Fluid Dynamics, but we also include chapters by experts in Geophysical Fluid Dynamics and Plasma Physics.

The chapters are grouped into different sections covering the main areas of our subject. The opening section contains two essays; the first provides a comprehensive introductory survey, while the second focuses on the history and pre-history of the relevant theoretical ideas. The next section presents the fundamental observational results, which underpin all of the theory that follows. Section III contains three chapters on purely hydrodynamic aspects of the tachocline, covering its structure, turbulence, and differential rotation. Magnetic fields enter next, in Section IV, with two chapters on the magnetic confinement of the tachocline and one on the influence of rotation on magnetohydrodynamic turbulence. This is followed by a section containing three chapters reviewing instabilities driven by rotational shear, magnetic fields and buoyancy. Section VI is concerned with the generation of large-scale magnetic fields by dynamo action, whether in the turbulent convection zone or, more likely, in the tachocline itself. The book then concludes with an overview that summarizes current controversies and points to future progress. There is inevitably some overlap between the different chapters and, given the lack of certainty about the nature of the tachocline, authors have not been restrained from expressing contradictory views. While we have tried to maintain some degree of uniformity regarding notation, we have not attempted to impose complete uniformity, nor have we prevented authors from using their preferred choices of electromagnetic and other units.

We thank all those who have helped to make this book possible. We are grateful to the staff of the Isaac Newton Institute, particularly Tracey Andrew and Christine West, for making the workshop possible as part of the overall programme at the Institute. We are also indebted to the members of the Scientific Organising Committee, Pascale Garaud, Douglas Gough and Jean-Paul Zahn.

D.W.H.
R.R.
N.O.W.

Part I

Setting the scene

1

An introduction to the solar tachocline

Douglas Gough

1.1 Preamble

The task that I have been assigned is to set the scene for the discussions that follow: to present my view of the principal issues that had confronted us before the meeting when trying to understand the dynamics of the solar tachocline. Most of what I write here is enlarged upon, and in some cases superseded by, the chapters that follow, in which references to most of the original publications can also be found. Nevertheless, I trust that it can serve as a useful elementary introduction to the subject, setting it into its wider astronomical context.

The tachocline is interesting to astrophysicists for a variety of reasons, the most important being (i) that it couples the radiative interior of the Sun, where nearly 90% of the angular momentum resides, to the convection zone, which is being spun down by the solar wind, (ii) that it controls conditions at the lower boundary of the convection zone, and is therefore an integral component of the overall rotational dynamics of the convection, and (iii), perhaps most relevant to the interests of the greater proportion of the participants of the workshop, it is now generally recognized as being the seat of the solar dynamo. It plays some role in shaping the evolution of the Sun, and it must be taken into account when interpreting the helioseismological diagnostics of the solar structure.

It is therefore perhaps useful first for me to make a few remarks about some properties of the Sun that are pertinent to the dynamics of the tachocline, and to agree on a practical definition, or at least a description, of what we even mean by the tachocline. I shall then discuss the two existing dynamical descriptions of the tachocline, and the extent to which I consider them to represent, or not, the response of the Sun to the predominant balance of forces in what in reality are circumstances much more complicated than those to which the idealized theories really apply. This raises many theoretical issues, some of which are supported either directly or indirectly by experiment (either physical or numerical) or by

The Solar Tachocline, D. W. Hughes, R. Rosner and N. O. Weiss.
Published by Cambridge University Press.

astronomical observation; more might also be so supported in the not-too-distant future, others not.

My discussion is not a balanced account of the various (disparate) views of the community; it is peppered with my own opinions, with many of which the reader may take issue. In so doing, I hope that it will stimulate productive thought that will advance our understanding of this topical subject.

1.2 Some basic properties of the Sun

The Sun is a star on the Main Sequence: that is to say, it is in a state of hydrostatic and thermal balance in which energy that is being produced in a hot central core from thermonuclear transmutation of hydrogen into helium is being transported down a temperature gradient through the surrounding envelope, and is finally radiated into space from the photosphere, the visible surface of the Sun. This is generally regarded as a state of one of the simplest phases of the evolution of a star. It is certainly the most well studied. It is also the longest phase, at least before the star finally condenses into a degenerate configuration such as a white dwarf or a neutron star, and so most ordinary (non-degenerate) stars are currently in their Main-Sequence phase. The duration of this phase, which ends when the supply of hydrogen fuel is exhausted from the centre of the core, depends on the mass of the star; for the Sun it is about 1.0×10^{10} years. The solar age is about 4.6×10^9 years, so the Sun is nearly half-way through its Main-Sequence life.

The Sun's core (the region in which, say, 95% of the thermonuclear energy is generated) extends to about 20% of the radius $R_\odot$ of the photosphere, and contains about 35% of the total mass of the star. The surrounding envelope is divided into two principal regions: a quiescent radiative region in which radiant heat diffuses down the temperature gradient, extending to a radius, r_c, of about $0.71R_\odot$, and an overlying turbulent convective region which extends to the photosphere. These two regions merge in – some would say are separated by – a thin boundary layer, which is the tachocline, on which I shall enlarge very shortly. But before doing so I should perhaps attempt to avert misunderstanding by drawing attention to a potential source of confusion brought about by some newcomers to the field who call the radiative envelope the core. I hope that any inadvertent occurrence of that misnomer in this book will be subliminally ignored. I shall sometimes use the term 'radiative interior' to denote the entire radiative region beneath the convection zone, encompassing both the radiative envelope and the core.

At this point it is perhaps useful to mention a few more timescales. The thermal diffusion time $\tau_{\rm th}$ through the radiative envelope, namely $r_c^2/\pi^2\overline{\kappa}$, where $\overline{\kappa} = 7.3 \times 10^5\ \text{cm}^2\ \text{s}^{-1}$ is an appropriate average (the square of the harmonic mean, over radius r, of the square root) of the thermal diffusivity κ, is 1.2×10^7 years; the viscous timescale, in the absence of boundary layers, is about 10^{12} years, substantially greater than the age of the Universe; and the thermal relaxation time of the convection zone, namely the characteristic time over which the zone would return to thermal equilibrium after a putative global perturbation to conditions at its base, is 10^5 years. The timescale for Ohmic decay of a large-scale magnetic field pervading the radiative interior is of order 10^{11} years. All four timescales are much greater than the characteristic times over which most of the dynamical processes discussed in this volume operate, and so for many purposes the phenomena directly associated with them can be ignored. On the other hand, the characteristic global dynamical timescale, which is given by the acoustic travel time from the centre to the surface of the Sun, is about 1 h, which is much shorter; therefore the radiative envelope and the core remain quite precisely in hydrostatic equilibrium on the timescales of major interest here.

The (sidereal) rotation period of the essentially uniformly rotating radiative interior of the Sun is about 27 days, corresponding to the angular velocity $\Omega_0 = 2.7 \times 10^{-6}\,\mathrm{s}^{-1}$. This is comparable with the characteristic turnover time of the large convective eddies in the lower reaches of the convection zone. The angular velocity in the convection zone varies only weakly with depth, and declines gradually from $2.9 \times 10^{-6}\,\mathrm{s}^{-1}$ at the equator to about $2.0 \times 10^{-6}\,\mathrm{s}^{-1}$ at the poles.

It should be realized that the convection zone contains only about 2% of the mass of the star. Therefore any reasonable redistribution of matter in the convection zone brought about by convection dynamics has very little impact on the overall weight of the envelope pressing down on the core. Therefore the core evolves inexorably on its own timescale of 10^{10} years, untrammelled by the machinations of the dynamically active outer envelope. I should mention also the Eddington–Sweet timescale, τ_{ES}, the circulation time of large-scale meridional flow through the radiative envelope that is associated with the baroclinicity induced by rotation, and which is enabled by thermal diffusion; it is the thermal diffusion time divided by the square of the rotational Froude number $Fr = 2\Omega_0/\overline{N}$ (where $\overline{N}$ is the value of the buoyancy frequency characteristic of the body of the radiative envelope, about $2.5\times10^{-3}\,\mathrm{s}^{-1}$), namely $\tau_{\mathrm{ES}} = Fr^{-2}\tau_{\mathrm{th}} \simeq 2.5 \times 10^{12}$ years. There is also a contribution to the circulation from spin-down resulting from the extraction of angular momentum from the convection zone by the solar wind via the external magnetic field, which is currently slowing down the Sun on a timescale of $\tau_{\mathrm{sd}} \simeq 10^{10}$ years, a time comparable with, although a little longer than, the solar age.

There are other timescales more pertinent to the subject-matter of this volume that will emerge in due course. Let me mention here just two. The first is the global Alfvénic time τ_{A}. This is the time it takes for an Alfvén wave to traverse the quiescent radiative interior of the Sun, and is characteristic of the period of a magnetically restored global torsional oscillation. It depends, of course, on the (harmonic) mean intensity of the large-scale magnetic field, which is unknown. Presumptions in the literature of its characteristic value vary widely, from zero, a most unlikely value, to the order of megagauss. Suffice it to say that if one adopts an intermediate value, say about 2 kG, a value characteristic of, though somewhat lower than, the fields observed in sunspot umbrae, then τ_A is about 22 years, a timescale which is central to issues discussed in this volume. It is therefore evident that no theory of the solar cycle can be considered complete unless it addresses the dynamical role of the radiative envelope. The second timescale is the internal thermal or rotational equilibration time of the convection zone: the time it takes for the convection zone to propagate (rather than merely to diffuse) a thermal or angular-momentum perturbation towards an internal equilibrium without the entire convection zone necessarily getting back into balance with its surroundings. That time is about one year, again not very different from the timescale of the solar cycle.

To conclude, I record in Table 1.1 characteristic values of various physical variables in the tachocline. They are evaluated at a radius $r = 0.70R_\odot$. The density, ρ, pressure, p, sound speed, c, and acceleration due to gravity, g, were obtained seismologically; the remaining, non-seismic, variables were inferred from a solar model (Model S) of Jørgen Christensen-Dalsgaard by adjusting it appropriately to be consistent with the seismic variables, and taking the relative hydrogen abundance by mass to be $X = 0.737$, as in the model. Under these conditions one can compute diffusion coefficients. There is considerable diversity amongst the values that one finds quoted in the literature; here I adopt what I consider to be the most reliable estimates: I evaluate the magnetic diffusivity η and the ion contribution ν_i to the kinematic viscosity from the formulae of Lyman Spitzer, using Georges Michaud and Charles Proffitt's more recent estimate of the Coulomb logarithm ($\ln\Lambda = 2.5$); I evaluate the photon-transport contribution ν_r to the viscosity from the formula of L. H. Thomas; I evaluate the helium–hydrogen diffusion coefficients from Michaud and Proffitt's extraction from the work of Paquette, Pelletier, Fontaine and Michaud. About 10% of the total kinematic viscosity $\nu = \nu_\mathrm{i} + \nu_\mathrm{r}$ and all of the thermal diffusivity κ come from photon transport; the magnetic diffusivity comes entirely from particle transport. From these coefficients one deduces a Prandtl number $\nu/\kappa \simeq 1.9\times10^{-6}$ and a magnetic Prandtl number $\nu/\eta \simeq 6.6\times10^{-2}$ characteristic of the tachocline. One can extrapolate the diffusion coefficients downwards through the radiative envelope using the approximate scaling laws $\eta \propto T^{-3/2}\ln\Lambda$, $\nu_\mathrm{i} \propto T^{5/2}/\rho\ln\Lambda$, $\nu_\mathrm{r} \propto T^4/\rho^2\hat{\kappa}$, $\kappa \propto T^3/\rho^2\hat{\kappa}$, $\chi \propto T^{5/2}/\rho\ln\Lambda$, with $\Lambda \propto \rho^{-1/2}T^{3/2}$, and k_p and k_T constant; $\hat{\kappa}$ is the opacity, which in the radiative interior satisfies roughly (to within 15%) the empirical law $\hat{\kappa} \propto \rho^{0.4}T^{-3}$. The Soret coefficients depend on chemical composition, and are each approximately proportional to $(1 - X^2)(3 + 5X)/(3 + X)$; they increase inwards through the core by about 50%. Beneath the tachocline the buoyancy frequency N increases gradually

Table 1.1. *Properties of the tachocline at $r = 0.70R_\odot$*

density	ρ	0.21	$\mathrm{g\,cm^{-3}}$
pressure	p	6.7×10^{13}	$\mathrm{g\,cm^{-2}s^{-2}}$
temperature	T	2.3×10^{6}	K
sound speed	c	2.3×10^{7}	$\mathrm{cm\,s^{-1}}$
opacity	$\hat{\kappa}$	19	$\mathrm{g^{-1}\,cm^{2}}$
gravitational acceleration	g	5.4×10^{4}	$\mathrm{cm\,s^{-2}}$
density scale height	H_ρ	$0.12R_\odot$	
pressure scale height	H_p	$0.08R_\odot$	
adiabatic exponent	γ_1	1.665	
buoyancy frequency	N	8×10^{-4}	$\mathrm{s^{-1}}$
magnetic diffusivity	η	4.1×10^{2}	$\mathrm{cm^2\,s^{-1}}$
kinematic viscosity	ν	2.7×10^{1}	$\mathrm{cm^2\,s^{-1}}$
thermal diffusivity	κ	1.4×10^{7}	$\mathrm{cm^2\,s^{-1}}$
helium diffusion coefficient	χ	8.7	$\mathrm{cm^2\,s^{-1}}$
pressure Soret factor	k_p	2.9	
temperature Soret factor	k_T	2.6	

The density and pressure scale heights are defined as $H_\rho = (-\mathrm{d}\ln\rho/\mathrm{d}r)^{-1}$ and $H_p = (-\mathrm{d}\ln p/\mathrm{d}r)^{-1} = c^2/\gamma_1 g$, where $\gamma_1 = (\partial \ln p/\partial \ln \rho)_s$ is the first adiabatic exponent, the partial thermodynamic derivative being taken at constant specific entropy s. The buoyancy frequency is defined by $N^2 = g(H_\rho^{-1} - gc^{-2})$; it rises with depth from essentially zero at the base of the convection zone to about 1×10^{-3} at the top of the tachopause. The (upwards) helium diffusion velocity v_{He} through hydrogen is given by $v_{\mathrm{He}} = -\chi(\mathrm{d}\ln Y/\mathrm{d}r - k_p\mathrm{d}\ln p/\mathrm{d}r - k_T\mathrm{d}\ln T/\mathrm{d}r)$, where Y is the helium abundance, whose value in the tachocline is 0.245. The diffusion velocity is defined such that $v_{\mathrm{He}}Y$ is the mass flux of helium through hydrogen. The total radius, mass and luminosity of the Sun are $R_\odot = 6.96 \times 10^{10}$ cm, $M_\odot = 1.99 \times 10^{33}$ g, and $L_\odot = 3.84 \times 10^{33}$ erg s^{-1}.

with depth to a maximum of about $2.9 \times 10^{-3}\,\mathrm{s^{-1}}$ at $r \simeq 0.1R_\odot$, and then declines to zero at $r = 0$.

1.3 Solar spin-down

Perhaps the most serious problem to be faced by any protostar is how to dispose of almost all of its angular momentum and magnetic field as it collapses to its eventual hydrostatic state. A globular region of mass $1M_\odot$ of dense interstellar gas, say with a density of 100 hydrogen atoms per cubic centimetre, rotating with an angular velocity equal to half the mean vorticity of the galactic rotation (namely, $1 \times 10^{-8}\ \mathrm{y}^{-1}$, Oort's second constant) and pervaded by a not atypical 3 μG magnetic field, would, if it could conserve angular momentum and magnetic flux as it contracted to solar dimensions, end up rotating some 3×10^5 times faster than the Sun, and would possess a global magnetic field in the radiative interior of about 1TG, exceeding the solar value by perhaps a similar factor; the disrupting Lorentz force would be comparable with the binding gravitational force, and the centrifugal force would exceed the gravitational force by a factor in excess of 10^4. There has been a fair amount of intellectual effort expended on the angular momentum issue, on which I shall not dwell here, except to note that when the Sun first attained hydrostatic equilibrium, having lost almost all of its angular momentum, it was still rotating considerably faster than it is today. From that time on the solar-wind torque, acting on a timescale comparable with, although at present apparently somewhat longer than, the Main-Sequence age, has slowed down the convection zone, which, as I pointed out earlier, equilibrates on the very short timescale of about a year. A consequent issue that has exercised the minds of stellar physicists quite considerably is the extent to which the spin-down of the convection zone has been transmitted to the radiative interior. In the early days opinions varied widely, but now the matter has essentially been settled by helioseismology. I raise it here because it was in that context that the first idea of a tachocline emerged.

There was considerable debate in the 1960s and 1970s on the rotation of the solar interior, brought about partly by Bob Dicke's claim that the surface of the Sun is some six times more oblate (implying a 2.5-fold increase in ellipticity) than one would have expected had the Sun been rotating more-or-less uniformly throughout. The import of the claim was that the external gravitational equipotentials deviate from sphericity by more than 100 times expectation, and thereby destroy the precise agreement between observations of the rate of precession of the perihelion of the orbit of the planet Mercury and the prediction by General Relativity. (The reason why the Sun's surface and the gravitational equipotentials do not deform proportionately is that in near-uniform rotation only 4% of the deviation from sphericity of the surface layers arises from the asphericity of the gravitational equipotentials, the rest being the direct centrifugal distortion arising from the observed rotation of those layers.) Dicke argued that because the timescale for viscous diffusion of angular momentum through the quiescent radiative interior is substantially greater than the age of the Sun – indeed, as I mentioned earlier, it is greater than the age of the Universe – the core is still spinning with angular velocity comparable with the high value it had when the Sun arrived on the Main Sequence, and is therefore still substantially flattened by the centrifugal force, inducing oblateness in the gravitational field. Against that view, Ed Spiegel was the standard-bearer for realism, pointing out that, as for any issue with non-uniformly rotating fluids, spin-down must be considered in a dynamically consistent way, and that statements based on diffusion alone are highly misleading. (I should take this opportunity to remind the reader that that is so also for issues concerning the tachocline today.) After some preliminary toy-model studies with Derek Moore and Francis Bretherton, Spiegel addressed the influence of the strongly stable stratification on the baroclinic angular-momentum-transporting meridional flow induced by spin-down, and conceived of a growing sequence of thin Holton shear layers which transported angular momentum principally by advection, and which advanced from the base of the convection zone to the core, spinning down the entire Sun. Although the advance occurred on a timescale much less than the age of the Sun, it was nonetheless quite slow by dynamical standards, so perhaps Spiegel's original appellation 'tachycline' was not entirely appropriate.

It appears today that the existence of the tachocline is not directly dependent on spin-down, and is instead driven principally by the stresses maintaining the latitudinal differential rotation of the convection zone against a rigidly rotating envelope below. This conclusion derives from the helioseismological measurements of the variation of angular velocity in the solar interior (in addition to observational estimates of the spin-down rate), which it is appropriate now to describe.

1.4 The rotation of the Sun today

I should point out immediately, to avoid possible misunderstanding, that by angular velocity I mean the azimuthally averaged value, any deviations from that value being regarded as zonal flow. I must also point out that, except when I state explicitly to the contrary, all descriptions of the angular velocity Ω actually refer to an average of values at equal latitudes in the northern and southern hemispheres, which is what global helioseismology tells us. Observers almost invariably talk about rotation rate, whose value they quote typically in nanohertz. For brevity, values of Ω are sometimes quoted in nanohertz, but what is meant, of course, is values of $\Omega/2\pi$.

It has long been known from direct visual observation of tracers that the surface layers of the Sun rotate differentially, with angular velocity which is quite accurately described by the three-term expression

$$\Omega_s(\theta) = \Omega_e(1 - \alpha_2\mu^2 - \alpha_4\mu^4), \tag{1.1}$$

where $\mu = \cos\theta$, θ being colatitude, and Ω_e, α_2 and α_4 are constants. Helioseismological inferences from the rotational degeneracy splitting of acoustic modes obtained originally and separately by Tim Brown, Ken Libbrecht and Jesper Schou, and their colleagues, have demonstrated that this kind of latitudinal variation persists almost unaltered down to the base of the convection zone, although, as has been emphasized recently by Peter Gilman and Rachel Howe, through the main body of the convection zone, at latitudes below about 70°, the angular-velocity contours depicted by Schou and his collaborators are more closely described as being inclined at a constant angle, about 30°, to the axis of rotation, except, of course, very near to the equator. At the base of the convection zone there is an abrupt transition to almost uniform angular velocity throughout the radiative envelope, with a value Ω_0 intermediate between the extremes of Ω_s. The angular velocity in the core is uncertain, although there is some evidence that its average is somewhat less than that in the surrounding envelope, a property which was not expected by anyone who had contemplated solar spin-down before helioseismology. The values of the constants defining Ω_s in the subsurface layers, at radius $r = 0.995R_\odot$, obtained seismologically by Schou and his colleagues, are $\Omega_e/2\pi \simeq 455\,\text{nHz}$, $\alpha_2 \simeq 0.12$, $\alpha_4 \simeq 0.17$; at $r = 0.75R_\odot$ the latitudinal variation of the angular velocity can be represented by the same expression with $\Omega_e/2\pi \simeq 463\,\text{nHz}$, $\alpha_2 \simeq 0.17$ and $\alpha_4 \simeq 0.08$ (implying a specific angular momentum, integrated over the sphere, the same as at $r = 0.995R_\odot$). Similar values are given by other investigators. The radiative interior rotates at a rate $\Omega_0/2\pi \simeq 430\,\text{nHz}$. The transition shear layer at the base of the convection zone – the object of study in this volume – is too thin to be resolved by current seismic data, but fits made by Sasha Kosovichev, Paul Charbonneau and others of simple functional forms have suggested values of its thickness Δ ranging from about $0.02R_\odot$ to $0.05R_\odot$. Most of the estimates rest on the assumption that the base of the convection zone is spherical. Any deviation from sphericity, for which there is some weak seismological evidence, smears a spherical view. Therefore the estimates of the thickness of the shear layer should probably be considered to be upper bounds.

To put our study into its context, I shall elaborate a little on the seismologically inferred angular velocity. These extra details do not all bear directly on the tachocline, but they must surely be accounted for in any comprehensive theory of the solar cycle. It is often said that the form $\Omega_s(\theta)$ persists unaltered through the convection zone, implying that the angular velocity contours are radial. As I have already pointed out, it has been shown, by work of nearly a decade ago, that throughout much of the convection zone the contours are more nearly uniformly inclined by about 30° from the axis of rotation; and equatorwards of latitude 20° or so there is a tendency for them to be even more nearly aligned with the axis of rotation. It should be realized, however, that in this equatorial region Ω deviates only slightly from being constant, so the slopes of the isotachs are not accurately determined. Recently, Howe and her colleagues, using more extensive, more highly resolved data, have concluded that the angle of inclination of the isotachs is closer to 25°. Near the poles the inclination of the isotachs from the axis of rotation continues to increase with latitude, and in the lower reaches of the convection zone the angular-velocity contours become more nearly horizontal: the angular velocity increases more rapidly with depth at constant latitude towards the base of the convection zone, as though there the stresses imposed by the more rapidly rotating radiative envelope extend further into the convection zone than do the corresponding stresses near the equator. In addition, there is another shear layer, entirely within the convection zone, and not far beneath the photosphere: the rotation rate rises with depth at constant latitude to a maximum at radius $r \simeq 0.94R_\odot$ some 20 nHz greater than the corresponding photospheric value, and then declines gradually to roughly its surface value. This is so equatorward of latitudes 50° or so. At higher latitudes seismological inferences are less reliable, although it appears that the general trend continues, but with an additional thinner shear layer of opposite sign immediately beneath the surface associated with which is a local minimum in rotation rate at $r \simeq 0.995R_\odot$, some 5–10 nHz slower than the corresponding surface rate. Also at high latitudes is an intriguing deviation of Ω from the parametrized function $\Omega_s(\theta)$; there is an abrupt decrease by 20 nHz or more in the rate of rotation of the surface layers in the vicinity of latitude 75°, extending to a depth of about $0.05R_\odot$. It is not possible at present to measure the rotation poleward of 80°, but I guess the regions of slow rotation extend all the way to the poles, and thereby, perhaps not accidentally, are coincident with the regions from which the fast solar wind blows.

I have already mentioned that the tachocline may not be spherical. By fitting parametrized shear-functions to the helioseismic data, Charbonneau and his collaborators inferred that the tachocline is prolate, with a likely ellipticity of about 0.25 (corresponding to a prolateness $(r_{\rm pole} - r_{\rm eq})/r_{\rm eq} \simeq 0.03$), although with considerable uncertainty. How much this result is contaminated by the greater spreading of the tachocline shear into the convection zone in polar regions is unclear. However, the result is not inconsistent with an earlier finding that the base of the convection zone, determined from the sound-speed stratification, appears to be similarly prolate with an ellipticity of about 0.20. There is no even half-convincing evidence of a static spatial variation of any other tachocline property.

Superposed on the basic pattern of angular velocity described in the previous paragraph are subphotospheric zonal bands of alternately fast and slow rotation which have been observed by Howe, Frank Hill, Rudi Komm, Christensen-Dalsgaard, Michael Thompson, Schou and others, the surface manifestation of which are the so-called torsional oscillations discovered earlier by Bob Howard and Barry LaBonte. The bands are about 15° wide and penetrate at least $0.15R_\odot$ into the convection zone; they are more-or-less symmetrically placed about the equator, and, according to Sergei Vorontsov and his colleagues, they migrate equatorwards from latitudes of about 42° at a rate which causes the angular velocity at any given latitude to oscillate with a period of about 11 years. At higher latitudes the bands appear to migrate towards the poles. Sunspots, whose locations also migrate equatorwards as the sunspot cycle progresses, but from latitudes of only 30°, are found very roughly at latitudes at which the potential vorticity associated with the angular velocity has a maximum. Various other features have been reported, but I conclude this brief description just with an observation that may be pertinent to the tachocline, and which has been dubbed the tachocline oscillation: it is an oscillation in angular velocity near the equator, discovered by Howe and her colleagues, with an amplitude of about 6 nHz and a period of 1.3 years. It was observed both immediately above and immediately beneath the base of the convection zone, the two regions being separated in depth by about $0.1R_\odot$ and oscillating in antiphase with each other, implying that the radial shear has an antinode at the location of the tachocline. I should point out that Sarbani Basu, Schou and Antia have failed to reproduce this finding in a convincing light, and have challenged it. By contrast, I might add that we have found a hint of a third region of the oscillation in Ω, with its antinode about $0.1R_\odot$ beneath the lower of the two found previously, with a similar amplitude and in antiphase with it, so the oscillation may not be a property of the tachocline alone.

1.5 The solar tachocline

It is generally believed that the differential rotation of the convection zone results from a combination of the effects of the anisotropy of the Reynolds stresses, asphericity of the heat flux, and angular-momentum transport by large-scale meridional flow. These processes adjust on timescales of a year to a decade, which are short compared with what is generally believed to be the timescale for angular-momentum redistribution in the radiative envelope below. But they have presumably been operating over the lifetime of the Sun, giving time for their influence to have been transmitted to the radiative envelope. So why is it that the radiative envelope rotates uniformly in the face of the differentially rotating convection zone? And why is the transition layer so thin? These questions were first addressed seriously in a seminal paper by Ed Spiegel and Jean-Paul Zahn, who argued that shear-driven turbulence in the stably stratified envelope immediately beneath the convection zone is almost two-dimensional, providing little stress to transport angular momentum vertically. They tacitly assumed that, in contrast to the turbulence in the convection zone above, the Reynolds stresses are horizontally isotropic, and act in the manner of viscous diffusion so as to oppose latitudinal gradients in Ω. They argued convincingly, subject to that assumption, that the outcome would be more-or-less in accord with the findings of helioseismology. Indeed, with not implausible values of vertical and horizontal turbulent viscosity coefficients, they obtained a thin shear layer, which they called the tachocline (from the Attic Greek *tachos* = speed, in preference to *tachys* = fast, + *klino* = to cause to bend or slope). Their model predicts that the angular velocity in the uniformly rotating radiative envelope is $\Omega_{0SZ} = 0.90\Omega_e$. Had Spiegel and Zahn ignored the dynamics of the meridional circulation associated with the baroclinicity (which, of course, they never would), thereby having a model with only pure rotational motion controlled by homogeneously viscous angular-momentum transport, they would have obtained $\Omega_{0\mathrm{visc}} = 0.96\Omega_{\mathrm{e}}$. This highlights the kind of error that can be made by a purely diffusive description (not to mention erroneous timescales), which in this case produces only 40% of the full equatorial angular-velocity jump across the tachocline of the model. Solar observations imply $\Omega_0 \simeq 0.93\Omega_{\mathrm{e}}$.

In their analysis, Spiegel and Zahn ignored any reaction of the tachocline stresses on the rotation of the convection zone, a simplification which is no doubt a very good first approximation. As I described in the previous section, there is helioseismological evidence that the reaction is not entirely negligible, however, and that the region of shear has spread upwards into the convection zone, particularly in the polar regions. The dominating dynamics of that spreading is very different from the dynamics in the stably stratified shear layer beneath, so we reserve the term tachocline for only the latter, in accord with Spiegel and Zahn's original appellation.

A word or two about the overlying convection zone is not out of place here. Because the heat capacity of the fluid in the main body of the convection zone is high, by the standards of the flux of heat that the Sun demands, convective velocities are very subsonic and, more pertinently, the mean stratification is very close to being adiabatic, the relative deviation from adiabaticity probably not exceeding 10^{-4} or so (in the lower reaches of the convection zone it is as small as 10^{-6}), except, of course, very close to what one might regard as the boundaries of the convection zone. The largest-scale eddies in particular have quite a high degree of coherence, so their overall dynamics is determined not just locally. Therefore the mean stratification in the convecting region need not be unstable everywhere, particularly near its boundaries. Here I adopt the pragmatic approach, which, I hasten to add, is not universally accepted, of regarding the entire region in which the mean stratification is almost adiabatic (together with the highly superadiabatic upper boundary layer) as constituting the convection zone, whether the stratification be superadiabatic or not. It is the region in which the motion is sufficiently vigorous to isentropize the fluid. It is a region that can be identified seismologically.

Beneath that region is the overshoot layer, in which maybe tongues of fluid from the convection zone trespass into the very stably stratified tachocline beneath, bringing with them their magnetic field, sometimes entraining quiescent fluid and mixing entropy and chemical species, but perhaps more often simply undergoing a reversal in the direction of their flow and returning to the convection zone after little or no mixing, causing the interface between the well mixed convecting layer and the relatively quiescent radiative layer beneath to undulate. Those tongues are not space-filling, so the mean stratification of the region into which they penetrate is intermediate between being adiabatic and being that of a quiescent fluid in radiative equilibrium; they provide a smoothing of the mean stratification in the vicinity of the base of the convection zone (not a sharpening of the discontinuity, as some purveyors of one-dimensional solar models have maintained), a smoothing for which Nick Ellis and I once sought a seismic diagnostic from low-degree g modes in the days before it was realized that reliable unambiguous internal-g-mode detection is not imminent. There have been a wide variety of estimates of the extent of the overshoot layer, some being greater than what we now believe to be the thickness of the tachocline. It appears to me to be most likely that the overshoot layer is rather thinner than the tachocline, but there remains some room for doubt. At the very bottom of the overshoot region the direct convective motion is too slow to remain adiabatic, and therefore transports almost no heat, although it might be a not insignificant agent for mixing magnetic field and chemical species. This is also a region where the more vigorous direct motion gives way to oscillatory wave motion, which itself has the potential for transporting heat, angular momentum, chemical species and magnetic field. The physics is very complicated, and it is being debated what role that region plays in the overall dynamics of the tachocline.

1.6 On the basic dynamics of the tachocline

If we start the discussion by recognizing that the tachocline is basically a thin, stably stratified, rotational shear layer, uniformly rotating at its base, we must conclude immediately that the imbalance of forces that must necessarily be present in such a configuration acts in such a way as to drive an angular-momentum transporting meridional circulation, towards the axis of rotation at latitudes where the body of the tachocline rotates faster than its base, away from the axis where it rotates more slowly. I should emphasize that the circulation is driven principally by the stresses associated with the differential rotation of the convection zone, and not significantly by the substantially weaker effects of spin-down as some have claimed. The high degree of stabilization in the vertical, expressed by the buoyancy frequency being very much higher than the top-to-bottom angular-velocity difference across the tachocline, tries to constrain that flow to be almost horizontal almost everywhere. I shall presume the motion also to be basically axisymmetrical, as has been assumed in most dynamical studies to date. As was pointed out by Spiegel and Zahn, conditions in the tachocline are such that the balance of forces in the latitudinal direction is cyclostrophic (which in this context Spiegel and Zahn called heliostrophic, the solar analogue of geostrophic), the pressure gradient being almost balanced (in a rigid frame of reference rotating with the characteristic angular velocity of the fluid) by the Coriolis force. I should emphasize that this balance is indeed a balance, a balance that is more-or-less permanently sustained, provided that the tachocline flow is not seriously unstable; although driven by a force imbalance which induces baroclinicity, the meridional tachocline circulation is not, strictly speaking, a baroclinic instability (or any other instability), as some people have described it in the past. I should emphasize also that the baroclinicity is not itself the fundamental driving force. The baroclinicity is a slave to the rotation and the associated anisotropy of the Reynolds stresses and the buoyancy associated with the asphericity of the heat flow in the convection zone, and is transmitted to the tachocline principally via the vertical hydrostatic balance of forces and advection by the meridional flow. The meridional circulation is driven predominantly by the stresses in the convection zone (and elsewhere) that provide the azimuthal force to produce the differential rotation; the stresses are transmitted to the tachocline to force a differential azimuthal flow which is deflected polewards, near the equator, or equatorwards, near the poles, by (in a rotating frame of reference) the associated Coriolis force. This is called gyroscopic pumping. As I said earlier, those convective stresses are more powerful than the stresses that can be set up in the tachocline, so to a first approximation, at least, one can safely adopt Spiegel and Zahn's tactic of ignoring any back reaction to the tachocline in the convection zone.

The meridional tachocline circulation must necessarily have a vertical component. In view of the very highly stable stratification, that flow can penetrate into the radiative envelope only if it does so slowly enough for thermal diffusion to cancel the opposing buoyancy. It can penetrate into the convection zone quite freely. Spiegel and Zahn demonstrated that if one starts from a radiative envelope that is rotating uniformly, and that is not pervaded by a large-scale magnetic field, then after a time t the penetration depth is of order $(t/\tau_{\mathrm{ES}})^{1/4} r_{\mathrm{c}}$, where r_{c} is the radius of the base of the convection zone. So even had the Sun arrived on the Main Sequence rotating uniformly, by now the meridional flow generated in the convection zone, and the angular momentum it advects, could have penetrated perhaps a quarter or more of the way to the centre of the Sun, producing a region of shear some ten or more times thicker than the tachocline is observed to be. This penetration is a robust result, despite a recent invalid claim to the contrary by Gilman and Mark Miesch. It follows that the tachocline is not an intrinsically transient diffusive phenomenon. There must be some agent operating either in or immediately beneath the tachocline to impose a degree of horizontal rigidity. Moreover, it is likely that the tachocline is in a more-or-less steady state, save for a gradual evolution in response to the spin-down of the entire Sun.

What I have said so far is probably not very controversial. Where opinions diverge is on the issue of what the agent that imposes the rigidity might be. I shall summarize my no doubt biased view of those opinions, on the whole refraining from discussing the purely diffusive studies which ignore the dynamics of the meridional flow, for, as in the case of spin-down, without appropriate caution they can lead to faulty conclusions.

As I said earlier, Spiegel and Zahn invoked layerwise almost two-dimensional horizontally isotropic shear turbulence in the tachocline, which they expected to occur on account of the large Reynolds number of the horizontal shear. The Reynolds stresses exerted by the turbulence were presumed to act locally in the manner of a viscosity, the viscosity coefficient being much greater for horizontal shear than for vertical shear. The study begs the question of whether or not the shear can sustain turbulence that is sufficiently vigorous to quench the very shear that drives it. That raises the issue of the mode of instability of the rotational shear, about which I shall remark briefly below. But more important is the objection raised by Michael McIntyre and me about the nature of the angular-momentum transport in layerwise two-dimensional turbulence. Because potential vorticity – namely $Q = \rho^{-1}\boldsymbol{\omega} \cdot \nabla s$, where $\boldsymbol{\omega}$ is vorticity measured in an inertial frame (ρ is density and s is specific entropy) – is conserved in dissipationless flow, turbulence tends to transport, and, in regions where it is vigorous, homogenize, perhaps in an apparently diffusive manner, potential vorticity, not angular velocity, which drives the mean flow away from rather than towards a state of uniform rotation. This property is observed in the Earth's atmosphere; and it has been simulated numerically by Peter Haynes, in plane geometry, and demonstrated in spherical geometry, in the weakly nonlinear limit, by Pascale Garaud. But in addition there is often strong coupling to internal waves, principally inertia-gravity waves and Rossby waves, which transport angular momentum far from the site of their generation to where the angular momentum can be returned to the mean flow either via viscous or thermal dissipation or by nondissipative, necessarily nonlinear, interactions. McIntyre has emphasized often that this wave transport is not merely incidental to the turbulence but is an essential ingredient of it, because a local reduction in a potential-vorticity gradient is not angular-momentum preserving, so the waves are absolutely necessary for carrying the angular momentum away. It follows that, unlike pure diffusion, the process of redistributing angular momentum by layerwise two-dimensional turbulence is usually not local. In the Earth's atmosphere there is often observed a mid-latitude band in which potential-vorticity mixing is strong, sandwiched between more nearly laminar polar and equatorial regions with which it communicates partly via waves. It is perhaps interesting to note that if there were such a region in the tachocline that matched smoothly onto a uniformly rotating polar cap, then the angular velocity in that mixed region would increase equatorwards: in the jargon of solar physicists, the turbulence would drive an equatorial acceleration.

At this point a digression on wave transport by gravity waves is not out of place. It was an issue of some interest in the late 1970s in the early days of helioseismology, particularly in view of reports of detections of the infamous large-scale 160 min oscillation of the Sun, an obvious gravity-mode candidate. It had already been found theoretically that some large-scale gravity modes (standing gravity waves) were likely to be self-excited, driven by the fluctuations in the nuclear reactions they induce, in the manner suggested by Eddington as a potential source of excitation of the pulsations of classical variable stars, although a solar gravity mode with a period as long as 160 min is too dissipative to be one of those modes. This was also a time when solar spin-down was a topical issue, and angular-momentum transport by gravity waves seemed a plausible candidate for spinning down the core. Microscopic dissipation processes were obviously much too slow to enable any large-scale gravity mode to extract enough angular momentum directly from the core to spin it down in the Sun's lifetime; for such modes to act as an effective spin-down agent the core would need to be turbulent, which seemed to be not out of the question at the time. (Core turbulence would homogenize the chemical composition, but helioseismology had not yet ruled that out.) Wojtek Dziembowski subsequently showed that three-mode coupling to pairs of resonant small-scale daughter modes damps such gravity modes so effectively that they are unlikely to grow to sufficient amplitude to transport an interesting amount of angular momentum. Because microscopic diffusion coefficients in the Sun are so tiny (to be more precise, the Reynolds and Péclet numbers characteristic of the oscillations are so large), the damping of the daughter modes is small enough to permit them to be driven by the large-scale parent mode to amplitudes so high that they readily extract energy from their parent and dissipate it, microscopically, at a high rate themselves, thereby keeping the parent at very low amplitude – a result which might at first seem counterintuitive. The process appeared to require very precise resonances to be maintained between the three modes over a characteristic growth time (a frequency resonance within about τ_κ^{-1} over a thermal diffusion time τ_κ of the mode – about a million years), but the gravity-mode spectrum is so dense that appropriate daughter pairs would have easily been found had the background state of the Sun been independent of time. It was subsequently shown by Chris Jordinson that continuous resonance is actually not necessary, so even though such resonances are readily broken by the changing structure of the Sun as it evolves on the Main Sequence (and as it responds to the activity cycle), new resonances come into play to replace the broken ones without changing the likely limiting amplitude of the parent. It would appear, therefore, that large-scale gravity modes do not partake in the spin-down process.

One is thus led to enquire of the role of the gravity waves that are excited by convective overshooting. Because of the enormous mismatch of the natural timescales of the two regions – the inverse buoyancy frequency N^{-1} is about 15 min at the top of the radiative envelope, compared with $l/v \sim 1$ month (l and v being characteristic length and velocity scales of the dominant eddies) in the lower reaches of the convection zone – the resonant gravity waves that match the characteristic timescale and horizontal lengthscale of the convection have such high vertical wavenumbers that they dissipate after propagating only a few kilometres beneath the convection zone, a distance so small that the waves could be no more than froth on the boundaries of the overshooting motion. The peak of the spectrum of waves that penetrate substantially into the radiative envelope, like the peak in the acoustic-mode spectrum, is therefore at a frequency rather higher than v/l. But gravity waves near the peak are so far off resonance that their amplitudes, which can be estimated by balancing the pressure fluctuations in the waves against the momentum flux of the overshooting motion, are normally expected to be too small to be dynamically interesting. However, the amplitudes of those waves are notoriously difficult to estimate, not least because the properties of the convective motion in the Sun are uncertain. So there is perhaps some chance that the wave flux is much higher than is normally expected, and that that flux could provide a kind of 'shearing rigidity' to the solar envelope, and act as the carrier for spin-down. To be sure, a spectrum of gravity waves would not lead to a state of uniform rotation, for prograde modes in an initial shear dissipate faster than retrograde modes, and so tend to enhance that shear; but when that property was first mooted the helioseismological evidence for uniform rotation was not yet available.

The beautiful and now classic experiment of Alan Plumb and Angus McEwan, demonstrating spontaneous gravity-wave-induced symmetry breaking that generates a shear from an initially uniformly rotating state, had not yet been published. At the very least, it proved that a broad spectrum of gravity waves enhances shear, even if that process was not instantly accepted as the mechanism that drives the quasi-biennial oscillation (QBO) of the Earth's atmosphere, as it is today. It is interesting that much more recently, despite these arguments, Pawan Kumar and Eliot Quataert, and Zahn, Suzanne Talon and José Matias simultaneously published papers claiming that convection-driven gravity waves are the agent that constrains the rotation of the Sun's radiative interior to be uniform. Although it was immediately obvious, in the light of this discussion, that those claims are wrong, I mention them because interestingly the gravity waves that penetrate deeply enough to be dynamically interesting were estimated in both investigations, by apparently different reasoning, to carry a hundred or so times the angular-momentum flux that previous pressure-balance arguments had yielded. This is a crucial discrepancy, for if the new estimates were correct they would imply that gravity waves play a significant role in the dynamics of the tachocline.

More recently, Talon, Kumar and Zahn have resurrected the idea that convectively excited gravity waves play an important role in transporting angular momentum right through the Sun, invoking a presence of turbulence throughout the radiative interior to provide an effective diffusivity which they set arbitrarily to be 10^5 times the radiative momentum diffusivity (i.e. kinematic viscosity) in order to enable the waves to dissipate angular momentum at an interesting rate. The implied turbulent diffusion timescale for the core is about 3×10^8 years, substantially longer than the radiative thermal diffusion timescale and therefore not upsetting the thermal stratification directly, but considerably shorter than the age of the Sun (and, of course, the characteristic microscopic diffusion time of helium through hydrogen), implying that the profile of chemical composition in the core has been smoothed to a degree that would be difficult to reconcile with (actually, it is generally considered to be ruled out by) helioseismology.

Faced with the realization that the purely fluid dynamical processes that are able to operate in the tachocline do not lead to a uniformly rotating radiative interior – a personal realization, I hasten to add, which is not yet accepted by everyone – McIntyre and I were forced to the conclusion that the rigidity required to oppose the tendency to shear can be provided only by the Sun's internal relic magnetic field. We did not *invoke* a magnetic field, as some commentators mistakenly reported; all Main-Sequence stars are expected to harbour the largest-scale remnants of the magnetic field that pervaded the interstellar gas from which they condensed. Those remnants would be predominantly dipolar, comprising, as Roger Tayler and, much more recently, Jonathan Braithwaite and Henk Spruit have emphasized, both poloidal and toroidal components. Nor was there anything novel in the concept that a magnetic field can hold the interior rigid, a concept which has been around since the time of Alfvén and Ferraro, and which we took for granted (notwithstanding Ferraro's law; we tacitly presumed that any shear across field lines is likely to set up Alfvén waves along field lines which would dissipate via phase mixing, except possibly near O-type neutral points where the Alfvén time around neighbouring field loops is almost constant). What was new is the conclusion that it necessarily has to be the magnetic field that holds the interior rigid.

Our view of the basic dynamics has much in common with Spiegel and Zahn's: a baroclinic meridional flow in cyclostrophic balance, descending from the convection zone in polar and equatorial regions, and converging at mid-latitudes where it returns to the convection zone in a region of little, or as I shall argue soon, essentially no vertical shear. The principal difference is that in our idealized model we ignored any turbulent transport in the main body of the tachocline, having instead a thin magnetic boundary layer at downwelling latitudes producing a sharp tachopause separating the tachocline from the uniformly rotating radiative interior. At those latitudes the magnetic field is essentially horizontal, and is prevented from diffusing into the tachocline by advection by the downward flow. This is no doubt a gross oversimplification, but it does highlight what is perhaps the dominant dynamics. The upwelling region is much more complicated, and we refrained from speculating on the details. My view is that the upward flow drags with it the magnetic field, which penetrates the convection zone and so provides the means to generate a

Lorentz torque to couple the radiative interior to the convection zone for spin-down. In most of that region the vertical shear has been quenched by the Maxwell stresses. At its latitudinal extremities the poloidal magnetic field is sheared to create toroidal field; the shear is susceptible to a thermal-diffusively moderated magnetorotational instability polewards of the shear-free regions where Ω^2 decreases away from the rotation axis, which can develop into weak three-dimensional turbulence generating Maxwell stresses and, to a lesser degree, Reynolds stresses that force the flow towards a state of locally uniform rotation and so keep the shear in check out to the transition to the essentially field-free downwelling region. There is yet no helioseismological evidence for such an extended shear-free zone, although such evidence is currently being sought. I should add that we also have little evidence for the latitudinal extent of that layer, should the layer exist. Some preliminary numerical simulations by Garaud and myself suggest that it might be small, comparable with the thickness of the tachocline. If that be so, it will be extremely difficult, if not impossible, to detect it seismologically.

McIntyre and I provided quantitative estimates of the tachocline structure only in the downwelling zones, far from its extremities where the flow is more complicated. However, we were able to predict a relation between tachocline thickness and the strength of the magnetic field at the tachopause. We predicted also a ventilation time τ_v by the meridional flow of order 10^6 years; this is essentially the same as that implied by Spiegel and Zahn's analysis, and is short compared with the timescale for gravitational settling of heavy elements (some 10^{12} years). It is therefore a consequence of the theory that, granted that the relic internal magnetic field is now dominated by its most slowly decaying, dipolar, component, magnetic confinement of the meridional flow weakens towards the magnetic axis (which probably, although not necessarily, is common with the axis of rotation), perhaps permitting a polar pit in which fluid from the convection zone descends to depths substantially beyond the mean extent of the tachocline (although that would require a megagauss magnetic field to sustain it, which is very much more intense than the estimate we made based on purely laminar flow). It was suggested that this may be deep enough for lithium to be destroyed by nuclear burning, so decreasing the photospheric lithium abundance to its low observed level. Another consequence of the ventilation is that the tachocline region immediately beneath the convection zone, which in standard solar-evolution theory has suffered a degree of element segregation by gravitational settling, is completely homogenized with the convection zone right down to the tachopause, producing a sound-speed anomaly (relative to a standard tachocline-free theoretical solar model) which Julian Elliott, Takashi Sekii and I calibrated seismologically to be $0.02R_\odot$ thick (under the assumption that it is spherical). If one accepts the model, then this provides the most accurate measure of the tachocline thickness Δ, because the sound-speed variation is seismically resolved much more finely than the angular velocity. I should emphasize that it is a calibration of a tachocline model, and not of a mere assumption as reviewers have been apt to say. Moreover, it is important not to forget that it does depend critically on the details of the theoretical modelling of the stratification of the Sun, which recent atmospheric chemical-abundance observations imply are not yet quite correct. In particular, the calibration relies heavily on the degree of gravitational settling calculated (by Christensen-Dalsgaard, Proffitt and Thompson) in the standard model. Moreover, it ignores material transport by residual motion beneath the tachopause, such as might possibly be produced by gravity waves driven in the vicinity of the lower boundary of the convection zone whose amplitudes are too low to have direct dynamical consequences; if the region of material transport were extensive, then $0.02R_\odot$ should certainly be regarded as an upper bound to Δ. The calibration implies a boundary–layer magnetic field of 1–10 G.

An important quantity that McIntyre and I did not predict is the angular velocity Ω_0 of the rigidly rotating radiative envelope. That requires an assessment of the overall angular-momentum balance of the tachocline, which our preliminary piecemeal discussion, without further elaboration, was unable to provide. I should emphasize that the Gough–McIntyre tachocline relies crucially on the downward pumping of the radiative-envelope magnetic field to prohibit its outward diffusion, preventing it from threading the tachocline except in the shear-free upwelling zones.

The two models of the tachocline dynamics have set the scene for further detailed study. Garaud has adopted a global view, taking an initial tachocline-free state of an idealized rotating solar model, and studying its dynamical evolution in two dimensions. As with all numerical simulations, she was unable to work with realistically low diffusion coefficients. But under some circumstances, starting with a dipolar field aligned with the axis of rotation, she found that much of the field was swept aside to produce a mid-latitude field-free shear layer beneath the convection zone. Her model was left also with a field concentration about the axis which penetrated into the convection zone, providing a torque to couple the radiative interior directly with the slowly rotating polar regions of the convection zone. Whether that field concentration would survive in a three-dimensional simulation (with realistic diffusion coefficients) remains to be seen, particularly if the magnetic axis (if there is one) is inclined from the axis of rotation. In an interesting simulation of a turbulent wave field, Miesch has sought to establish whether tachocline turbulence could act in such a way as to inhibit latitudinal shear, as Spiegel and Zahn had presumed. Miesch considered a spherical shell of fluid between impenetrable, isothermal, horizontal-stress-free boundaries, initially rotating uniformly with angular velocity Ω_0. He supplied the vertical component of the vorticity equation with a steady axisymmetric body source term whose strength diminished with increasing depth, to mimic mean forcing of differential rotation by overshooting convection, which he took to extend to the base of the shell; he also added an unsheared distribution of random sources, rotating with angular velocity Ω_0, either to the vertical component of the vorticity equation, to generate Rossby waves, or to the equation for the horizontal divergence of the velocity, to generate gravity waves. He carried out a detailed analysis of the spectrum of the turbulence, and related the direction of angular-momentum transport to the gradient of the mean zonal flow, labelling downgradient transport as being diffusive. The outcome of the simulations most pertinent to the aim of the investigation was that, not surprisingly, wave drag from the unsheared random sources reduced the latitudinal shear that had been induced in the mean flow by the steady component of the forcing, particularly in the lower layers of the fluid where the strength of that forcing was low. Miesch, perhaps too naively, claimed that that result supports Spiegel and Zahn's presumption that horizontal two-dimensional turbulence reduces latitudinal shear. One should realize, however, that by imposing random wave sources in a rigidly rotating frame an external body force was being applied to the fluid. Indeed, Miesch recognized that his representation suffered shortcomings, accepting in particular that more realistic forcing scenarios need to be considered. It would be interesting to see the result of having the sources of the waves provided solely by internal, albeit artificial, stresses within the fluid.

Gravity-wave transport beneath the convection zone has also been considered by Edgar Knobloch and, in quasi-linear theory, by David Fritts, Sharon Vadas and Øyvind Andreassen, who described how wave transport can contribute not only to a modification of the angular velocity, but also to the driving of rectified meridional flow which can advect light elements in particular perhaps down to levels deep enough for them to undergo nuclear transmutation. This process is important not only for solar physics, but also for understanding the light-element abundances of all moderate-mass Main-Sequence stars. A characteristic ventilation time of the wave-driven flow that penetrates deeply enough for lithium to have been destroyed was estimated by Fritts, Vadas and Andreassen to be between 2×10^4 and 2×10^5 years, rather shorter than the timescale of 10^6 years of the flow discussed by McIntyre and myself, although rather uncertain because it depends directly on the poorly determined value of the wave momentum flux. Nevertheless, the physical processes are certainly operating at some level. More recently, Talon, Kumar and Zahn have reconsidered the possibility of a QBO-like oscillation, correcting their previous error and extending the scope of their discussion to include the influence of rotation and a large-scale magnetic field on the propagation of the waves. To estimate the gravity-wave amplitudes they used the formalism of Peter Goldreich, Norman Murray and Kumar, originally developed for acoustic modes, as had Kumar and Quataert before them, quoting similar, although somewhat different, results. With a broad spectrum of gravity waves they succeeded in generating a shear layer beneath the convection zone with a characteristic scale of $0.1R_\odot$, although at first they failed to produce a temporal oscillation. They argued that in reality the shear would be destroyed by merging with the convection zone, analogously to the process in Plumb and McEwan's experiment, thereby permitting cyclic behaviour with perhaps a solar-cycle timescale. Following similar work by Eun-jin Kim and Keith McGregor, who studied a model with just a single pair of gravity waves, they succeeded in a subsequent paper, adopting a somewhat higher wave flux, in generating a QBO-like oscillation, this time deeper in the star where the residual wave flux is lower and the oscillation period correspondingly greater than in the tachocline; the oscillation they generated had a period of about 300 years. They concluded that not only would the waves induce this QBO-like oscillation, but they would also establish an almost uniform rotation profile throughout the Sun on a timescale of only 10^7 years. This timescale, and the oscillation period, are inversely proportional to the flux of angular momentum carried by the waves, as Plumb and McEwan had shown, so whether the process in the tachocline is actually pertinent to the solar cycle depends critically on the veracity of Kumar, Talon and Zahn's (apparently high) wave-amplitude estimates. Numerical simulations of gravity-wave generation currently being carried out by Tami Rogers and Gary Glatzmaier may resolve this matter.

It is perhaps useful to add a few words about the effect of a putative dynamo field diffusing from the convection zone into the tachocline. I refer to the work of Emese Forgács-Dajka and Kristof Petrovay, who purport to have demonstrated that such a field suppresses the imprint of the convection-zone shear onto the radiative envelope beneath, an imprint which Spiegel and Zahn had shown to be inevitable in the absence of highly anisotropic tachocline turbulence (and an internal magnetic field). Forgács-Dajka and Petrovay considered a model in which an oscillating poloidal magnetic field is presented to the radiative zone at the base, $r = r_c$, of the convection zone, the radiative zone being forced by viscous stresses to rotate differentially with the convection zone at $r = r_c$. They found, under the assumption that the motion driven by viscosity and mediated by Maxwell stresses is purely azimuthal, that in certain circumstances a thin shear layer could be produced. At face value this study was essentially purely diffusive (thereby falling into the category of discussions that I promised not to discuss), but I mention it because it has been the subject of a fair amount of informal discussion of late. Moreover, in a subsequent paper Forgács-Dajka and Petrovay imposed a specific meridional flow, finding that it had little qualitative effect on the rotation, and suggesting that perhaps a dynamically consistent model might behave similarly. It is important to realize, however, that the dynamo magnetic field that suppressed the shear was itself presumed to be unsheared by the convection zone. This property was achieved by asserting that $B_\phi = 0$ at $r = r_c$, a boundary condition that was imposed without comment. The field that diffuses into the radiative interior of this model therefore acts as an almost rigid restrainer through which the flow must diffuse, and it is hardly surprising that with suitable choices of field strength and artificial turbulent diffusivity the shear in the model is suppressed in a layer whose thickness can be made to be comparable with that of the tachocline; it appears to me, therefore, that the required outcome was essentially written into the formulation of the problem at the outset. Would it not have been more realistic to have presented the radiative interior with a magnetic field that had been distorted by the differential rotation of the convection zone? In that case one might expect the penetration of the shear to be enhanced, not suppressed, by the magnetic field. It seems to me that the real issue here is not whether an undistorted magnetic field can suppress shear, but is to understand the difference between the manner in which the convection-zone turbulence influences the field and the manner in which the tachocline turbulence does: can dynamo action generate a component of the azimuthal field to annul the component generated by the shear? The observed rigid rotation of the magnetically constrained coronal holes and the existence of active longitudes evince that such a process might not be impossible.

I conclude this section by recalling that the tachocline dynamics is determined predominantly by the convection-zone stresses that maintain the latitudinal differential rotation. Although spin-down also plays a role, its importance today is negligible. That is indicated by the fact that the spin-down time $\tau_{\rm sd}$, which is about 10^{10} years, is much longer than the characteristic ventilation time $\tau_{\rm v}$ of the large-scale tachocline flow. The latter is of the order of 10^6 years. That is a robust result. It can be expressed in terms of the observed differential rotation, the buoyancy frequency, the thermal diffusivity (actually the thermal diffusion time) and the (observationally inferred) thickness of the tachocline, and when expressed in terms of these quantities it is independent of the details of the dynamical balance that causes the tachocline flow to be confined; the formula for $\tau_{\rm v}$ is the same whether the principal agent confining the shear is anisotropic viscosity, as Spiegel and Zahn have suggested, or the interior magnetic field, as McIntyre and I maintain. Of course, to be sure of this conclusion one must actually compare the magnitudes of the terms in the governing equations, the details of which would be out of place here. One might nevertheless wonder whether spin-down could have driven a tachocline flow at least early in the Sun's Main-Sequence lifetime, when the angular velocity was much greater and the spin-down time much shorter than they are today. That is not completely out of the question, particularly if one can imagine the Sun to have arrived on the Main Sequence with an angular-velocity distribution through its interior that was far from its current form. But, in my opinion, that is unlikely, because the distribution of angular momentum through the Sun would surely have adjusted as the Sun evolved from the Hayashi track to the Main Sequence, on a timescale somewhat greater than even today's tachocline ventilation time, in response to forces not unlike those at work today.

1.7 Tachocline instability and the solar dynamo

There has been a considerable amount of discussion of putative instabilities arising in the solar tachocline. The first I should mention is the two-dimensional non-axisymmetric inviscid shear instability arising from the latitudinal variation of the angular velocity, which I presume, as have others, was principally in the minds of Spiegel and Zahn as the mechanism driving their shear-destroying turbulence. It is Rayleigh's swirling-flow version of what is now commonly called the inflexion-point instability, a necessary condition for which is that the potential vorticity Q be stationary somewhere; Fjørtoft's extension, applied to simple flows like the rotation of the Sun, is tantamount to saying that the stationary value be a maximum in $|Q|$. If the boundaries of the flow are sufficiently far (in units of the characteristic scale of variation of Q) from that maximum, then the flow actually is unstable. Dziembowski and Kosovichev studied the stability of angular-velocity variations of the form given by Equation (1.1), and found that for values of α_2 and α_4 characteristic of the base of the convection zone, the flow might just be unstable. As I have mentioned already, Garaud subsequently studied the weakly nonlinear development of the flow, and showed it to produce a flattening of the maximum of $|Q|$. That resulted in a barely perceptible modification to Ω, one that is not contradicted by helioseismology. It is not unlikely, therefore, that the tachocline is at most marginally unstable. The perturbation to the angular velocity associated with the potential-vorticity flattening can be described as a pair of azimuthal jets either side of the latitude 60° of the maximum in $|Q|$, a prograde jet poleward, whose velocity maximum, according to Garaud, is located at latitude 65°, and a weaker retrograde jet centred about latitude 52°. The difference in the velocities of the jets is a consequence of the spherical geometry, and is such that the angular-momentum deficit in the retrograde jet is largely compensated by the excess angular momentum in the prograde jet. Thus, the shear-driven small-scale flow, which in reality one might expect to be layerwise two-dimensional turbulence, does not so obviously demand the existence of waves to remove angular momentum as it does in the case of turbulence imposed far from any maximum in $|Q|$, the situation studied by Haynes which I mentioned in the previous section. I should mention also that Charbonneau, Mausumi Dikpati and Gilman subsequently presented a similar result, although, like Garaud, not quite in the terms I have used to describe it here.

Dikpati and Gilman have also investigated a mildly three-dimensional version of the instability in the 'shallow-water' approximation. As one would expect, when the stable density stratification of the fluid layer is strong, the result agrees with the two-dimensional analysis. But as the stability of the density stratification weakens, an otherwise stable shear flow becomes unstable to a new kind of motion. That motion has non-zero kinetic helicity, which Dikpati and Gilman suggested, notwithstanding the global nature of the instability, could produce a positive local α-effect to drive the solar dynamo.

Another obvious instability to investigate is that which might arise from the vertical shear. For adiabatic perturbations the flow should be linearly stable if the Richardson number $Ri = N^2/S^2$, where N is the buoyancy frequency and S is the rate of shearing, exceeds $1/4$. Evry Schatzman, Zahn and Pierre Morel have pointed out, however, that the motion in the tachocline can be of such a small scale that radiative thermal diffusion has the potential of ironing out the stabilizing buoyancy force. They cited analysis by Dudis and by Zahn which showed that under such conditions the Richardson number should be replaced by the product of the Richardson number and the Péclet number vl/κ characterizing the unstable eddies, and that then the vertical tachocline shear could actually be weakly unstable close to the equator. This instability could, therefore, have some influence on the equatorial tachocline dynamics; but it is unlikely to have a serious effect on the overall tachocline structure.

A magnetic field can have a profound influence on the stability characteristic of a flow, and on the nonlinear development of the instabilities. Most pertinent to the tachocline is perhaps the magnetorotational instability, whose importance to astrophysics was first stressed by Steve Balbus and John Hawley. The instability arises in shearing flows in which Ω^2 decreases outwards, and is particularly important for redistributing angular momentum in accretion discs. In a convectively stable tachocline, where the angular velocity increases outwards everywhere on horizontal surfaces, the instability can develop only on timescales long enough for thermal diffusion to annul the constraint imposed by the stabilizing buoyancy. It tends to develop into weak three-dimensional turbulence which transports angular momentum down the gradient of angular velocity, thereby reducing shear; it is not the dynamically rapid adiabatic instability that operates unhindered by negative buoyancy in accretion discs. It is prone to operate particularly at the poleward interface between the putative vertical-shear-free upwelling zone in the solar tachocline and the adjacent downwelling region. The turbulence would tangle and strengthen the magnetic field, isotropizing and possibly strengthening its elasticity, so that, as Gordon Ogilvie and Michael Proctor have shown, the turbulent fluid becomes on a large scale more like a visco-elastic medium. That can add to the torque that couples the convection zone to the radiative envelope in spin-down. It is not out of the question that, poleward of the shear-free zone, the magnetic boundary layer in the tachocline model proposed by McIntyre and myself is also magnetorotationally unstable, or unstable to the Tayler kink instability, as Spruit has recently argued. The ensuing turbulence would increase the effective diffusion coefficients, the magnetic diffusivity by a larger factor than the thermal diffusivity, and so increase the thickness of the magnetic boundary layer. It would also substantially increase the value of the intensity of the relic magnetic field in the radiative envelope that is required to maintain the tachopause at its observed level, and it would decrease the ventilation time. But it too would probably not produce a dramatic qualitative change in the overall tachocline structure.

Instabilities driven by magnetic buoyancy are potentially important to tachocline dynamics, if the field strength is great enough. They are an essential ingredient of many solar dynamo theories. Buoyant fluid containing concentrations of magnetic flux emerging into the convection zone from the mid-latitude upwelling region of the tachocline could be caused to rise against the restraining tension. It must hardly be mere fortuitous coincidence that the upwelling zone, situated where there is little or no vertical tachocline shear, is inferred by helioseismology to be located at the very same latitude as that at which sunspots emerge at the start of a new activity cycle.

McIntyre and I argued that the downwelling regions of the tachocline are likely to be essentially free from magnetic field. This presupposes that the tachocline is not turbulent. As Garaud had shown, if the polarity of the magnetic field in the lower layers of the convection zone alternates with the solar cycle, having, in the long term, essentially zero mean, there is negligible field penetration into a quiescent region beneath. A tachocline meridional flow having a ventilation time of 10^6 years, or even 2×10^4 years, does not materially alter that conclusion. However, if the dynamics of the dynamo process is genuinely wholly stochastic, the field would not average precisely to zero. To be sure, convective overshooting into the stable layers beneath must bring field down with it, but, as simulations by Nic Brummell, Tom Clune and Jüri Toomre suggest, the overshoot layer may be rather thinner than the tachocline, although one must be aware that the smallest dynamically important scales may not have been resolved adequately for modelling the most deeply penetrating overshooting motion; however, for practical dynamical purposes the overshooting layer can probably be viewed simply as a diffuse transition between the tachocline and the convection zone. The model that McIntyre and I outlined starts beneath that transition. The situation is rather different for the model proposed by Spiegel and Zahn, however; in that model one would expect the turbulence to entrain the overshooting field, in consequence becoming more three-dimensional and enhancing the down-gradient vertical transport of angular momentum, which would act in a direction to oppose the suppression of latitudinal shear by the horizontal Reynolds stresses. The outcome would be to increase the prediction of the angular velocity Ω_0 of the radiative envelope, possibly leaving it significantly closer to the value observed. The shear would stretch and strengthen the field into predominantly azimuthal bands, which eventually become buoyantly unstable and rise into the convection zone at the start of a new activity cycle. The general effect of a toroidal magnetic field on the susceptibility of the rotational shear to instability is thus a pertinent issue. According to Paul Cally, and David Hughes and Steve Tobias, the direct effect of the Maxwell stresses tends to be stabilizing, whereas the magnetic buoyancy is destabilizing. Which of the two opposing influences dominates depends on the details of the equilibrium configuration.

That the tachocline is pervaded by a predominantly azimuthal magnetic field appears to be the view that is now most commonly held. It prompted Gilman, Dikpati and Peter Fox to undertake an extensive study of the stability of configurations of this genre, which resulted in a series of papers reporting that large-scale instability occurs for a wide variety of magnetic profiles, in a manner not dissimilar to the Tayler kink instability, and the magnetorotational instability of accretion discs with toroidal magnetic fields that had been discussed earlier by Ogilvie and Jim Pringle. It was also argued that the tension in the field would squeeze the tachocline to produce the observed prolateness, although a full global equilibrium model of the Sun with a prolate tachocline was not constructed.

It appears now to be generally accepted that the solar dynamo resides principally in the tachocline, following arguments put forward by Spiegel and Nigel Weiss. I shall not attempt to review the plethora of models that have been proposed, but instead commend the reader to the excellent review by Tobias.[1] Suffice it to say that the appellation 'solar dynamo' is a technical term that refers to the sum of magnetohydrodynamical processes commonly called dynamo action that twist and strengthen the magnetic field to maintain it against Ohmic decay, and in so doing cause the polarity of the dipole component of the field to reverse approximately every 11 years. Whether that process is truly self-sustaining, or whether it relies for its existence on being continually fed by the seed field that is being dredged up by the upwelling tachocline flow, and is therefore running down as the global relic field in the radiative interior decays, is not of concern to the majority of solar-dynamo theorists, perhaps because it cannot be unambiguously checked observationally. It does not even seem to command interest simply as an academic issue amongst those whose principal motivation is explaining the variations of the external magnetic field of the Sun. It is an important issue to astronomers, however, because it raises the possibility of the evolution of stellar magnetic fields having a dependence on time that is separate from the time-varying rotation and the structural evolution of the convection zone. I once interested Nic Brummell in the matter for a while. However, I failed to persuade him to carry out any numerical simulations to try to distinguish between dynamo-like flows that do and do not require continual refuelling, because Nic was funded to simulate dynamos; evidently it would have been imprudent of him to spend a significant amount of supercomputer time on

[1] In *Fluid Dynamics and Dynamos in Astrophysics and Geophysics*, ed. A. M. Soward, C. A. Jones, D. W. Hughes & N. O. Weiss (London: CRC Press, 2005), p. 193.

non-dynamos, for that would jeopardize his chances for further funding in the future.

It is always amusing to play Devil's advocate on matters upon which there is almost universal agreement, and healthy too for the opposing advocates in cases such as this where we lack a complete robust theory. And so I have occasionally espoused heretical alternatives to the solar-cycle mechanism. One such possibility, on which I worked for a while with Phil Goode, was a revival of the idea apparently proposed by Walén that the cycle is controlled by a magnetically restored torsional oscillation of the radiative interior. That would require a global field of intensity comparable with that found in the umbrae of sunspots, at first sight a not unreasonable value. If that were the case, then one would naturally expect the oscillation to be basically periodic, the fluctuations in cycle period observed at the surface presumably being produced during the stochastic journey of the information through the convection zone. The timing statistics of such a process are different from those of a turbulent dynamo, in which the very restoring process contains a stochastic element, and perhaps the difference could be detected in the sunspot record. I think it is true to say that unfortunately the record is too short to distinguish between the possibilities, although I hasten to add that Dicke, who had also performed a statistical analysis, had more confidence in the outcome than I, claiming that it provides evidence in favour of the torsional oscillation. One of the difficulties encountered by such a model is to find a mechanism to drive the oscillation against Ohmic dissipation enhanced by phase mixing. McIntyre and I entertained the idea that it could be driven by differential dissipation of gravity waves, in the manner of the QBO, but McIntyre has declared that he is now convinced that magnetorotational instability would kill the oscillation. He is probably right. But I shall sit on the fence for a while. Could the 1.3-year tachocline oscillation be excited similarly? And should it too be expected to be similarly quenched by the magnetorotational instability? Some would say that one should not spend too much time worrying about such questions until it is clearer whether that oscillation actually exists.

Finally, I should point out that if the tachocline really does harbour the solar dynamo there should be helioseismologically observable consequences. Putative magnetic-field concentrations, which must necessarily be aspherical, and the associated density perturbations split the frequencies of otherwise degenerate components of a seismic oscillation multiplet, thereby providing a diagnostic of solar asphericity. One might hope that in the near future a magnetic signature in the seismic frequencies produced by a latitudinal variation of the wave propagation speed of magnitude comparable with a (substantial) fraction of the mean (spherically averaged) tachocline sound-speed anomaly would be detectable, which translates, perhaps optimistically, into a magnetic field intensity of order 5×10^5 G. Unfortunately, frequency perturbations produced by a magnetic field cannot be distinguished from perturbations produced by any other agent, aside from horizontal flow, so other, theoretical, considerations would need to be brought to bear in order to be able to infer whether it is actually a magnetic field that is responsible for the signature. To be sure, a magnetic field distorts the eigenfunctions differently from a horizontal density variation, but it will probably be a very long time before that can be measured in modes that penetrate as far as the tachocline. Therefore, at least for the time being, we shall have to be content with calibrating models and looking for temporal variation. Evidence for 11-year variations in structure and rotation in the vicinity of the tachocline has been sought by several investigators, but none has yet convincingly been found. The search continues.

Acknowledgments

I am very grateful to all those with whom I have discussed the dynamics of the tachocline, most notably Pascale Garaud, Michael McIntyre, Ed Spiegel and Jean-Paul Zahn.

2

Reflections on the solar tachocline

Edward A. Spiegel

Solar activity takes place in narrow bands of latitude that move like solitary waves from mid-latitudes toward the solar equator. This behaviour points to the existence of a thin layer in the Sun that may serve as a waveguide. With its grand minima, the cycle is intermittent in a way that does not occur in the simplest chaos models. To be useful as a primitive model of the cycle, a differential equation should be of high enough order to display such strong intermittency. These and other features of solar fluid dynamics led to the adumbration of an intermediate shear layer between the convection zone and the radiative core. This layer, like the weather layers in planetary atmospheres, produces coherent structures – sunspots and perhaps vortices. Similar layers may play a role in stellar activity in cool stars other than the Sun and perhaps even in hot stars if their atmospheres are turbulent.

2.1 The maculate Sun

Rotation and turbulence in stars are significant for an understanding of stellar evolution and for the fluid dynamics of accretion discs. We can watch these processes most closely in our own Solar System. Observations of the Sun, the giant planets and the earth reveal coherent structures whose study has been one of the most exciting adventures in the mathematical science of the twentieth century. (At a meeting in the Newton Institute, we ought to recall this.) To put it simply, a *coherent structure* is a dynamical object that lasts much longer than we might have expected on the basis of simple dimensional arguments. The discrete vortex tubes seen in Jupiter's atmosphere are good examples – the Great Red Spot has been observed since the telescope was invented. But here I am more concerned with the relatively dark spots on the Sun that mark the locations where discrete magnetic flux tubes protrude from the photosphere.

The space-time diagram of the locations of spots called the (Maunder) butterfly diagram shows how the bands of solar excitation converge from the mid-latitudes

The Solar Tachocline, D. W. Hughes, R. Rosner and N. O. Weiss.
Published by Cambridge University Press.

toward the equator every 11 years or so. The narrow extent of these activity belts in their stately motion toward the equator suggest that we look for a layer in the Sun that can serve as a wave guide. Such a layer can be thought of as analogous to what planetary scientists call a weather layer or even to the oceanic thermocline. This perception led to the name *tachocline* (Spiegel & Zahn 1992). In adopting this name we attempt to reduce the problem of the solar cycle to a previously unsolved problem. We may usefully recall that, in discussing the atmospheres of the outer planets, Ingersoll (1990) wrote 'How deep do the zonal winds extend? Is there a level below which the fluid rotates uniformly? These are fundamental questions, but there are no simple answers'. He mentions various possibilities such as that 'the winds could be confined to a thin weather layer above cloud base with the interior in solid rotation ... '.

The difficulty is that we really do not understand why vortices form in rotating turbulent fluids. Nor do we know why vortex tubes are formed on Jupiter while magnetic tubes appear in the Sun. That the solar atmosphere is ionized, while Jupiter's is not, is probably relevant, but that is not an explanation. It is worth reflecting on this difference but I want to do some recollecting before I begin reflecting. Let me then first describe the erratic path that led to my being asked to reflect on the tachocline by the organizers of the workshop.

2.2 Braking the Sun's rotation

I have alluded to planetary vortices to underscore the connections of our subject to well studied processes of geophysics and planetary physics. A possible connection of the internal rotation of the Sun to the verification of Einstein's theory of gravity was also debated for a short period some 40 years ago. There were other, less pressing issues that involved the problem of solar rotation as well. Stars like the Sun are observed in galactic clusters like the Pleiades and the Hyades. These young counterparts of the Sun rotate more rapidly than the Sun does (Kraft 1967). Thus they offer evidence for believing that the rotation of solar-type stars decreases with time. This is probably caused by torques exerted by magnetic fields that are drawn out by stellar winds, as suggested by Schatzman (1962; see Mestel 1999). The lithium abundances in solar-type stars also decrease with time. Hence this topic has implications for mixing in stars, with possible consequences for stellar evolution and perhaps also certain issues arising in cosmology.

Let J be the Sun's angular momentum, R its radius and M its mass. (I omit the $\odot$ subscript so that these quantities may also refer to those properties for any suitable object.) Let us assume at first that the angular velocity, Ω, is roughly constant throughout the body of the Sun. Given a solar model, we can compute the coefficient α in $J = \alpha MR^2\Omega$. If there were no magnetic fields involved, the rate of loss of angular momentum would be $\dot{M}R^2\Omega$. However, because the solar wind pulls the magnetic field out with it, the departing mass rotates at roughly the surface angular velocity out to a limiting (Alfvén) radius, R_C, as foreshadowed by the (Ferraro) law of isorotation. (I shall try to adhere to the McIntyre dictum of calling things by descriptive names rather than by the names of people who worked on them. Perhaps I should not even call it the McIntyre dictum.)

Beyond R_C the field is too weak to maintain the near constancy of angular velocity and the material can be considered to have left the Sun. Hence we estimate the loss rate of angular momentum as

$$\frac{dJ}{dt} = \dot{M}R_C^2\Omega. \tag{2.1}$$

We may write this loss rate as a slowing down of the solar rotation:

$$\frac{d\Omega}{dt} = \frac{\dot{M}R_C^2}{\alpha MR^2}\Omega. \tag{2.2}$$

The rotational lifetime implied in this formulation cannot be expected to be constant. The mass loss rate depends on coronal heating which is a function of the magnetic field and that in turn depends on the current rotation rate. Though dimensional estimates of the dependence of the magnetic field strength on rotation rate exist, such as the pleasantly simple one of Cowling (1969), they are not of much help in estimating $\dot{M}$, which ultimately depends on coronal heating rates. This is unknown so, for qualitative purposes, let us suppose this lifetime ultimately depends on Ω as a power law. We then have

$$\dot{\Omega} = -\beta\Omega^{n+1}, \tag{2.3}$$

where β and n are constants. Thus, we find (Spiegel 1968) that

$$\Omega = \frac{\Omega_0}{(1 + t/\tau)^{1/n}}, \tag{2.4}$$

where Ω_0 is the rotation rate when the star reaches the Main Sequence and

$$\tau = (n\beta\Omega_0^n)^{-1}. \tag{2.5}$$

The ages of the Pleiades, Hyades and the Sun are respectively 50, 500 and 5000 megayears and the surface rotation rates of the solar-type stars are 19, 9 and 2 km/s (Kraft 1967). A rough fit of this formula to those data suggests that $\tau \approx 10^8\,y$ and $n \approx 1$. The simpler approach of fitting a power law to those data with no τ suggests $n = 2$ (Skumanich 1972). In either case, one finds that the half-life of the solar rotation is comparable to the Sun's age.

So far, I have presumed that the effects of the loss of angular momentum from the solar surface are transmitted to the solar interior in a short time. This had been the standard assumption for some time as it had generally been believed that the differential rotation resulting from a slowing down of the outer layers leads to instability. However, Dicke (1964) proposed that the rotational instability could not overcome the stable density stratification below the convection zone. Hence, he argued, the radiative core of the Sun still rotated at about the rate it had on arriving at the zero age Main Sequence. This claim took no account of the possibility of doubly diffusive instability, although Townsend (1957) had discussed the role of radiative transfer in reducing the stabilizing effect of density stratification in parallel shear flow: 'The criterion for turbulent motion ... is usually expressed as a critical value of the Richardson number ... However, the derivation of this criterion neglects radiative transfer ... its inclusion will cause a reduction in the magnitude of the buoyancy forces which are responsible for the inhibition of turbulent motion'.

Both the Richardson criterion and Townsend's modification of it provide only necessary conditions for instability but, in the cylindrical case studied by Yih (1961), necessary and sufficient conditions emerge from the linear theory. Yih's study of the instability of Taylor–Couette flow showed how the Rayleigh criterion for instability of swirl in a nominally stable density gradient is promoted by thermal diffusivity. He wrote that 'The destabilizing effect of thermal diffusivity ... is almost exactly the same as [in convective] instability ... ' and referred to the so-called salt fountain (Stommel *et al.* 1956). Yih's work paralleled Stern's (1960) results on doubly diffusive convection, which also referred to the salt fountain that has provided a paradigm of this type of diffusive destabilization. (For an account of the convective work see Spiegel (1969) and the papers in Brandt & Fernando (1995).) The study of doubly diffusive instability of differential rotation in the spherical case by Goldreich & Schubert (1967) and Fricke (1968) had a much more decisive influence on astrophysical work. These latter authors noted a further instability caused by variation of Ω along the rotation axis, developed more fully by Lebovitz & Lifschitz (1993).

The discussion concerning Dicke's suggestion of a rapidly rotating solar interior heated up when he and Goldenberg announced that they had measured a significant solar oblateness. Their observations were reported at the Texas Conference on Relativistic Astrophysics in New York in December of 1967. I was in Cambridge then at the invitation of Dennis Sciama who had gone off to attend the conference. When Sciama returned to Cambridge, he handed me the preprint of their paper (Dicke & Goldenberg 1967), of which I had not yet heard, and assigned me to give a seminar that same day in which I was to criticize the paper from *both* the theoretical and observational point of view. The reason for this sense of urgency was that it was being proposed that a solar oblateness was evidence for a quadrupole term in the solar mass distribution. That would imply an inverse cube term in the solar gravitational force field. In turn, this would cause some precession of the perihelion of Mercury and so would spoil the agreement between the prediction of general relativity and the result of celestial mechanics for the precession rate.

This work came at the same time that the question of the depletion of lithium in solar type stars became current. Bob Kraft had organized a meeting on *The Sun Among the Stars* in 1966 that was, I suspect, a forerunner of the meetings named for what became known as the solar–stellar connection. It does not take a very high temperature (by stellar standards) to destroy lithium, and it was not easy to come up with a process that would mix lithium down from the outer layers at just the right rate (Spiegel 1968). Schwarzschild used to put it that anything connected with convection would be too fast if it worked at all and much too slow if it did not work. The arguments that were used against Dicke's model were helpful in this respect. They implied that there was rotational turbulence below the convection zone in the Sun that offered a possible resolution of the lithium problem. A modern treatment of this question has recently been presented by Brun *et al.* (1999).

2.3 Solar spin-down

The solar wind torques have a direct influence on the outer layers of the Sun all the way down to the roots of the surface fields. Even if those roots are not very deep, the convective turbulence will quickly spread the effects to the bottom of the convection zone. Then what happens? The problem of how the loss of solar angular momentum in the solar wind is passed on to the rest of the Sun has not been fully resolved. The process will likely involve magnetic fields but, in the early 1970s, many people working on the problem granted, for the sake of argument, Dicke's claim at the peak of the discussion that there was no magnetic field in the solar interior. If there were any field, it would be enhanced by the proposed differential rotation and, by opposing, end it. So there was a nonmagnetic interregnum with only a few interruptions that lasted from the 1950s, when Alfvén and Cowling battled in the pages of the *Monthly Notices* over Alfvén's theory of how the solar cycle arose in a strong dipole field in the solar core, to the present when Gough & McIntyre (1998; McIntyre 2002) have restored the internal field in order to maintain a rigid rotation in the solar interior.

Guidance for the nonmagnetic case of solar spin-down came from the paper of Greenspan & Howard (1963) on impulsive spin-up and the even more closely related paper by Bondi & Lyttleton (1948) on the slowing of the Earth's rotation. Both of these relied on Ekman pumping, as did the suggestion that such a process occurred in the Sun (Howard *et al.* 1967).

In the conventional treatment of spin-down, in either the gradual or the impulsive case, the slowing down of a solid boundary sets up a boundary layer of thickness $\delta = \sqrt{\nu/\Omega}$, where ν is the kinematic viscosity and Ω is the current angular velocity.

The resulting mismatch of rotation rates between the boundary layer and the fluid interior establishes a pressure gradient that drives flow between the two parts of the fluid. This induces a large-scale circulation in the body of the fluid that redistributes the vorticity and leads to spin-down of the interior fluid. The timescale for the process is the geometric mean between the rotation period and the viscous time of the fluid, or d/δ times the rotation period, where d is the depth of the fluid. But a more powerful driving occurs if there is a relatively deep layer at the boundary that has a much higher viscosity than the interior fluid (Bretherton & Spiegel 1968). In the case of the Sun, the convection zone is such a layer even without the extra boost from its (mildly) unstable entropy gradient. The higher viscosity would be turbulent viscosity and, since the convection zone is much thicker than any supposed (Ekman) boundary layer, spin-down is greatly speeded up by this mechanism.

As we did not know how to treat a turbulent flow like that in the convective zone, we simply represented the turbulent viscosity by a Darcy frictional law. Even then the problem was difficult and we cravenly assumed that the Darcy damping time is much shorter than the rotation period. The implied analogy to a porous medium allowed a quick if qualitative check of the idea in the lab. In the 1969 GFD Program at Woods Hole, Joseph Buschi tried to test the notion of convective pumping more directly by the experiment of spinning up a layer with penetrative convection (the ice water version) in a rotating frame. A numerical study of rotating penetrative convection at high Rayleigh number would be of value in confirming these notions.

For a brief description of the effects of stable stratification, we may use the model of a rotating layer of fluid stratified by a constant gravitational force antiparallel to the rotation vector. If the dissipation is small enough to be neglected in the body of the fluid, the spindown of a *homogenous* fluid is as just described. A disturbance made at some depth in the fluid with horizontal extent h will be felt all along a column parallel to the rotation axis as implied by the Taylor–Proudman theorem. The effect of a stable stratification is to truncate this (Taylor) column to a stub whose height is

$$\ell = \frac{2\Omega h}{N}. \tag{2.6}$$

Here the buoyancy frequency is given by

$$N^2 = \frac{g}{c_p}\frac{\mathrm{d}S}{\mathrm{d}z}, \tag{2.7}$$

where S is the specific entropy, c_p is the specific heat, assumed constant, and z is the vertical coordinate. The introduction of stratification confines the rotational control to an ever decreasing range of depths as N/Ω is increased. (That ratio is now around one thousand in the Sun.)

In GFD, only the stably stratified case, $S' > 0$, is normally considered and one usually asks the complementary question of how wide a region must be before a disturbance in a layer of depth ℓ can be controlled by the rotational constraint, as it is called. This width is known as the (Rossby) deformation radius, $N\ell/(2\Omega)$. A corresponding wavenumber appears in the theory of rotating convection, as can be seen in a careful reading of a paper by Cowling (1951). In that case the meaning seems much clearer: for wavenumbers smaller than the (Cowling) wavenumber, rotation prevents convective instability in the absence of dissipation.

Since the spin-down of a non-dissipative, stratified medium can directly affect only a sublayer of depth ℓ, the process takes only ℓ/d of the full spin-down time. We spent some time working these things out (during what Andrew Ingersoll called The Great Oblate Debate) as did geophysicists such as Holton (1965), though for other reasons. The problem was not a difficult one given the usual idealizations, though complications caused by boundaries did raise some subtle issues. In the solar case, the disturbance comes down from the convection zone over a horizontal distance of order $2R$. Therefore the short initial phase of spindown affects only a sublayer of thickness $2R\Omega/N$.

When lecturing on solar rotation, I used to illustrate these processes with a cave drawing like that in Figure 2.1 (Spiegel 1972). I called the sublayer that is spun down by direct convective pumping the tachycline. Since I did not expect such a layer to be observed in the then foreseeable future, I felt free to choose that infelicitous name as a joke that is too convoluted to be recalled here. Fortunately,

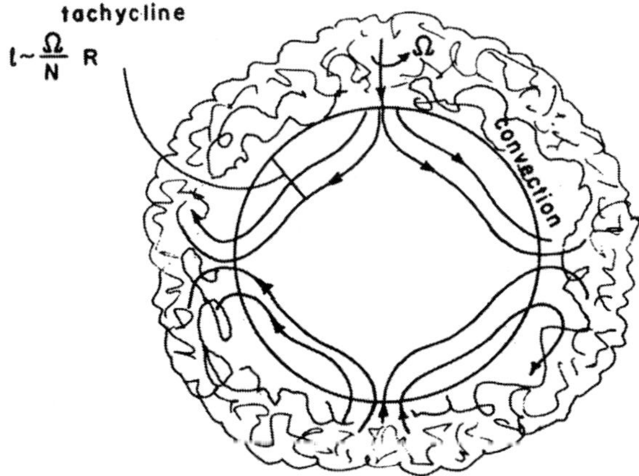

Figure 2.1. An early cartoon of solar spin-down currents.

since the tachycline is transient, this terminology allows it to be distinguished from its more permanent relative, the tachocline, as Zahn and I (1992) later called the quasi-stationary velocity transition layer.

Ultimately, dissipation cannot be ignored in this process. If the spun-down layer is not unstable, the rest of the radiative core will be spun down dissipatively on the Eddington–Sweet time. This is $(N/\Omega)^2$ times the thermal time of the core. That thermal time is normally called the Kelvin–Helmholtz time in astrophysics while the Eddington–Sweet time is the thermal time scale of a region of radius RN/Ω.

Since the spun-down sublayer is likely to be unstable and even turbulent, it is not obvious how the process will evolve. It may be that a turbulent tachocline will produce further pumping and spin down a second sublayer. This, in its turn, will become turbulent and so on. That version of the process would give rise to a rotationally layered interior flow reminiscent of the layering seen in thermohaline convection. This suggestion could be checked by numerical simulations with the means that are currently available. But all this has to do with a transient process that seems to have done its job by now.

2.4 Solar intermittency

The notion of a quasi-static sublayer, or tachocline, in fact arose (in my own case) in attempting to model the intermittency of the solar cycle. In the early 1960s, Derek Moore and I were studying the origin of the solar oscillations that had been reported by Leighton *et al.* (1962) . We considered the possibility that the oscillations were driven by a convective overstability like those known in magnetic and rotational convection. (In fact, sound waves can be convectively overstable in the manner of those other convective instabilities (Spiegel 1964). However, it is believed that acoustic instabilities are too weak to account for the observed oscillations.) To understand convective overstability, we constructed what we intended as a generic model of the process (Moore & Spiegel 1966). We wrote the dynamical equations for a fluid element moving vertically under the influence of a restoring force in an unstably stratified medium. The restoring force, represented by a nonlinear spring, was meant to represent a magnetic force or a stabilizing molecular weight gradient. This led to an equation of the form

$$\ddot{x} = -\frac{\partial V}{\partial x} - \mu \dot{x}, \tag{2.8}$$

where the potential V is a quartic function of (amplitude) x and is linear in a control parameter, p. (We normally omitted the friction term back then but, with time, I came to feel the need for it.) The key to the overstability in the model is to let p vary slowly in time. We made the model autonomous by introducing an equation

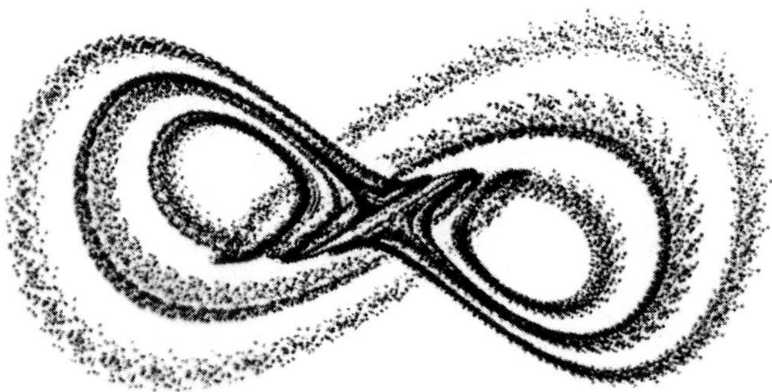

Figure 2.2. A projected orbit of a thermally excited nonlinear oscillator.

for the slow variation of p with the form

$$\dot{p} = \epsilon g(x, p), \tag{2.9}$$

where ϵ is a small positive constant and g is a simple function, linear in p. For a suitable range of the parameters, numerical solutions of the system are aperiodic, as in Figure 2.2, an orbit projected onto the $x - \dot{x}$ plane as computed (much) later by Leonard Smith. We naturally speculated that many phenomena of astrophysics whose aperiodic behaviour had been ascribed to some *dei ex machina* were simply exhibiting the sort of natural dynamics we were seeing.

Later, when Eddy (1976) re-examined the historical record of sunspots and reported that indeed, as Maunder had noted, there were very few spots during the life of Newton, the plot thickened. We had not seen any such behaviour in the simple third-order systems that were by then thought to be representative chaotic systems. We wondered whether a more complicated system might give rise to the kind of intermittency that was seen in the grand minima of solar activity. As a means to model this, we imagined that some secondary dynamo process in the Sun could be driving the solar cycle. Its chaotic interaction with the main convective dynamo could then produce the grand minima. Such a dynamo process ought to operate in a thin layer, given the narrow (if time dependent) activity zones. Since we had already worked on the rotational sublayer involved in solar spindown, we naturally considered the possibility that such a layer could persist at the bottom of the convection zone. The activity of this layer could nonlinearly interfere with the main convective dynamo process. This leap of faith provided a working hypothesis to guide the construction of an intermittency mechanism.

In dynamical systems theory, around that time, the term intermittency was coming to mean vacillation between different kinds of behaviour (Pomeau & Manneville 1990). However, I had in mind the meaning intended by Batchelor & Townsend (1949) in describing the structure of turbulence. The plan was to couple two dynamo models into a single functioning machine and use some property of this object to quantify the level of solar activity. For the component dynamo models, I turned to the nonlinear system that Malkus (1972) had derived to describe his modification of the Bullard disc dynamo model. His system had the same general form as our aperiodic oscillator, but differed in significant details. It turned out to be equivalent to the Lorenz (1963) system, which I did not learn about till the early 1970s, so sparse were the people working on such things then.

I made a symbiotic system of two of those disc dynamos, that is, of two Lorenzian systems. A dynamo model representing the main convective process was coupled to a second dynamo model taking the role of the tachocline. For this, I enlarged Equation (2.8) to allow V to depend on a second amplitude y and set

$$\ddot{y} = -\frac{\partial W}{\partial y} - \xi \dot{y}, \tag{2.10}$$

with $W = W(x, y, p)$ and $V = V(x, y, p)$. To bring about symbiosis, I let them share the parameter p and set $g = g(x, y, p)$. In the conservative limit (ϵ, μ and ξ all $= 0$), this system becomes a two-dimensional nonlinear oscillator whose potential is one of Thom's catastrophes. The set of three equations forms a system of fifth order. For suitable parameter choices, it has an unstable, invariant manifold, $y = 0$. In that manifold there lives a third-order system that is equivalent to the Lorenz system.

For non-zero initial y, in a certain finite parameter range, the system leaves the invariant manifold and wanders about in the five-dimensional state space until it encounters a stable manifold of the invariant manifold. Then it is brought back to small $|y|$ where it skims along the invariant manifold only to fly off again after a while. The first run with that model produced the intermittency illustrated in Figure 2.3. It was so easy to get it that I was (and am – see Hardenberg *et al.* 1997) persuaded that the behaviour is robust. (However, I have yet to make a case for the suggestion that it has something to do with intermittency in turbulence, cf. Spiegel 1981.) Though the detailed behaviour of the model did not resemble the observed variation of the sunspot number, the general behaviour seemed to support the supposition that a secondary dynamo process in the Sun could be at work in the solar cycle.

When Nathan Platt, a GFD fellow, took an interest in the intermittency model (Platt 1990), we worked with Tresser to develop the process, which we called on/off intermittency (Platt *et al.* 1993a). Our attempt to model the time variation of the sunspot number as an intermittent dynamical system worked reasonably well (Platt *et al.* 1993b). Naturally, detailed prediction is not possible for a chaotic process that the cycle resmbles. (The data are not good enough to establish that it really is chaotic (Spiegel & Wolf 1987).) We assumed that a (suitably defined) distance from the invariant manifold measured the level of activity (y^2 worked well) and we saw

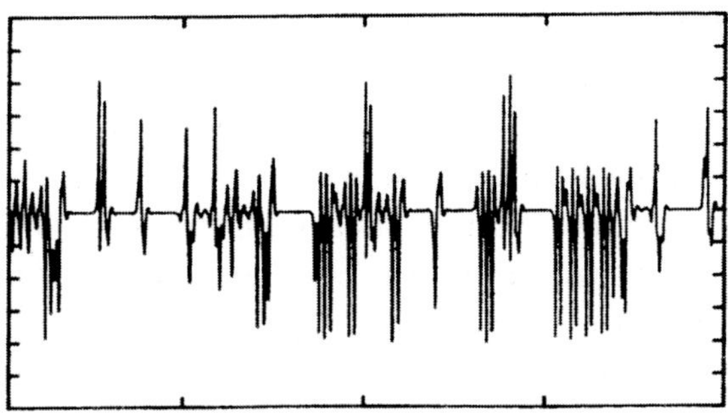

Figure 2.3. On/off intermittency: y vs t.

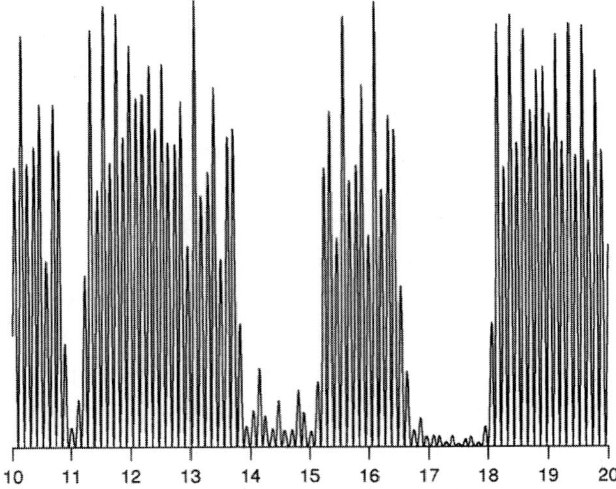

Figure 2.4. The improved model of the solar cycle (see the text).

grand minima somewhat like those observed. There was almost no activity ($y^2 \approx 0$) during the grand minima and I thought this was a good thing. Later, I learned that careful study of historical records showed that there was some spot activity during the Maunder minimum (Ribes & Nesme-Ribes 1993; Beer *et al.* 1998). A slight modification of the model made with Claudia Pasquero and Antonello Provenzale fixed that problem and led to results of the kind shown in Figure 2.4. Also, Axel Brandenburg and I tried putting the on/off process into his dynamo simulation and this produced a reasonable looking butterfly diagram. (We produced a manuscript about this that is, or was, on his web page.)

Several (groups of) people, such as Jones *et al.* (1985), have made dynamical models designed to mimic the solar activity cycle. In her Laureate thesis, Claudia Pasquero (1996) assessed their successes in modelling the solar activity cycle. We tried to write a paper about this with Provenzale, but found that it is not so easy to report on such things. However, one of Pasquero's conclusions that seemed to be indirect evidence for a secondary dynamo process was that the on/off intermittency model provides a reasonable qualitative representation of the solar intermittency (see also Schmitt *et al.* 1996). Still, as Dirac said, 'Just because the results happen to be in agreement with observation does not prove that one's theory is correct'.

2.5 Into the 1990s

Ocean circulation is thought to be driven by wind stresses on the ocean surface. There is a thermal boundary layer in the upper ocean below which lies the abyss. In the Sun, the convection zone pumps fluid into (and out of) a velocity sublayer that couples the convection zone to the radiative core. Seen in this coarse-grained manner, the two situations are somewhat analogous (with the convection zone playing an oversized analogue of what the oceanographers call the mixed layer) and, as mentioned, it is this loose analogy that is behind the naming of the tachocline. I have spent enough time in Woods Hole to have formed the impression that there is no generally agreed upon reason why a thermocline in the ocean should remain thin. I have asked many oceanographers why the thermocline does not spread and, as often as not, the answers I have received were rebutted by others that I asked. The different answers I have gotten make me think that few would believe in the thermocline if one had not been observed. Before its detection by helioseismology (Brown *et al.* 1989), there were only indirect reasons for postulating a long-lived, thin tachocline. Once the tachocline was detected by observation, it was possible to look at the problem more confidently.

In 1990, Jean-Paul Zahn and I gave a course in stellar fluid dynamics at Woods Hole. There I learned that Jean-Paul had worked to understand the newly observed shear layer below the convection zone. In his lectures, he gave his own slant to the subject of shear dispersal in a stratified medium (Zahn 1990). He and Brian Chaboyer applied his ideas to the mixing of lithium. In his report on the summer's work, Chaboyer (1990) wrote 'For the youngest clusters [Charbonneau *et al.* 1989] studied, no Li depletion is observed, contrary to what would be expected from the Eddington–Sweet circulation. In order to account for this fact, Charbonneau *et al.* stated: "*one must investigate mechanisms that could reduce the expected transport through meridional circulation.*" An obvious candidate for such a mechanism is horizontal diffusion, which hinders the meridional advection'. (Italics not mine.) In other words, strong horizontal mixing weakens the inhomogeneity and so inhibits the mixing of elements in the radial direction (Chaboyer & Zahn 1992). This idea works to inhibit the radial spreading of the thermocline as well (Spiegel & Zahn 1992).

In fact, anisotropic advection as a result of stratification is also in line with oceanographic thinking. Pedlosky (1969) had written that 'In order to consider a model of geophysical relevance the viscosity and diffusivity of temperature must be considered as parametrizations of small-scale turbulent mixing processes. It is natural therefore to allow that the transport coefficients need not be the same for horizontal as well as vertical fluxes of momentum and heat. The theory will be developed with the ratio of the horizontal to vertical mixing coefficients as a free parameter ... '. Pedlosky's model could serve as a reasonable description of the tachocline, though it was meant to be geophysical. However, since those words were written, much work has been done on turbulence in thin rotating layers. By now the words geostrophic (and quasi-geostrophic) turbulence resound through geophysical fluid dynamics, bringing a more sceptical view of the notions of eddy viscosity with them. By making the vertical diffusion almost negligible, we may have been stacking the decks, but the results have provided a useful starting point for discussion of the place of origin of the solar cycle.

There was also other, indirect evidence of the tachocline's role in the solar cycle. Galloway & Weiss (1981) remarked that a strong toroidal field of some thousands of gauss tucked into a layer some tens of thousands of kilometres thick would carry as much magnetic flux as protruded out of the Sun during sunspot maximum. This suggests that the solar cycle is to some extent a sequence of rearrangements of the field in the tachocline whose stretching compensates for the minor dissipative losses. Parker (1979) gives a nice description of the basic physics of the breakup of a magnetic layer and the nonlinear development is vividly demonstrated in the computations of Cattaneo & Hughes (1988) that give an idea of the nature of such rearrangements. Weiss and I (1980) invoked such a rearrangement process driven by instability in the transition layer, and not just for its contribution to the emergence of the fields that make spots. We suggested that slight variations in the solar constant could be caused by the roughly decennial disruption of the sublayer. They may do this but we rather overestimated the strength of the effect (Gough 1981; Gilliland 1982).

Some objected to the idea that global changes in the tachocline could occur in tens of years when its thermal time is more like a million years. However, a short timescale is possible if the disturbances are not spherically symmetric. There are many timescales in the turbulence driving the process and I expect the tachocline to respond on the timescale that it can most readily accommodate. In setting up such a thermal impedance match we need to find the characteristic adjustment time of the tachocline. This is a thermal analogue of the spin-up time that I call the warm-up time.

In spin-up, the viscous effects operating in the boundary layer and the circulation in the fluid interior act in series to give a total timescale that is the geometric mean of the two characteristic times. In warm-up, the vertical passage of a disrupting influence, such as a sheet of magnetic field, makes an internal boundary layer that locally disrupts the hydrostatic equilibrium. This operates in series with thermal diffusion in the layer and, as in spin-up, the timescale for the process is the geometric mean of the times of the basic contributing mechanisms (Spiegel 1987). The tachocline is thermally compliant to *non-radial* disturbances on a timescale of a decade or two.

So end my discursive recollections of the reasons that led me to believe in the tachocline before I had any real right to. Fortunately, much of this reasoning is no longer needed now that we can be confident that the tachocline is there. What we ought to do now is to better explain why it is there.

2.6 From recollections to reflections

By the time this volume appears, the ideas of the contributors shall certainly have developed beyond those expressed in the talks (in November 2004) on which their chapters are supposedly based. Some of the work I shall now mention as desirable may have already been carried out. So much the better. I will nevertheless risk superfluity and attempt to carry out my assigned reflections. After all, that is what such meetings are for. Still, those who do not like standup astrophysics should perhaps avert their eyes at this point.

I suppose that the foremost issue that we ought to confront is the nature of the flow that produces the tachocline. The tachocline studies that I am aware of take the differential rotation of the convection zone as given, say by observation, and use this as a kinematic boundary condition to compute the resulting flow. This is like what oceanographers do when they use the curl of the wind stress to induce motions in the mixed layer and the thermocline. But, just as the oceanographers are enlarging their view of things, we need to move toward analysis in which the convection zone, the tachocline and the radiative interior are treated as a single system. The flow coupling these regions will affect the large-scale motions of all of them. The motions in the tachocline driven by convective pumping can induce horizontal temperature inhomogeneity on a large scale. In its turn, the convection zone may respond to local warming with updrafts that become part of the convective pumping mechanism. Accurate simulations of such processes may not be possible in realistic parameter ranges without drastic approximations, yet some reduced version of this kind of calculation is now possible with the available means and is likely to be illuminating.

If it is true that the solar cycle originates in the tachocline, there can be little doubt that the process is hydromagnetic. This is suggested by the observed correlation of the solar-surface rotation rate with solar activity (Ribes & Nesme-Ribes 1993). The rough image is that the plunging plumes of large-scale (non-Boussinesq) convection at very high Rayleigh numbers bring down magnetic fields from the convection zone. These torrents may be seen in films such as are shown by Malagoli *et al.* (1990). The magnetic fields raining onto the tachocline are stretched out into a shearing toroidal structure. We then have a magnetic boundary layer reminiscent of the boundary layers in ordinary shear flow and this analogy may guide us to what happens next. Horseshoe vortices form in the boundary layers of laboratory shear flow in a way whose understanding is developing apace (Waleffe 2003). Magnetic horseshoes (or hairpins, if you prefer) should arise similarly in the tachocline and float to the solar surface, twisting as they go as in what Elsasser used to call Parker's bathtub mechanism. The twisting may help the arches to maintain their discrete identities rather than being splayed out like the horseshoes in shear turbulence as they enter the body of the turbulent fluid.

The return of the magnetic field to the convection zone in horseshoe flux tubes takes place mainly in the spot latitudes while the field rains down in other latitudes. Why this mechanism leads to a relatively ordered cycle rather than a jumbled magnetic climatology is not evident and I will not try your patience with reference to another analogy at this point. I have already strayed beyond what has been learned from the fluid dynamics equations. My excuse, if one is needed, is that I feel that we can enhance our understanding of the full process by studying isolated features such as those I have just described. We can also grasp at clues that seem to have no obvious explanation. For example, I am subliminally aware of reports that large spot groups recur at particular longitudes. How can such places be marked? It seems unlikely that the markers could be the fields themselves since we would see more or less permanent spots in those places.

Since a rotating, turbulent layer like the tachocline may well be expected to produce vortices like those that form in the weather layers of planets, might there be semipermanent vortices like Jupiter's red spot in the tachocline? Then, every time the wave of magnetic excitation in the tachocline (Proctor & Spiegel 1991) passes by, the field would be wrapped up and extruded into the convection zone as in Weiss's (1966) flux expulsion calculations or some that Steve Meacham performed to test these notions. This process does not seem likely to lead directly to spot groups and I mention it as a possible trigger for a more promising mechanism, like horseshoes. It also recalls the question that I touched on at the beginning: vortex formation versus magnetic flux tube formation. The two processes seem mutually exclusive as we see in Jupiter and the Sun, though there is an occasional solar exception (Akasofu 1985). We may learn more about this when it becomes possible to see the differential rotation in the tachocline with some precision. Though the differential rotation on the solar surface does not at all resemble the analogous flows in the atmospheres of the giant planets (Dowling & Spiegel 1990), that of the tachocline may well resemble what we see in those atmospheres. In any case there seems to be good reason to simulate the passage from the fluid dynamical to the MHD case and to study the crossover from vortex formation to flux tube formation. Observationally, the transition should appear at some intermediate mass in the brown dwarf sequence.

Another possible transition that I find interesting is the disappearance of obvious cyclic activity in fully convective stars. For those objects there would be no tachocline and solar-type cyclic activity should not be detected. This is a prediction (by now, a postdiction Joe Patterson tells me) that we are led to by tachoclinic studies. Yet, there may be a compensation for this loss.

There is a relatively thin shear layer at the top of the convection zone (Basu & Antia 2001) that I have discussed with Kumar Chitre. Would stars with larger convective structures than the sun form deeper shear layers of this kind and produce spots in them? Such a layer at the top of a convection zone is not fed by plunging plumes, but it might produce some sort of coherent structures nonetheless.

I must confess that I am much more enthusiastic about the prospects of extending these considerations to activity in the hottest stars. Here the situation is even more complicated than is sometimes appreciated. Those stars are rapid rotators and they have convective cores that are likely to produce convective dynamos. If the issues raised in this volume have any generality, we may expect to find some kind of transition layers at the tops of those convective cores. I suspect that this suggestion will raise no eyebrows, but the situation in the atmospheres of hot stars is even more difficult to anticipate.

Cassinelli (1985) has summarized the observational evidence for his belief that spots are formed on hot stars. This is a subject that needs a meeting of its own, but it is worth mentioning its possible relation to our present concerns. There are many timescales of variation in hot stars and an interesting case is that of the Hubble–Sandage variables, a.k.a. LBVs (Humphreys & Davidson 1994). Their various time scales include tens or hundreds of years. What could produce such characteristic times? Perhaps the source of this activity can be traced to Struve's well known discovery of (what he called) macroturbulence. While those stars do not satisfy the usual convective instability criterion, there is no shortage of possible instability mechanisms that one might invoke to rationalize the intense (apparently supersonic) velocity fields that cause the line broadening Struve reported.

Hot young stars rotate rapidly and they are likely to be subject to baroclinic instability. They are also subject to instabilities induced by radiative forces (e.g. Spiegel & Tao 1999). For this, and possibly other reasons, they pulsate as well. A strong enough pulsation, in its turn, can cause convection through parametric instability (Poyet & Spiegel 1979) and this will lead to photon bubbles (as briefly described in Spiegel 2005). We may also go overboard and consider that the conjectured transition layer at the edges of the convective cores could produce magnetic bubbles that twist off and rise up to produce photomagnetic cauldrons at the stellar surfaces. If transition layers form at the bottoms of these turbulent atmospheres, cycle timescales of variation on the order of decades or perhaps centuries might seem reasonable to one who does not know precisely how to determine cycle times.

Finally, we should recall that, during its first thirty million years or so on the Main Sequence, the Sun had a convective core. With a rotation period of a day, a healthy convective dynamo ought to have operated. What are the consequences? If we could catch a star in that early state, we might observe large fluctuations in luminosity and asteroseismologists would certainly have a fine time. I do not expect us to be so lucky very soon. Still, we may think about the fate of that (presumed) early magnetic field. Will it survive all this time or will it have found its way out by some form of instability? If it is still there, it may be just what Gough & McIntyre (1998) want to render the radiative core rigid. (As I am about to send this off I am downloading a paper by Brun and Zahn destined for *A & A*. They rediscuss the issues raised by Gough and McIntyre, so we shall still have much to discuss even after reading the present volume.) At any rate, if a large-scale field is still in the core, we would need to think about whether it can not only keep the tachocline thin but how the tachocline can reconnect at the rate needed to produce a variation of 11 years. So it seems that in this end is our beginning.

As you have no doubt noticed, many of my reflections are aimed at suggesting that we look at tachoclinic issues in a wider context that includes other stellar types and other solar epochs. Both observationally and theoretically, we are likely to learn the most if the stellar, solar, planetary and geophysical studies are kept in close touch. In my own case, I have learned much from the many collaborators I have mentioned here and from a large number of friends and colleagues from many disciplines who have kindly offered opinions and insights. Above all, it has been a great pleasure to have editors who actually edit and do it well.

References

Akasofu, S.-I. (1985). *Planet. Space Sci.*, **33**, 275.

Basu, S. & Antia, H. M. (2001). *Mon. Not. Roy. Astron. Soc.*, **324**, 498.

Batchelor, G. K. & Townsend, A. A. (1949). *Proc. R. Soc. Lond.*, A**199**, 238.

Beer, J., Tobias, S. & Weiss, N. (1998). *Solar Phys.*, **181**, 237.

Bondi, H. & Lyttleton, R. A. (1948). *Proc. Camb. Phil. Soc.*, **44**, 345.

Brandt A. & Fernando, H. J. S. (eds.) (1995). Double-diffusive convection, *Geophys. Monograph* **94** Washington D.C.: American Geophysics Union.

Bretherton, F. P. & Spiegel, E. A. (1968). *Astrophys. J.*, **153**, L77.

Brown, T. M., Christensen-Dalsgaard, J., Dziembowski, W. A. *et al.* (1989). *Astrophys. J.*, **343**, 526.

Brun, A. S., Turck-Chièze, S. & Zahn, J.-P. (1999). *Astrophys. J.*, **525**, 1032.

Cassinelli, J. P. (1985). In *The Origin of Nonradiative Heating/Momentum in Hot Stars* (NASA 2358), ed. A. B. Underhill & A. G. Michalitsanos, p. 2.

Cattaneo, F. & Hughes, D. W. (1988). *J. Fluid Mech.*, **196**, 323.
Chaboyer, B. (1990). In *Stellar Fluid Dynamics*, ed. R. Salmon & B. DeRemer (WHOI-91-03, Woods Hole), p. 267.
Chaboyer, B. & Zahn, J.-P. (1992). *Astron. Astrophys.*, **253**, 173.
Charbonneau, P., Michaud, G. & Proffitt, C. R. (1989). *Astrophys. J.*, **347**, 821.
Cowling, T. G. (1951). *Astrophys. J.*, **114**, 272.
Cowling, T. G. (1969). *Observatory*, **89**, 217.
Dicke, R. H. (1964). *Nature*, **202**, 432.
Dicke, R. H. & Goldenberg, H. M. (1967). *Phys. Rev. Lett.*, **18**, 313.
Dowling, T. E. & Spiegel, E. A. (1990). *Ann. N. Y. Acad. Sci.*, **617**, 190.
Eddy, J. A. (1976). *Science*, **192**, 1189.
Fricke, K. (1968). *Zs. Astrophys.*, **68**, 317.
Galloway, D. J. & Weiss, N. O. (1981). *Astrophys. J.*, **243**, 945.
Gilliland, R. L. (1982). *Astrophys. J.*, **253**, 399.
Goldreich, P. & Schubert, G. (1967). *Astrophys. J.*, **150**, 571.
Gough, D. O. (1981). In *Variations of the Solar Constant*, ed. S. Sofia. New York: G.I.S.S., p. 185.
Gough, D. O. & McIntyre, M. E. (1998). *Nature*, **394**, 755.
Greenspan, H. P. & Howard, L. N. (1963). *J. Fluid Mech.*, **22**, 449.
Hardenberg, J. G., Paparella, F., Platt, N. *et al.* (1997). *Phys. Rev. E*, **55**, 58.
Holton, J. R. (1965). *J. Atmos. Sci.*, **22**, 402.
Howard, L. N., Moore, D. W. & Spiegel, E. A. (1967). *Nature*, **214**, 1297.
Humphreys, R. M. & Davidson K. (1994). *Pub. Astron. Soc. Pacific*, **106**, 1025.
Ingersoll, A. P. (1990). *Science*, **248**, 308.
Jones, C. A., Weiss, N. O. & Cattaneo, F. (1985). *Physica*, **14D**, 161.
Kraft, R. P. (1967). *Astrophys. J.*, **150**, 551.
Lebovitz, N. & Lifschitz, A. (1993). *Astrophys. J.*, **40**, 603.
Leighton, R. B., Noyes, R. W. & Simon, G. W. (1962). *Astrophys. J.*, **135**, 474.
Lorenz, E. N. (1963). *J. Atmos. Sci.*, **20**, 131.
McIntyre, M. E. (2002). In *Meteorology at the Millennium*, ed. R. P. Pearce. UK: Academic Press & Royal Meteorology Society, p. 283.
Malagoli, A., Cattaneo, F. & Brummell, N. H. (1990). *Astrophys. J.*, **321**, L33.
Malkus, W. V. R. (1972). *Eos, Trans. Amer. Geophys. Union*, **53**, 671.
Mestel, L. (1999). *Stellar Magnetism*. Oxford: Clarendon Press.
Moore, D. W. & Spiegel, E. A. (1966). *Astrophys. J.*, **143**, 871.
Parker, E. N. (1979). *Cosmical Magnetic Fields*. Oxford: Clarendon Press.
Pasquero, C. (1996). *Modelli di variabilità solare* (Tesi di Laurea, Facultà di Scienza M.F.N., Università di Torino).
Pedlosky, J. (1969). *J. Fluid Mech.*, **56**, 401.
Platt, N. (1990). In *Stellar Fluid Dynamics*, ed. R. Salmon & B. DeRemer (WHOI-91-03, Woods Hole, MA), p. 320.
Platt, N., Spiegel, E. A. & Tresser, C. (1993a). *Phys. Rev. Lett.*, **70**, 279.
Platt, N., Spiegel, E. A. & Tresser, C. (1993b). *Geophys. Astrophys. Fluid Dyn.*, **73**, 147.
Pomeau, Y. & Manneville, P. (1990). *Commun. Math. Phys.*, **74**, 1889.
Poyet, J.-P. & Spiegel, E. A. (1979). *Astron. J.*, **84**, 1918.
Proctor, M. R. E. & Spiegel, E. A. (1991). In *The Sun and the Cool Stars*, ed. D. Moss, G. Rüdiger & I. Tuominen. New York: Springer-Verlag, p. 117.
Ribes, J. C. & Nesme-Ribes, E. (1993). *Astron. Astrophys.*, **276**, 549.
Schatzman, E. (1962). *Ann. d'Astrophysique*, **25**, 18.
Schmitt, D., Schüssler, M. & Ferriz-Mas, A. (1996). *Astron. Astrophys.*, **311**, L1.

Skumanich, A. (1972). *Astrophys. J.*, **171**, 565.
Spiegel, E. A. (1964). *Astrophys. J.*, **139**, 959.
Spiegel, E. A. (1968). In *Highlights of Astronomy*, ed. L. Perek. Dordrecht: Reidel, p. 261.
Spiegel, E. A. (1969). *Comm. Astrophys. Space Phys.*, **1**, 57.
Spiegel, E. A. (1972). In *Physics of the Solar System*, ed. S. I. Rasool (NASA SP-300), p. 61.
Spiegel, E. A. (1981). *Ann. N.Y. Acad. Sci.*, **357**, 305.
Spiegel, E. A. (1987). In *The Internal Solar Angular Velocity*, ed. B. R. Durney & S. Sofia. Dordrecht: Reidel, p. 321.
Spiegel, E. A. (2005). In *Mechanics of the 21st Century*, ed. W. Gutkowski & T. Kowalewski. Dordrecht: Springer, p. 365.
Spiegel, E. A. & Tao, L. (1999). *Phys. Rep.*, **311**, 163.
Spiegel, E. A. & Weiss, N. O. (1980). *Nature*, **287**, 616.
Spiegel, E. A. & Wolf, A. N. (1987). In *Chaotic Phenomena in Astrophysics*, ed. J. R. Buchler & H. Eichhorn, *Ann. N. Y. Acad. Sci.*, **497**, 55.
Spiegel, E. A. & Zahn, J.-P. (1992). *Astron. Astrophys.*, **265**, 106.
Stern, M. E. (1960). *Tellus*, **14**, 747.
Stommel, H., Aarons, A. B. & Blanchard, D. (1956). *Deep-Sea Res.*, **3**, 152.
Townsend, A. A. (1957). *J. Fluid Mech.*, **4**, 361.
Waleffe, F. (2003). *Phys. Fluids*, **15**, 1517.
Weiss, N. O. (1966). *Proc. R. Soc. Lond., A***293**, 310.
Yih, C.-S. (1961). *Phys. Fluids*, **4**, 806.
Zahn, J.-P. (1990) In *Stellar Fluid Dynamics*, ed. R. Salmon & B. DeRemer (WHOI-91-03, Woods Hole), p. 52.

Part II
Observations

3

Observational results and issues concerning the tachocline

Jørgen Christensen-Dalsgaard & Michael J. Thompson

The region near and just below the solar convection zone is characterized by a strong shear in rotation rate, between the latitudinally differential rotation in the convection zone and the nearly uniform rotation of the radiative interior. This so-called *tachocline* is also a region of substantial uncertainty in the modelling of solar structure, where convective overshoot and rotationally induced mixing may affect the thermal and compositional structure. Helioseismology led to the identification of the rotational shear and has provided fairly detailed information about the properties, structure and rotation of the tachocline, although unavoidably at somewhat limited resolution. Here we briefly discuss the techniques used in the helioseismic analyses and review the results of such analyses, as a background for the modelling of the properties of the tachocline and its effects on the generation of the solar magnetic field.

3.1 Introduction

As will be abundantly evident from other articles in this volume, knowledge of the solar internal rotation is essential for understanding solar magnetic activity, as it is for understanding important aspects of solar structure and evolution. Before the advent of helioseismology little was known about solar rotation below the surface, beyond the indication, from the surface latitudinal differential rotation, that it was non-uniform. Modelling of the evolution of solar rotation, from an assumed earlier state of rapid rotation, indicated that the Sun might still have a rapidly rotating core, (e.g. Dicke 1964; detailed modelling of the solar spin-down by Pinsonneault *et al.* 1989 also obtained a core rotating at several times the surface rate). Some constraints on the internal rotation were obtained from the observed solar surface oblateness (Hill & Stebbins 1975). Models of the interaction between rotation and convection in the convection zone, matching the surface differential rotation (e.g. Glatzmaier 1985; Gilman & Miller 1986), showed rotation that depended only on

The Solar Tachocline, D. W. Hughes, R. Rosner and N. O. Weiss.
Published by Cambridge University Press.

distance to the rotation axis, i.e. 'rotation on cylinders', in accordance with the Taylor–Proudman theorem (Pedlosky 1987). The resulting radial variation of the angular velocity within the convection zone appeared difficult to reconcile with dynamo models of the solar magnetic activity (e.g. Gilman 1986).

Already the first reliable helioseismic inference of solar rotation in much of the Sun, in the equatorial region (Duvall *et al.* 1984), showed marked deviations from these notions. In much of the radiative interior rotation was if anything slightly *below* the surface equatorial rate. Rotation of the core was poorly determined, with only barely significant evidence for a slight speed-up relative to the rest of the radiative interior. Early inferences of the latitudinal variation of rotation (Brown 1985; Duvall *et al.* 1986; Brown & Morrow 1987) indicated that the latitudinal variation was largely confined to the convection zone. More detailed results obtained from such analyses (Christensen-Dalsgaard & Schou 1988; Kosovichev 1988; Brown *et al.* 1989; Dziembowski *et al.* 1989) strengthened this conclusion and provided the first evidence that the transition between the latitudinally varying rotation in the convection zone and the nearly uniform rotation in the radiative interior takes place in a relatively narrow region, located near the base of the convection zone. Spiegel & Zahn (1992) presented an initial analysis of the dynamics of this region and named it the *tachocline*. Gilman *et al.* (1989) noted that this new insight into the properties of the solar internal rotation had important consequences for the understanding of the solar magnetic activity.

The dynamics of the tachocline is closely linked to the establishment of the solar internal rotation. It is generally assumed that the envelope rotation rate of lower-mass stars decreases as they age, owing to the loss of angular momentum in a magnetic stellar wind (see Mestel 1999); this is confirmed by the observed anti-correlation between age and rotation rate for solar-like stars (see Skumanich 1972). The angular-momentum loss presumably directly affects the outer convection zone; the effect on the radiative interior depends on the coupling of the interior to the convection zone, through the tachocline. Also, the properties of the tachocline play an important role in most current dynamo models aiming at explaining the generation and variation of the solar magnetic field. From the point of view of solar structure this is an interesting transition region with likely penetration, although to an unknown extent, of motion below the convection zone. Evidence for such motion comes from the reduction by a factor of around 150 in the abundance of lithium in the solar atmosphere, relative to the meteoritic abundance which presumably reflects the pre-solar nebula from which the Sun was formed (Anders & Grevesse 1989). This requires mixing, at some stage during solar evolution, from the convection zone down to a region where the temperature exceeds 2.5×10^6 K, substantially higher than the temperature at the base of the convection zone during Main-Sequence evolution (e.g. Christensen-Dalsgaard *et al.* 1992).

Although helioseismology is a powerful probe of solar internal properties, the achievable resolution and precision is unavoidably limited by the properties of the observed modes and the quality, however excellent, of the observations. As discussed elsewhere in this volume, modelling of the tachocline and its effects on the dynamo processes depends sensitively on the details of the thickness of the tachocline and its location relative to the base of the convectively unstable region, at a level which is barely resolved in current helioseismic inferences. Also, it is of evident interest to investigate to what extent the properties of the tachocline change with time, on solar-cycle or other timescales. Addressing these issues requires careful attention to the properties of the analysis of the data, including the inverse analysis, and the currently unavoidable limitations in the inferences must be kept in mind in the use of the results. Our goal with the present chapter is to provide a brief overview of the techniques used to obtain information about the solar interior, particularly the tachocline region, and discuss the results obtained so far, emphasizing their strengths and weaknesses. More extensive reviews on solar oscillations, helioseismology and the properties of solar structure and rotation have been given by, for example, Gough (1993), Christensen-Dalsgaard (2002), Thompson *et al.* (2003) and Miesch (2005).

3.2 Helioseismic techniques

The Sun oscillates simultaneously in many global resonant modes; through helioseismic analysis these provide observational constraints on the structure and dynamics of the solar interior. Because the Sun's structure is nearly spherical, the horizontal structure of each global mode is described to a very good approximation by a spherical harmonic $Y_l^m(\theta, \phi)$, where θ and ϕ are respectively the heliocentric colatitude and longitude. The integers $l \geq 0$ and m $(-l \leq m \leq l)$ are called the degree and azimuthal order of the mode. The description of the modes of a spherically symmetric star is completed by a third integer, the order n: the absolute value of n is approximately equal to the number of nodes in the radial direction in the mode's eigenfunction for, say, the radial displacement. The frequencies ω_{nlm} of the modes are increasing functions of n at fixed l and m. (We use both the angular frequency ω and the cyclic frequency $\nu = \omega/2\pi$ in this chapter.) Those global modes that have been unambiguously observed on the Sun are p modes, for which $n > 0$ and for which the primary restoring force is pressure, and f modes, for which $n = 0$ and which at high degree have the character of surface gravity waves.

The p-mode frequencies are sensitive primarily to conditions in the interior within an acoustic cavity. This is the region of the Sun in which the waves comprising the mode have the nature of propagating – as opposed to evanescent – waves. The acoustic cavity of the observed p modes extends from essentially the surface down to the radius r at which the horizontal phase speed $\omega r/(l+1/2)$ is equal to the adiabatic sound speed $c(r)$: this is the location of the lower turning point of the mode. Since c/r is an increasing function of depth, it follows that low-degree modes (small l) penetrate more deeply than high-degree modes, for fixed frequency. It is therefore the observed p modes with degrees smaller than about 40 that have direct sensitivity to the tachocline region, since it is these modes whose acoustic cavity includes the region of the tachocline. The f modes, as do deep-water waves, decay with depth on a scale of roughly $R_\odot/l$ and hence the observed f modes, with $l \gtrsim 100$, do not reach the tachocline region; however, they are important in providing constraints on the near-surface rotation. We note that for both types of modes the cavity has a latitudinal extent also, from the equator to the heliocentric colatitudes θ at which $\sin\theta = m/(l+1/2)$.

For a spherically symmetric star the frequencies would be independent of the azimuthal order m. Symmetry-breaking agents such as rotation, magnetic fields and structural asphericities raise that degeneracy. For the Sun the effect of rotation on the modes is adequately described for the present purpose by a first-order approximation:

$$\omega_{nlm} = \omega_{nl0} + m \int_0^{R_\odot} \int_0^{\pi} K_{nlm}(r,\theta)\Omega(r,\theta) r \mathrm{d}r \mathrm{d}\theta. \tag{3.1}$$

This effect is an odd function of m, and hence distinguishable from the effects of asphericities which do not distinguish between eastward- and westward-propagating waves and which thus give rise to frequency perturbations that are even functions of m. Here Ω is the solar internal angular velocity, radial coordinate r is the distance from the centre of the Sun, $R_\odot$ is the radius of the photosphere, and kernels K_{nlm} are functions of the spherically symmetric structure of the Sun. The kernels are generally assumed to be known functions to adequate accuracy, so that Equation (3.1) provides observational constraints on the unknown angular velocity $\Omega(r,\theta)$ inside the Sun. The kernels are north–south symmetric about the solar equatorial plane, so the data

$$d_{nlm} \equiv \frac{1}{2m}(\omega_{nlm} - \omega_{nl-m}) \tag{3.2}$$

are sensitive only to the north–south symmetric component of the internal rotation.

The kernels corresponding to different modes (n, l, m) have different sensitivities to the solar interior, in line with the variation of acoustic cavities between different modes. This variation in sensitivity permits the use of helioseismic inverse techniques to make inferences about the rotation as a function of position inside the Sun (see Gough 1985; Kosovichev 1999). One such technique is optimally localized averages (OLA): taking a linear combination of constraints (3.1) for different modes gives

$$\sum_{nlm} c_{nlm} d_{nlm} = \int_0^{R_\odot} \int_0^{\pi} \sum_{nlm} c_{nlm} K_{nlm} \Omega r \mathrm{d}r \mathrm{d}\theta, \tag{3.3}$$

where the coefficients $c_{nlm}(r_0, \theta_0)$ are chosen so that the averaging kernel

$$\mathcal{K}(r_0, \theta_0; r, \theta) \equiv \sum_{nlm} c_{nlm}(r_0, \theta_0) K_{nlm}(r, \theta) \tag{3.4}$$

has unit integral and is localized so that it has substantial amplitude near $r = r_0$, $\theta = \theta_0$ and is small elsewhere. In this case the left-hand side of Equation (3.3) may be regarded as an estimate of the localized average $\bar{\Omega}(r_0, \theta_0)$ of the rotation Ω in the vicinity of the target location (r_0, θ_0). The form of that average, often called the 'solution', is described by the averaging kernel. There are various ways in which the OLA coefficients may be chosen: see for example Pijpers & Thompson (1992).

By varying the target location, a map of the rotation rate in the solar interior can be built up. Its strict interpretation is always best understood in terms of the averaging kernels. However, a successfully localized averaging kernel generally has a peak near the target location, and it is possible to define measures of, say, the radial and latitudinal resolution achieved at that location in terms of the radial and latitudinal widths of that peak (Schou *et al.* 1994). Some typical averaging kernels for a solar rotation inversion are shown in Figure 3.1.

In reality, the data d_{nlm} contain noise: these errors propagate through to $\bar{\Omega}$. If the error in each datum is ϵ_{nlm} and this has standard deviation σ_{nlm}, then the standard deviation of the error in $\bar{\Omega}$ is

$$\left(\sum_{nlm} c_{nlm}^2 \sigma_{nlm}^2 \right)^{1/2}.$$

Since the solutions $\bar{\Omega}$ at different target locations, (r_1, θ_1) and (r_2, θ_2) say, are built from the same data d_{nlm}, the errors in the solutions at the two locations will be correlated. The error correlation is in principle straightforward to compute, given the coefficients $c_{nlm}(r_1, \theta_1)$, $c_{nlm}(r_2, \theta_2)$ and the standard deviations of the data errors (Howe & Thompson 1996).

The solution of any linear inversion of the rotational splitting data may similarly be written in the form of Equation (3.3). Likewise, averaging kernels can be defined as above and error propagation understood in terms of the weights given to the data. One such inversion technique, also widely used in helioseismic studies, is regularized least squares (RLS) inversion, in which the data are fitted by finding a solution function $\Omega(r, \theta)$ which minimizes the weighted sum of the chi-squared misfit to the data and a regularization term which could for example be large if the

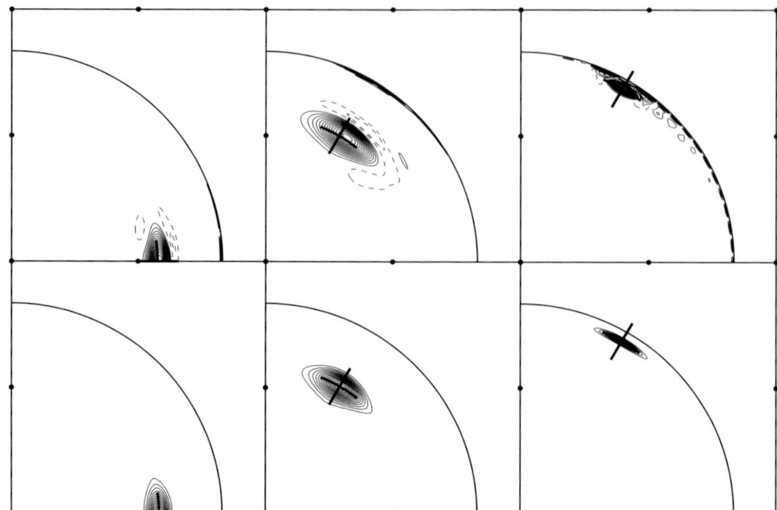

Figure 3.1. A few typical averaging kernels for an RLS inversion (top row) and for an OLA inversion (bottom row). The target location to which each kernel corresponds is indicated with a cross. The target locations are at the base of the convection zone at the equator (left column) and at latitude 60° (middle column), and at radius $0.95R_{\odot}$ and latitude 60° (right column). Solid curves indicate positive contour values, broken curves negative contour values.

second derivatives of the solution in the radial or latitudinal directions are large (Schou *et al.* 1994, 1998).

Such inversions have succeeded in mapping the rotation throughout much of the solar interior. However, the finite resolution of the inversions, as evidenced by the spatial extent of the averaging kernels (Figure 3.1), means that a rather sharp transition such as the tachocline will appear wider in the results than it is in reality (e.g. Thompson 1990). This effect is illustrated in Figure 3.2, where we show the results of inverting splittings data for an artificial rotation profile possessing a tachocline. The tachocline appears more spread out in the inversion results than in the original input profile. To correct for this effect of resolution, one may attempt to 'deconvolve' the averaging kernel from the solution in the vicinity of the tachocline (Charbonneau *et al.* 1999a), but this means that one must presume some particular parametrized shape for the rotation profile across the tachocline, which introduces a prejudice. Similarly, it is possible to fit the frequency splittings with a highly prescribed shape for the rotation profile: this is sometimes referred to as forward modelling. Alternatively, Corbard *et al.* (1998) devised a nonlinear inversion technique which was better adapted to handling sharp variations in the underlying rotation profile and did not smear out the tachocline to the same extent as the linear inversions do.

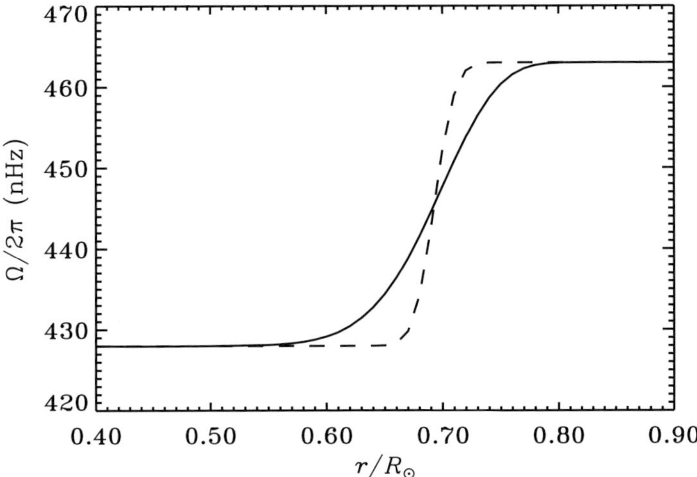

Figure 3.2. An illustration of the broadening of the apparent width of the tachocline caused by the finite resolution of one of our inversion methods. The dashed curve shows a cut through the latitudinally independent rotation profile used to generate a set of artificial data (for a mode set similar to that used in the inversion of MDI data shown in Figure 3.8). The solid curve shows an equatorial cut through a SOLA inversion of the artificial data.

Linear inversion techniques similar to those used to infer the solar rotation can be used to make inferences about the Sun's internal structure, for example the variation of adiabatic sound speed with radial coordinate. The dependence of the frequencies on the structure is however inherently nonlinear, and so to use these techniques one typically poses the problem of determining the solar structure on the assumption that the structure is a small perturbation about that of a known reference model. In the simplest but most common case in which one seeks to determine just the spherically symmetric component of the structure of the solar interior, the relevant data to use are the mean multiplet frequencies (i.e. averaged over all m values, for fixed n and l). Using the assumption that the solar interior is in hydrostatic equilibrium, the relative difference in frequency $\delta\omega/\omega$ between the Sun and the reference model can be expressed in terms of differences f_1 and f_2 in two seismically relevant properties of the structure, thus:

$$\left.\frac{\delta\omega}{\omega}\right|_{nl} = \int_0^{R_\odot} \left[K_1^{nl}(r)f_1(r) + K_2^{nl}(r)f_2(r)\right] \mathrm{d}r + \mathcal{F}_{\rm surf} \qquad (3.5)$$

(e.g. Gough & Kosovichev 1988; Dziembowski *et al.* 1990; Gough & Thompson 1991). The term $\mathcal{F}_{\rm surf}$ accounts for near-surface differences between the Sun and model: this is taken to have the form of a mode-weighted function of frequency. In Equation (3.5) f_1 and f_2 may be chosen so that for example $(f_1,f_2) = (\delta c/c, \delta\rho/\rho)$, where $\rho(r)$ is the density; or $(f_1,f_2) = (\delta u/u, \delta\gamma_1/\gamma_1)$, where $u = p/\rho$, $p(r)$ being the pressure, and $\gamma_1(r)$ is the first adiabatic exponent (see Section 3.3). With additional assumptions, such as that the equation of state of the solar plasma is known, the inversion may be posed in terms of properties that affect the frequencies indirectly, for example $(f_1,f_2) = (\delta c/c, \delta Y)$, where $Y(r)$ is the fractional abundance of helium.

The structure inversions are therefore somewhat more complicated conceptually than the inversions for rotation, since two unknown functions have to be determined and so too does the function accounting for different surface properties. However, this does not alter the principles we have discussed above. For details, see Dziembowski *et al.* (1990), Däppen *et al.* (1991), Basu *et al.* (1997), Kosovichev (1999), Rabello-Soares *et al.* (1999).

Sharp transitions at some depth in the Sun of some property may give rise to a characteristic periodic signal in the frequencies of modes that penetrate well beneath that depth (e.g. Gough & Thompson 1988; Vorontsov 1988; Gough 1990). Such a transition may occur in the temperature gradient or its derivative at the base of the adiabatically stratified convective envelope, or in the chemical abundances beneath the convection zone, and both of these would cause a similar transition in the sound speed. Specific approaches to detect and interpret such a characteristic signal have been developed, and are addressed further in Section 3.5.

In order to map the properties of the tachocline it is necessary to use observational frequencies (for radial structure) and frequency splittings (for rotation) for a set of modes whose lower turning points span the location of the tachocline and extend beneath it. Thus the most detailed inferences have used medium-degree data, from low degree (but not necessarily the lowest degrees) up to $l = 60$ or higher. The data have come from the observational instruments that resolve the solar disk and are thus sensitive to the medium-degree modes. Medium-degree data used to study the tachocline and the structure of the region include: the mode frequencies and splittings up to $l = 60$ observed in the summers of 1986, 1988, 1989 and 1990 from the Big Bear Solar Observatory (Libbrecht *et al.* 1990); those up to $l = 99$ measured since early 1994 with the LOWL instrument (Tomczyk *et al.* 1995); and data up to even higher l values observed since mid-1995 by the Global Oscillation Network Group (GONG) (Harvey *et al.* 1996) and since 1996 by the Michelson Doppler Imager (MDI) on board the SoHO satellite (Scherrer *et al.* 1995). Accurate data on the lowest-degree modes, crucial for constraining the rotation of the deep solar interior, have been obtained from spatially unresolved observations in the BiSON (Chaplin *et al.* 1996) and IRIS (Fossat 1991) ground-based networks and the GOLF instrument (Gabriel *et al.* 1995) on SoHO.

The frequency splittings are generally available in the form of so-called a-coefficients. These come from fitting the m-dependence of the frequencies within each multiplet by a polynomial in m:

$$\nu_{nlm} = \nu_{nl} + \sum_{j>0} a_j^{nl} \mathcal{P}_j^{(l)}(m). \tag{3.6}$$

The polynomials $\mathcal{P}_j^{(l)}$ have order j and have definite parity. The polynomials generally used now are those proposed by Ritzwoller & Lavely (1991) – see also Schou *et al.* (1994) for a discussion of their properties and their use in inversions.

The odd terms in this expression for the multiplet's frequencies are sensitive to the rotation. Thus the inversion for the internal solar rotation can proceed using just the odd a-coefficients. The a_1 coefficient is sensitive only to the spherically symmetric component of the rotation as a function of radius: the higher-order coefficients are sensitive to the rotation profile's latitudinal variations as well. The lowest three odd a-coefficients as measured by MDI are illustrated in Figure 3.3.

The mean multiplet frequency ν_{nl} is sensitive only to the spherically symmetric Sun, while the even a-coefficients sense structural asphericities (including centrifugal distortion of the Sun), second-order effects of rotation and the non-spherically symmetric effects of magnetic fields.

3.3 Solar internal structure and rotation

As discussed in Section 3.2, inferences of solar structure are most often obtained from analyses of differences between observed frequencies and those of a reference solar model, resulting in determination of averages of structural differences between the Sun and the model. A commonly used, although by now somewhat dated, reference model is Model S of Christensen-Dalsgaard *et al.* (1996). This used OPAL opacities (Iglesias *et al.* 1992) and equation of state (Rogers *et al.* 1996) and reaction rates predominantly from Bahcall & Pinsonneault (1995); the age of the present Sun, from the zero-age Main Sequence, is 4.6 Gyr. The model assumed a ratio $Z_s/X_s = 0.0245$ between the present surface abundances by mass, X_s and Z_s, of hydrogen and elements heavier than helium, respectively (Grevesse & Noels 1993). Diffusion and settling of helium and heavy elements were treated according to the formalism of Michaud & Proffitt (1993); helioseismic inferences have demonstrated the importance of including such effects (e.g. Christensen-Dalsgaard *et al.* 1993). A striking result of the settling of helium and heavier elements out of the convection zone is the establishment of sharp gradients in the abundances just beneath the convection zone (see Figure 3.4). The outer convection zone of the model extends to the distance $r_{cz} = 0.7115 R_\odot$ from the centre. Within the convection zone the stratification is essentially adiabatic, while below the convection zone it rapidly becomes substantially subadiabatic. This is illustrated in Figure 3.5a, in terms of the logarithmic gradient $\nabla = \mathrm{d}\ln T/\mathrm{d}\ln p$ of temperature T. In the convection zone

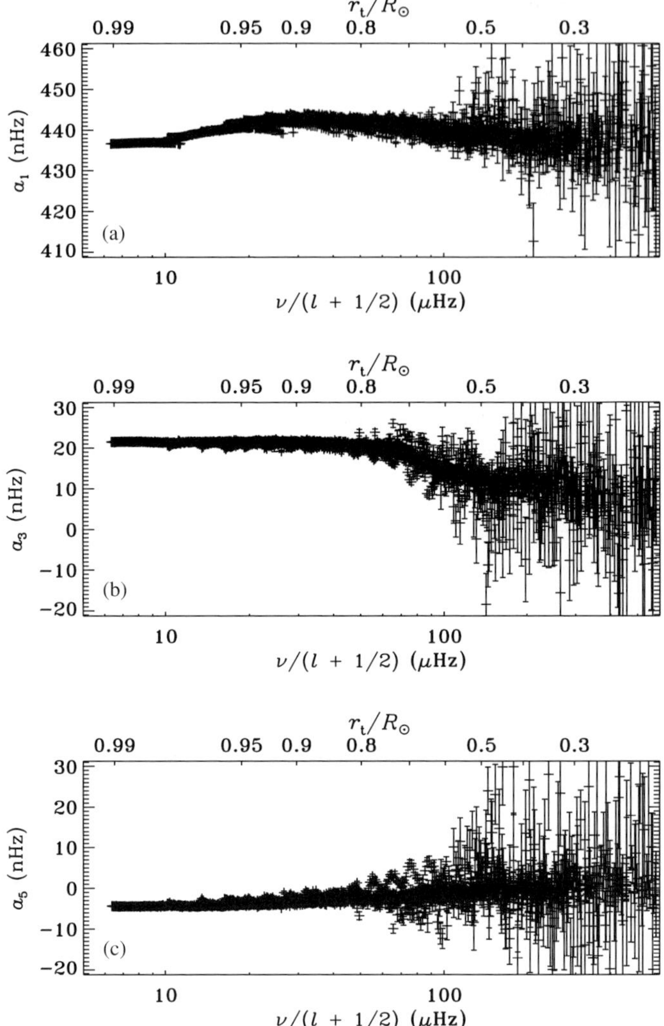

Figure 3.3. Odd a-coefficients (a) a_1, (b) a_3 and (c) a_5 from MDI data, with one-standard-deviation error bars. The bottom axis indicates the value of $\nu/(l + 1/2)$ for the modes, which maps onto the location $r_t/R_\odot$ of the lower turning point of the mode (top axis).

$\nabla \simeq \nabla_{\rm ad} = (\partial \ln T/\partial \ln p)_{\rm ad}$, the derivative being at constant specific entropy; since matter in the lower parts of the convection zone is well approximated by a fully ionized ideal gas, $\nabla_{\rm ad} \simeq 2/5$. Of greater relevance to the helioseismic analysis is the behaviour of the adiabatic sound speed c, given by

$$c^2 = \frac{\gamma_1 p}{\rho} \simeq \frac{\gamma_1 k_{\rm B} T}{m_{\rm u} \mu}, \tag{3.7}$$

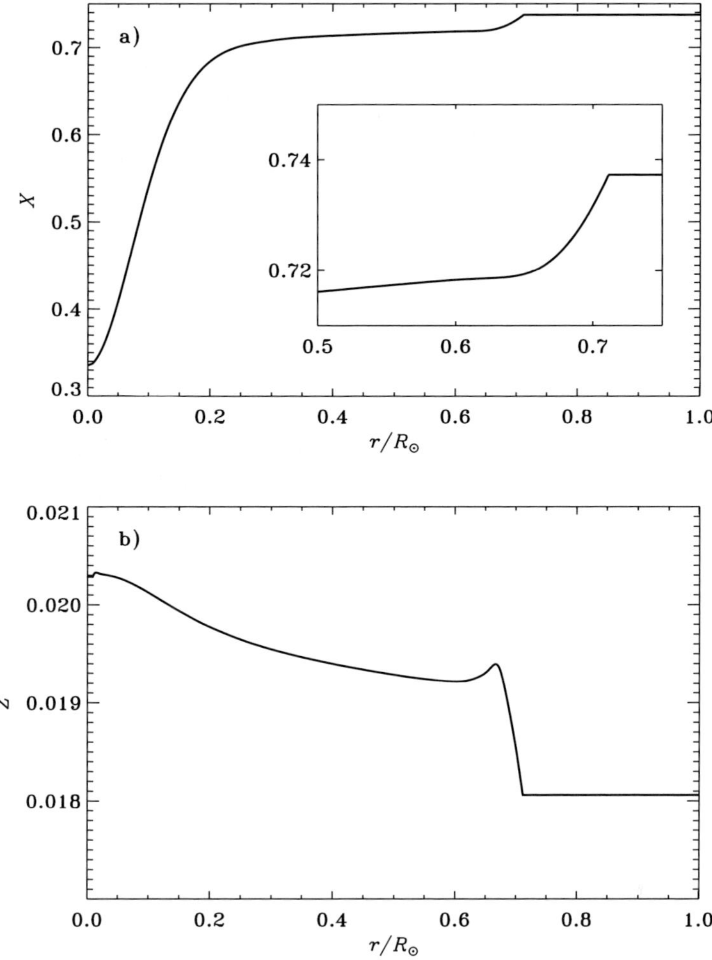

Figure 3.4. The mass fractions X of hydrogen and Z of elements heavier than helium in Model S of the Sun as a function of distance r to the centre, in units of the surface radius $R_\odot$ of the model. The insert in panel (a) shows a blow-up of the region near the base of the convection zone. The uniform composition in the convection zone and the effect of settling beneath the convection zone are evident.

where $\gamma_1 = (\partial \ln p / \partial \ln \rho)_{\text{ad}}$ is the adiabatic compressibility; the second approximation assumes the ideal gas law, k_{B} being Boltzmann's constant, m_{u} the atomic mass unit and μ the mean molecular weight. The gradient in c^2, corresponding to ∇, is illustrated in Figure 3.5b; in the ideal-gas approximation and neglecting the derivative of γ_1, it is given by

$$\frac{\mathrm{d} \ln c^2}{\mathrm{d} \ln p} \simeq \nabla - \frac{\mathrm{d} \ln \mu}{\mathrm{d} \ln p}. \tag{3.8}$$

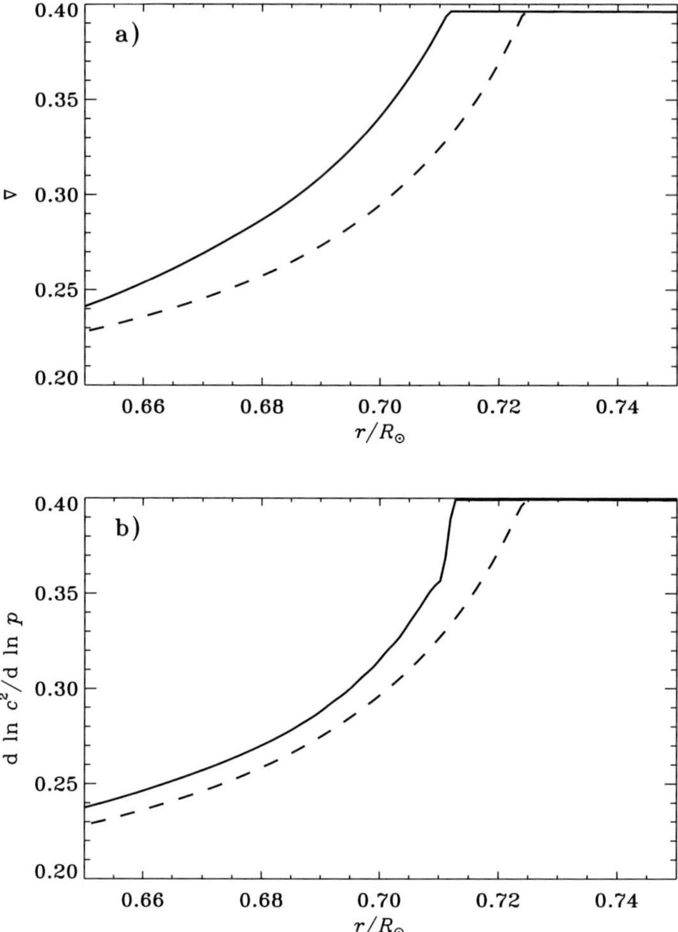

Figure 3.5. Temperature gradients $\nabla = \mathrm{d}\ln T/\mathrm{d}\ln p$ (panel a) and gradients of squared sound speed (panel b). The solid curves are for Model S, including settling of helium and heavier elements, while the dashed curves are for a similar model but without settling (and with a slightly shallower convection zone).

On comparing Figures 3.5a and 3.5b, it is evident that the gradient in the composition, reflected in μ, has a strong effect on the gradient in sound speed in this region. For comparison, the figure also illustrates a corresponding model without settling, and hence without compositional gradients in this region.

Figure 3.6 shows the relative difference in squared sound speed between the Sun and Model S, inferred from analysis of differences between a set of observed oscillation frequencies and those of the model (cf. Section 3.2). Analyses based on other datasets generally give similar results. Although the overall differences are relatively small, they are evidently highly significant. Particularly striking is the

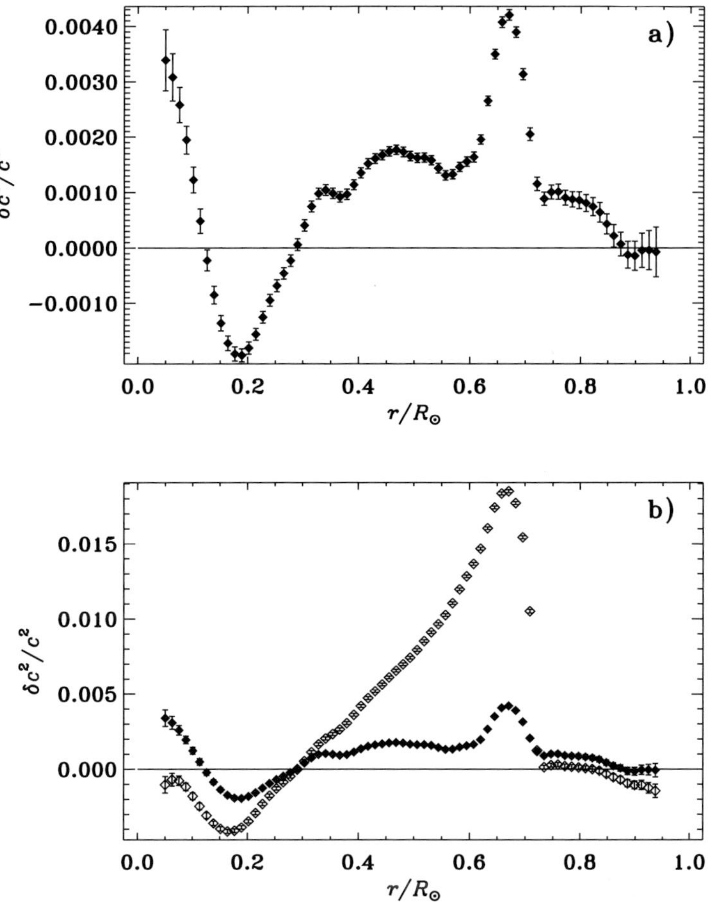

Figure 3.6. Inferred relative differences in squared sound speed against fractional radius, between the Sun and two solar models, in the sense (Sun) − (model). The results are based on inversion of a combined BiSON and LOWL dataset (Basu *et al.* 1997). In panel (a) the reference model is Model S of Christensen-Dalsgaard *et al.* (1996), with a surface composition characterized by $Z_s/X_s = 0.0245$. The error bars indicate one standard deviation in the inferred difference, based on the quoted errors in the observed frequencies. In panel (b) (Pijpers *et al.*, in preparation) the filled symbols show the same case as in panel (a), while the open symbols are for a model computed with reduced abundances of heavy elements (Asplund *et al.* 2005a), assuming a value of $Z_s/X_s = 0.0185$.

bump in $\delta c^2/c^2$ just beneath the convection zone, in the region of strong variation in the hydrogen abundance. We discuss this feature in more detail in Section 3.5.

As noted above (cf. Figure 3.5), the base of the convection zone is reflected in a rapid change in the sound-speed gradient. This was evident already in the first asymptotic sound-speed inversion by Christensen-Dalsgaard *et al.* (1985)

and led to a model-independent estimate of the depth of the convection zone. Christensen-Dalsgaard *et al.* (1991) made a careful analysis using various asymptotic inversion techniques, determining the radius at the base of the convection zone as $r_{cz} = (0.713 \pm 0.003)R_\odot$; a similar value was obtained by Kosovichev & Fedorova (1991). Subsequent analyses by Basu & Antia (1997) and Basu (1998) have confirmed this result, and substantially increased the precision. It should be noted, however, that the determination specifically refers to the location where the gradient begins to deviate substantially from being adiabatic; if, for example, overshoot beneath the convection zone, into the convectively stable region, were to result in a subadiabatic, but nearly adiabatic, zone with sufficiently efficient mixing to keep it chemically homogeneous[1], r_{cz} as inferred here would refer to the base of that zone.

Recently the determination of the composition of the solar atmosphere has been revised, resulting in substantial reductions in the inferred abundances of, in particular, carbon, nitrogen and oxygen (Allende Prieto *et al.* 2001; Asplund *et al.* 2004, 2005b; for a review, see Asplund *et al.* 2005a), and leading to a ratio Z_s/X_s of 0.0165, rather than the value of 0.0245 used in Model S. Since these elements make substantial contributions to the opacity in the radiative interior, the opacity is similarly reduced, leading to changes in the structure of the computed models and, in particular, a general reduction in the sound speed beneath the convection zone (Turck-Chièze *et al.* 2004; Bahcall *et al.* 2005a,b). This has a serious impact on the comparison with the helioseismic inferences. As an example, Figure 3.6b shows the inferred difference between the squared sound speed in the Sun and a model with revised surface composition[2]. It is evident that the revision has very substantially increased the discrepancy between the model and solar sound speed; in particular, the dominant discrepancy is no longer strongly localized near the base of the convection zone. This is accompanied by a decrease in the depth of the convection zone; for example, in a model with the revised composition Bahcall *et al.* (2004) found $r_{cz} = 0.726R_\odot$, which is certainly inconsistent with the helioseismically inferred value. An obvious, but perhaps aesthetically not very pleasing, way of reducing the discrepancy relative to helioseismic inferences would be to modify the intrinsic properties of the opacities in such a way as to compensate for the change in the composition (Basu & Antia 2004); Bahcall *et al.* (2005a) noted that an opacity increase of around 10% at temperatures between 2 and 5×10^6 K would be needed. It was pointed out by Antia & Basu (2005) and Bahcall *et al.* (2005c) that, rather than implicating the physics of the opacity calculation, the required opacity increase could be accomplished through an increase in the solar neon abundance by

[1] As, for example, in the model of Zahn (1991).

[2] Here a slightly earlier version of the revised composition than presented by Asplund *et al.* was used.

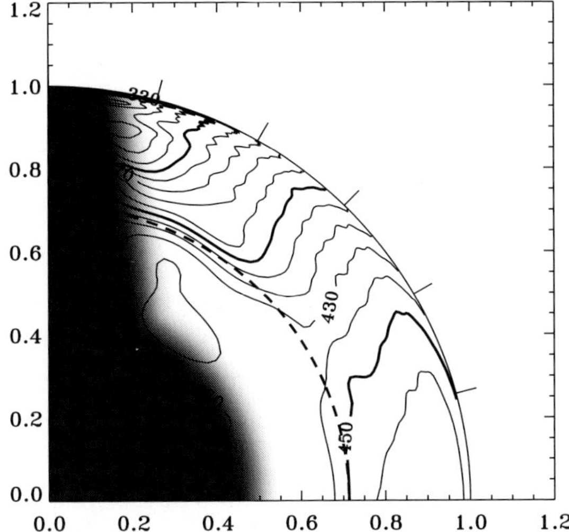

Figure 3.7. Inferred rotation rate $\Omega/2\pi$, based on OLA inversion of 144 days of MDI observations; some contours are labelled in nanohertz (nHz) and, for clarity, selected contours use a bolder line. Results are shown in a quadrant of the Sun, with the equator along the horizontal axis and the pole towards the top; the marks at the surface are in steps of 15° in latitude. The dashed circle marks the base of the convection zone. In the shaded region no reliable inference could be obtained from these data. (Adapted from Schou *et al.* 1998.)

a factor of 2.5–4. It remains to be seen whether such an increase is consistent with the, admittedly rather weak, observational constraints on the solar neon abundance. However, it is interesting that Drake & Testa (2005) very recently have found from X-ray observations that the neon-to-oxygen ratio in stars in the solar neighbourhood is substantially higher than the normally assumed solar value; if this were indeed to be representative also of the solar composition, the discrepancy between the solar models and the helioseismic inferences may well be substantially reduced (Bahcall *et al.* 2005c).

Figure 3.7 illustrates the helioseismically inferred rotation rate in the outer parts of the Sun; also, Figure 3.8 shows cuts at selected latitudes, illustrating the errors in the inferences. Similar results have been found in several independent analyses; however, there is some sensitivity, particularly at high latitude, to the choice of data set and, in particular, to the method used for the time-series analysis (Schou *et al.* 2002). The sharp distinction between rotation in the convection zone and in the radiative interior is evident. Within the convection zone there is strong differential rotation with latitude similar to that observed on the surface. It appears, however, that the lines of constant angular velocity, at least at low and moderate latitude, are not in the radial direction but show a roughly constant inclination relative to the rotation axis (Gilman & Howe 2003; Howe *et al.* 2005). Near the surface, for

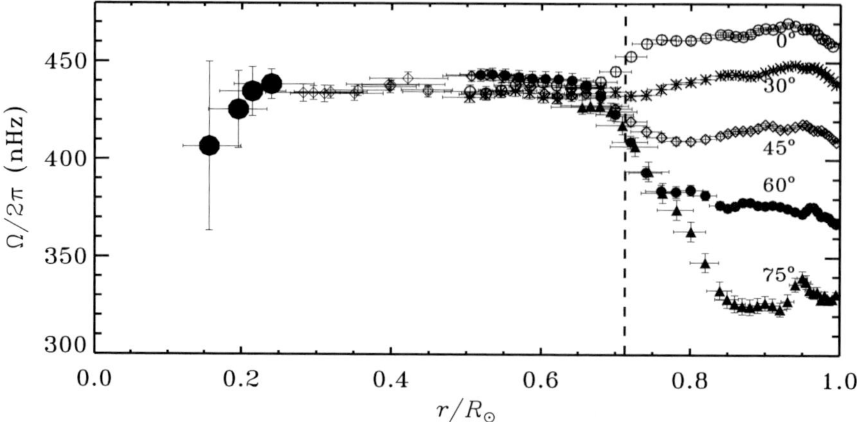

Figure 3.8. The inferred rotation rate $\Omega/2\pi$ as a function of fractional radius, at five solar latitudes: the equator, 30°, 45°, 60° and 75°. The vertical bars indicate one-standard-deviation errors, based on the quoted errors of the observations, while the horizontal bars provide a measure of the width of the averaging kernels (cf. Section 3.2) and hence the resolution of the inversion. The vertical dashed line marks the helioseismically determined base of the convection zone (Christensen-Dalsgaard *et al.* 1991). The results are from OLA inversions of MDI data in the outer region $r > 0.45R_\odot$ (from Schou *et al.* 1998) and of combined data from the LOWL instrument and the BiSON network, in the inner region $r \leq 0.45R_\odot$ (from Chaplin *et al.* 1999).

latitudes below $\sim$ 50°, there is a thin shear layer with rotation increasing with depth (e.g. Corbard & Thompson 2002); the maximal rotation rate, (469 ± 3) nHz, is found at the equator at a depth of roughly $0.06R_\odot$. Beneath the convection zone rotation is approximately uniform, at a rate corresponding to the surface value at a latitude of around 40°. The transition between these two regimes marks the tachocline.

The rotation rate down to the deep interior is illustrated in Figure 3.8. It is evident that the determination of the rotation of the deep interior is highly uncertain: very few modes reach this region; these have low degree and hence less well-determined rotational splittings d_{nlm} (cf. Equation 3.2); and the sound speed is high, reducing the time spent by sound waves in this region and hence the impact of the core rotation on the splitting. It is interesting, if not significant, that the rotation closest to the centre shows a slight reduction, relative to the bulk of the radiative interior; a similar trend was observed by Elsworth *et al.* (1995). Other recent observations have shown somewhat varying results (e.g. Charbonneau *et al.* 1998; Eff-Darwich *et al.* 2002; Couvidat *et al.* 2003; Fossat *et al.* 2003; García *et al.* 2004; Lazrek *et al.* 2004), although generally consistent with uniform rotation of the deep interior.

Another large-scale flow that has been invoked in some dynamo models is meridional circulation. This is not accessible to study by global helioseismology to leading order, but has been investigated using local helioseismic methods. Such studies indicate that the meridional circulation in the near-surface layers is generally poleward in both hemispheres, but the equatorward return flow has not so far been unambiguously detected (Giles *et al.* 1997, 1998; Haber *et al.* 2002; Zhao & Kosovichev 2004).

3.4 Dynamical properties of the tachocline

The existence of the tachocline is evident in inversion solutions for the solar internal rotation such as are shown in Figures 3.7 and 3.8. An early investigation that in effect sought to locate the tachocline was made by Goode *et al.* (1991), who performed a regularized least-squares inversion for rotation in which a discontinuity in the rotation rate was permitted: they found that the best fit to the data was obtained when the discontinuity was colocated with the base of the convection zone.

Subsequent investigations have shown that the rotational shear layer that constitutes the tachocline has a finite, non-zero thickness. The first quantitative results on the tachocline's location and thickness were obtained by Kosovichev (1996). He made a fit to the a_3 splitting coefficients in the Big Bear data: a_3 is the lowest-order coefficient to be sensitive to latitudinal variation in the rotation rate (cf. Figure 3.3). Since the rotation in the convection zone displays latitudinal dependence, whilst that in the radiative interior beneath the tachocline does not, this transition is evident when the a_3 coefficient is fitted with a forward model of the depth dependence of the latitudinal differential rotation. Kosovichev assumed a functional dependence $\Phi(r)$ for the transition in depth of the differential rotation of the form

$$\Phi(r) = \frac{1}{2}\left[1 + \mathrm{erf}\left(2(r - r_{\mathrm{c}})/w\right)\right], \tag{3.9}$$

where erf is the error function. This continuous step function varies from 0 to 1: it is centred on radial location $r = r_{\mathrm{c}}$ and has a characteristic width w. Quantitatively, over the width w, $\Phi(r)$ varies from 0.08 to 0.92 (see Figure 3.9). From the variation of a_3, Kosovichev found a fit for the parameters of the tachocline as $r_{\mathrm{c}}/R_\odot = 0.692 \pm 0.05$ and $w/R_\odot = 0.09 \pm 0.04$. This places the centre of the tachocline (and hence most of its extent) beneath the base of the convection zone (which from helioseismic estimates is at $0.713R_\odot$; cf. Section 3.3).

The functional form (3.9) introduced by Kosovichev has been adopted by various other investigators, for example to describe the variation of the rotation rate across the tachocline as a function of depth, at fixed latitude (e.g. Charbonneau *et al.* 1999a). Basu (1997) chose a different functional form,

$$\Phi_2(r) = \left[1 + \exp\left(-(r - r_{\mathrm{c}})/w_{\mathrm{B}}\right)\right]^{-1}. \tag{3.10}$$

Some care is needed in comparing results for the tachocline width when the two different functional forms have been used. A numerical fit of Equation (3.9) to

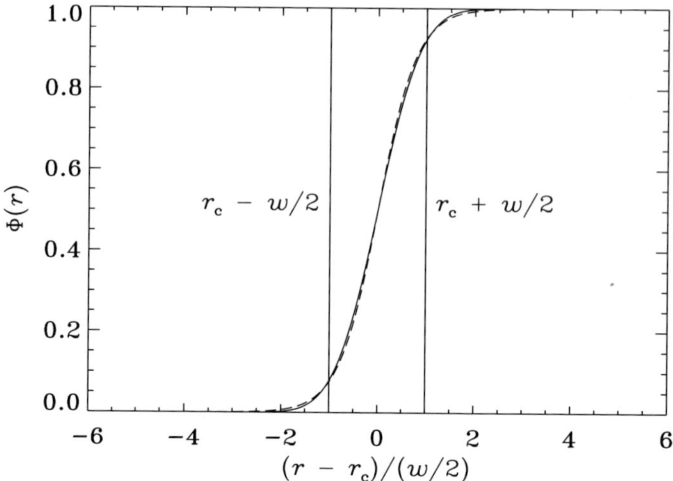

Figure 3.9. The function Φ introduced by Kosovichev (1996) and defined in Equation (3.9) (solid curve), together with the locations at a distance $\pm w/2$ either side of the centre location r_c (vertical lines). This region captures 84% of the variation of Φ, from 0.08 to 0.92. Also illustrated (dashed curve) is the function used by Basu and collaborators to characterize the tachocline, defined in Equation (3.10), where we have adopted $w_B = w/5$. With this identification the two functions coincide very closely.

Basu's function (3.10) indicates that the width directly comparable to w can be obtained by multiplying the value of w_B by a factor of 5, so $w = 5w_B$; this is illustrated in Figure 3.9. Neither functional form is evidently superior, since they are not physically motivated; but as a measure of width, w captures more of the transition than does the smaller w_B and so w probably accords better with what theorists might describe as the width of the tachocline.

Basu (1997) obtained values $r_c/R_\odot = 0.7050 \pm 0.0027$ and $w_B/R_\odot = 0.0098 \pm 0.0026$ (corresponding to $w/R_\odot = 0.049 \pm 0.013$), based on the behaviour of a_3 as measured by the GONG network. Charbonneau *et al.* (1998), also following Kosovichev in modelling a_3 but using LOWL data, obtained $r_c/R_\odot = 0.704 \pm 0.003$ and $w/R_\odot = 0.050 \pm 0.012$. Corbard *et al.* (1998), applying a nonlinear inversion method to LOWL data, obtained $r_c/R_\odot = 0.695 \pm 0.005$ and $w/R_\odot = 0.05 \pm 0.03$. These results all confirmed Kosovichev's finding that the centre of the tachocline lies beneath the seismically determined location of the base of the convection zone, but indicated that the tachocline thickness is at the low end of what Kosovichev reported.

An extensive analysis of the first two years of LOWL data by Charbonneau *et al.* (1999a), using both inversion and forward modelling, sought to quantify also whether the tachocline properties vary with latitude. At the equator they obtained $r_c/R_\odot = 0.693 \pm 0.002$ and $w/R_\odot = 0.039 \pm 0.013$. At 60° they found no significant

Table 3.1. *Location and thickness of the tachocline at different latitudes*

Latitude	$r_c/R_\odot$	$w/R_\odot$
0° (eq)	0.692 ± 0.002	0.033 ± 0.007
15°	0.691 ± 0.002	0.039 ± 0.007
45°	0.710 ± 0.002	0.052 ± 0.006
60°	0.710 ± 0.002	0.076 ± 0.010

Results as determined by Basu & Antia (2003), but given to lower precision than quoted in that paper. The values of the width w here have been obtained by multiplying the values of w_B given by Basu & Antia by a factor of five.

difference in the width of the tachocline, but found that the central location was further from the centre of the Sun by an amount $\Delta r/R_\odot = 0.024 \pm 0.004$, a significant difference. Thus they concluded that the tachocline is prolate. This result has been confirmed by the work of Antia *et al.* (1998) and Basu & Antia (2001).

The most extensive investigation of tachocline properties to date, using GONG and MDI data from 1995 to 2002, is that by Basu & Antia (2003). This essentially confirmed the results of Charbonneau *et al.* (1999a) but in addition reported a significant latitudinal variation of the tachocline width. The results are given in Table 3.1. They indicate that the tachocline is not only prolate but is also thicker at higher latitudes. Basu & Antia pointed out that their results are consistent with the location and thickness taking one value at low latitudes (beneath say 30°) and another value at high latitudes with essentially a discontinuity in between, rather than there being a smooth change in properties as the latitude changes. Of course it is hard to speak of the tachocline properties at mid-latitudes where the shear vanishes. The location of the tachocline at low and high latitudes with respect to the base of the convection zone is illustrated in Figure 3.10.

Attempts have been made to quantify any temporal variations in the solar internal rotation and in the tachocline properties. There is little evidence over the time that helioseismic observations have been made for any change in the rotation of the deep interior. In contrast, there are small but significant changes in the rotation profile in the convection zone on the timescale of the solar cycle. Weak but apparently coherent bands of faster and slower rotation superimposed on the near-surface rotation migrate during the solar cycle: these were first discovered in surface measurements of rotation (Howard & LaBonte 1980) and are termed 'torsional oscillations'. This is something of a misnomer, and the phenomenon is more likely to have

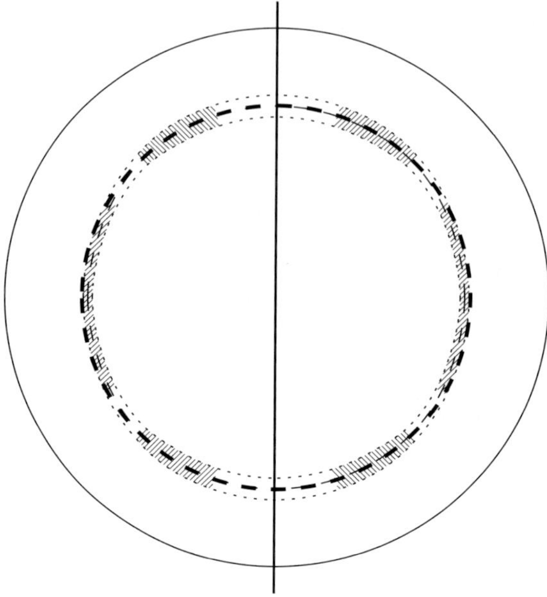

Figure 3.10. Schematic diagram showing the tachocline location and thickness corresponding to the values given in Table 3.1. Where the tachocline properties have been determined, a region of width w centred on r_c is shown shaded. The spherically symmetric location of the base of the convection zone at $r = r_{cz}$ is also shown (broken line). The tachocline is unshaded at mid-latitudes where it is poorly defined because the rotational shear vanishes, and at very high latitudes where it is not yet determined. The straight line indicates the rotation axis of the solar envelope.

a magnetohydrodynamic origin (see the discussion by Tobias Weiss in Chapter 13 of this book). At low and mid-latitudes the zonal bands of faster and slower rotation migrate toward the equator along with the bands of magnetic activity. These have been studied seismically by Kosovichev & Schou (1997), Schou *et al.* (1998) and Schou (1999). Howe *et al.* (2000a) quantified that the low-latitude migrating zonal bands extend down to about $0.92R_\odot$. There is also a high-latitude branch of the torsional oscillation which migrates poleward from about 60° and at the same time grows in strength. These flows are illustrated in Figure 3.11. Results of a nonlinear inversion technique used by Vorontsov *et al.* (2002) indicate that the high-latitude torsional oscillation variability extends over much of the depth of the convection zone and possibly down to its base.

There have also been reported variations of the rotation rate near the tachocline. Howe *et al.* (2000b), studying rotation inversions of GONG and MDI over a number of years, found evidence for a quasi-periodic variation of the rotation rate in equatorial regions above the tachocline at a radius of about $0.72R_\odot$, and an oscillation in antiphase with this in the radiative interior at a radius of about

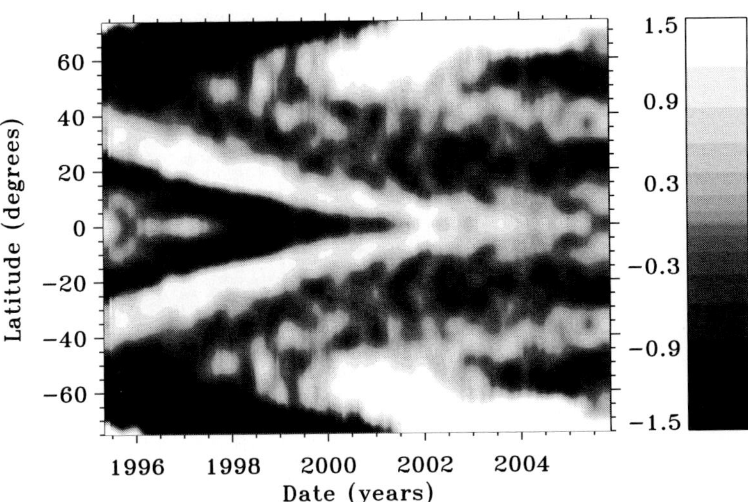

Figure 3.11. The evolution with time of the zonal flows at $0.99R_\odot$, inferred from RLS inversions of data from GONG and from the MDI instrument on SoHO, after subtraction of the time-averaged rotation rate. The results are presented as a function of time and latitude, the grey scale at the right giving the signal in nanohertz. Note that the plot is symmetrical about the equator, since global rotational inversion is sensitive only to the symmetrical component of the rotation rate. (Adapted from Howe *et al.* 2000a.)

$0.63R_\odot$ (see Figure 3.12). The amplitude of the oscillation reduces as one moves to 30° latitude. At higher latitudes, there is also evidence of an oscillation with a one year period, but the results are noisier and such an annual variation is possibly an artefact in the data. A one-year periodic signal does not explain the low-latitude 1.3-year oscillation, since the two would be clearly out of phase after the 4 years of observations used in the original Howe *et al.* (2000b) paper. A signal very like the 1.3-year oscillation has been seen also in analyses by Basu & Antia (2001) but they concluded that the variation was not significant. The oscillation has been sought by others in series of datasets that were one-year long but without finding the variation (Corbard *et al.* 2001; see also Eff-Darwich & Korzennik 2003); but it is evident that the amplitude of any 1.3-year variation would be greatly suppressed if sampled with a one-year cadence. The existence of a 1.3-year variation in the rotation rate at low latitudes near the tachocline certainly remains disputed.

Various theoretical models predict the existence of prograde zonal jets in the tachocline region. Firstly, if toroidal magnetic field exists in the region in a band confined in latitude, then the tendency of the field to slip poleward owing to magnetic curvature stress may be at least partially balanced by equatorward Coriolis forces from a prograde jet inside the magnetic band (Dikpati & Gilman 2001a, Rempel & Dikpati 2003). Secondly, a two-dimensional instability of the tachocline

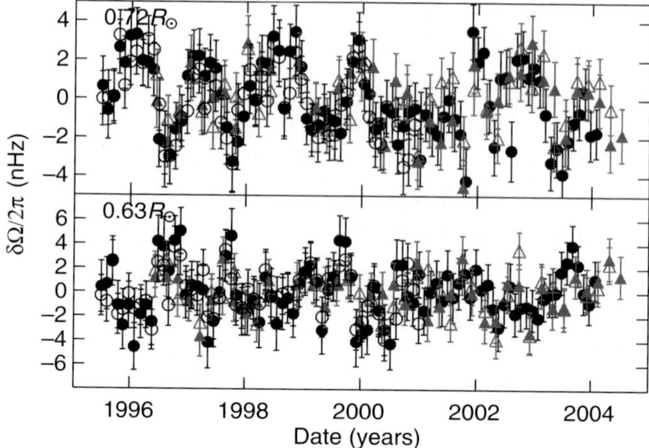

Figure 3.12. Residuals from rotation inversions at the equator and radii $r = 0.72R_\odot$ (top) and $r = 0.63R_\odot$ (bottom) after a temporal mean of the inversions has been subtracted from each. The results at $0.72R_\odot$ show some evidence for a quasi-periodic 1.3-year oscillation (Howe *et al.* 2000b) with the results at $0.63R_\odot$ being in antiphase, at least in the earlier part of the time series. Circles denote GONG results, triangles MDI results, filled symbols are for an RLS inversion, open symbols are for an OLA inversion.

differential rotation, or a combination of differential rotation and a toroidal field, can transport angular momentum from low latitudes to a narrow band of high latitudes, thus creating a jet there (Gilman & Fox 1997; Charbonneau *et al.* 1999b; Dikpati & Gilman 2001b). A preliminary study by Christensen-Dalsgaard *et al.* (2005) looked for such tachocline jets using GONG data from 1996 to 2003, and reported some evidence from both OLA and RLS inversions for a steady jet feature at a radius of about $0.72R_\odot$ and between latitudes 55° and 65° in all years except 1996. Since the feature does not migrate in latitude, Christensen-Dalsgaard *et al.* concluded that it provides some observational support for the second kind of model prediction discussed above, namely jets caused by a hydrodynamic instability. More work is, however, required to assess the significance of this observational finding, particularly in the light of error correlations between neighbouring points which can lead to the appearance of jet-like features in inversion solutions.

3.5 The structure of the tachocline region

From the point of view of tachocline dynamics, it is important to constrain the precise location of the region of transition of rotation relative to the base of the convection zone. The inferred radius $r_{\rm cz} = 0.713R_\odot$ of the base of the convection zone (cf. Section 3.3) places the centre $r_{\rm c}$ of the transition well below the convection zone at the equator, although with some slight overlap between the convectively unstable layer and the transition. At higher latitudes, assuming that $r_{\rm cz}$ does not change with latitude, the inferred prolate nature of the rotation transition and the possibly larger width of the transition at higher latitude (Table 3.1) would indicate that much of the transition takes place within the convection zone (cf. Figure 3.10). Of course, it may be wrong to suppose that the base of the convection zone, as defined by the transition to strong subadiabaticity, is indeed spherically symmetric.

As discussed in Section 3.3, the issues of the depth of the convection zone and the dynamics of this region are closely tied to the details of the behaviour of convective motion at the border between convective stability and instability. It is evident that motion cannot stop abruptly where the stratification becomes convectively stable. However, the properties and extent of such overshoot is highly uncertain[3]. Simple models (e.g. van Ballegooijen 1982; Schmitt *et al.* 1984; Zahn 1991) generally show a nearly adiabatic extension of the convection zone, corresponding to a negative convective flux, and with a transition to no motion and a radiative temperature gradient in a thin boundary layer. Rempel (2004) made a somewhat more realistic model in terms of downflows, possibly having a distribution of strengths and hence potentially providing a more gradual transition to the radiative stratification. More detailed hydrodynamical simulations are greatly complicated by the huge ratio between the thermal and dynamical timescales in the lower convection zone (e.g. Nordlund *et al.* 1996), although Nordlund *et al.* developed an interesting 'toy model' to elucidate some of the properties of this region. Extensive calculations, although still quite far from solar conditions, were carried out by Brummell *et al.* (2002), modelling convection in a rotating box at various angles between the rotation axis and the direction of gravity; they found substantial motion into the stable layer, although of insufficient strength to extend the nearly adiabatic region. Interestingly, the extent of the overshoot depended significantly on the angle between the rotation axis and gravity, reflecting the latitude of the corresponding region in the Sun being modelled (see also Julien *et al.* 1997). However, the extrapolation of these results to solar conditions is still somewhat uncertain.

Observationally, the structure of the region near the base of the convection zone affects the solar oscillation frequencies owing to the sharp variation in the sound-speed gradient in this region (cf. Figure 3.5). Such sharp variations give rise to oscillatory signatures in the frequencies whose properties reflect the location and nature of the sharp feature. Monteiro *et al.* (1994) expressed the effect of the base of the convection zone in terms of the variation $\delta\omega_p$ of the frequencies in the Sun, relative to a corresponding model where the transition had been smoothed; for

[3] Here we use overshoot or penetration indiscriminately for motion beneath the convectively unstable region. We note, however, that Zahn (1991) recommends the use of *penetration* for motion sufficiently vigorous to lead to a nearly adiabatic stratification and *overshoot* for weaker motion.

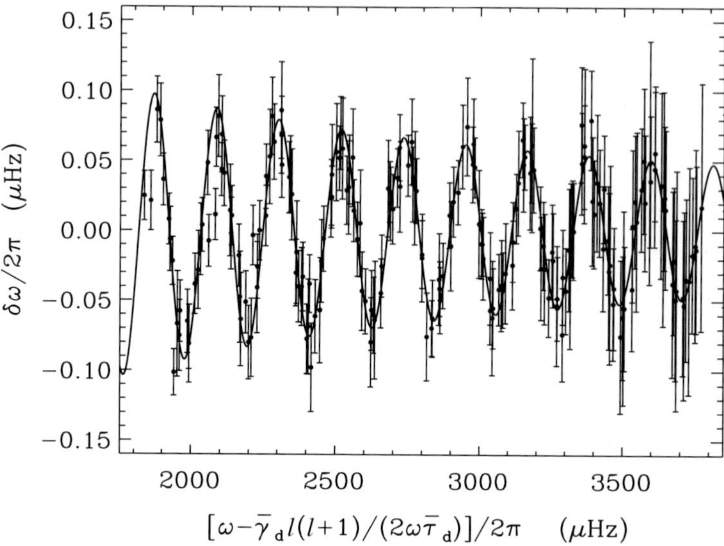

Figure 3.13. Oscillatory signal in observed oscillation frequencies, resulting from the rapid change in the sound speed at the base of the convection zone; observed frequencies from MDI were used in the analysis. The solid curve denotes the fit. The term in $\overline{\gamma}_{\mathrm{d}} l(l+1)$ corrects for the decreasing depth of penetration of the modes with increasing degree; for low-degree modes it is negligible.

low-degree modes, this can be approximated by

$$\delta\omega_p \sim A(\omega)\cos(2\omega\overline{\tau}_{\mathrm{d}} + 2\phi_0), \tag{3.11}$$

where the amplitude can be expressed as

$$A(\omega) = a_1\left(\frac{\tilde{\omega}}{\omega}\right)^2 + a_2\left(\frac{\tilde{\omega}}{\omega}\right), \tag{3.12}$$

in terms of a reference frequency $\tilde{\omega}$. Also, $\overline{\tau}_{\mathrm{d}}$ is essentially the acoustical depth $\int_{r_{\mathrm{d}}}^{R_\odot} \mathrm{d}r/c$ of the sharp feature, r_{d} being its distance from the centre, but including a correction for near-surface effects, and ϕ_0 is a phase which also depends on the near-surface structure of the Sun. The parameters $a_1, a_2, \overline{\tau}_{\mathrm{d}}$ and ϕ_0 can be obtained by fitting an expression of this form to the observed frequencies, after subtracting a suitable smooth function. An example of the resulting oscillatory signal, for observed solar data, is shown in Figure 3.13.

In Equation (3.12) the terms in a_1 and a_2 arise essentially from contributions to the oscillatory signals from discontinuities in the second and first derivatives of sound speed, respectively (see also Christensen-Dalsgaard *et al.* 1995). Thus determination of the amplitude provides a characterization of the transition between the convective and radiative regions, which can be calibrated by considering suitable solar models. As discussed above, simple models of convective overshoot (e.g. Zahn 1991) predict a slightly subadiabatic extension of the convection zone, followed by an abrupt change to the radiative gradient, thus essentially causing a discontinuity in the temperature and sound-speed gradients, and a relatively large amplitude of the oscillatory signal. From analysis of observations such as those shown in Figure 3.13, Monteiro *et al.* (1994) and Christensen-Dalsgaard *et al.* (1995) concluded that any overshoot of this nature would have to extend less than $0.1H_p$, $H_p \simeq 0.08R_\odot$ being the pressure scale height at the base of the convection zone. A similar limit was obtained by Basu *et al.* (1994), while Roxburgh & Vorontsov (1994) obtained a somewhat weaker limit using an approximate expression for the dependence of the amplitude on the extent of overshooting.

These analyses were all calibrated using models with no settling beneath the convection zone. As shown in Figure 3.5, the gradient in composition caused by settling has a strong effect on the sound-speed gradient and hence on the amplitude of the oscillatory signal. It was shown by Basu & Antia (1994) that even in the absence of any overshooting the signal amplitude in models with settling was significantly higher than the observed value. Thus it appears that the actual structure in the Sun is smoother than in such models.

In assessing these results it should be recalled that in all the analyses the frequencies reflect the spherically averaged structure of the solar interior. Also, since the frequencies are typically determined by analysing observations extending over several weeks or months, they involve substantial temporal averaging, over timescales comparable with or longer than the expected convective timescales near the base of the convection zone. Thus, even though individual overshooting convective plumes may reflect the structure assumed in the simple models with an abrupt braking after nearly adiabatic motion, the implied spatial and temporal average reflected in the oscillation frequencies may likely result in a smoother structure. Such averaging was indeed found in the numerical simulations by, for example, Nordlund *et al.* (1996) and Brummell *et al.* (2002), leading to a relatively smooth mean structure.

Departures from spherical symmetry of the solar structure are reflected in the even component of the dependence of frequencies ω_{nlm} on m (cf. Section 3.2). Thus, as pointed out by Monteiro & Thompson (1998), it is possible to make combinations of frequencies with the same (n, l) but different m that are sensitive predominantly to the structure at specific latitudes. This in principle enables a search for possible variations with latitude in the structure of the transition layer between the convective and radiative regions. Using the techniques of Monteiro *et al.* (1994) and Christensen-Dalsgaard *et al.* (1995) discussed above, Monteiro & Thompson demonstrated that it was indeed possible to see the effects of an overshoot region restricted to intermediate latitudes. However, a preliminary application to solar observations showed little if any significant variation. Further investigations in this direction are clearly warranted.

A striking feature of the inferred sound-speed difference shown in Figure 3.6a is the bump just beneath the convection zone, where the solar sound speed is higher than that of the model in a relatively restricted region. This essentially coincides with the region of sharp composition gradients (cf. Figure 3.5). If the abundance gradients were to be smoothed the hydrogen abundance would be locally increased, the mean molecular weight decreased, and the sound speed consequently increased in this region (cf. Equation (3.7)), bringing the model closer to the observations. A more quantitative determination of this effect can be obtained by inferring the composition profile from analysis of the oscillation frequencies. Using a technique developed by Gough & Kosovichev (1990), Kosovichev (1997) carried out an inversion specifically for the hydrogen abundance. An alternative technique, developed by Shibahashi & Takata (1996), Antia & Chitre (1998) and Takata & Shibahashi (1998), is based on inferring the hydrostatic structure of the Sun from helioseismic analysis and subsequently determining the thermal and composition profile such that this structure results, given the constraints of the equations of stellar structure and the observed surface luminosity. In both cases the resulting hydrogen profile is significantly smoother beneath the convection zone than the profile, illustrated in Figure 3.4, resulting from model calculations including settling. For both techniques the analysis depends on the assumed physics of the stellar interior, although in a careful analysis Takata & Shibahashi (2003) showed that, at least in the case of the latter technique, the result is relatively robust.

Such smoothing of the composition profile might plausibly result from motion induced by the rotational gradient in the tachocline region or by circulation or instabilities associated with the spin-down of the solar interior from an assumed state of rapid initial rotation. In models of solar evolution following the evolution of the solar internal rotation, Chaboyer *et al.* (1995) found in some cases a hydrogen-abundance profile with a substantially less steep gradient beneath the convection zone than in the non-rotating model[4]. Brun *et al.* (1999, 2002) considered models without and with rotationally induced mixing in the tachocline region; their models without mixing yielded sound-speed differences similar to those illustrated in Figure 3.6a, while with rotational mixing the bump beneath the convection zone was essentially eliminated. Elliott *et al.* (1998) and Elliot & Gough (1999) assumed that the composition was fully mixed in the tachocline region and determined the width of this region by fitting to the bump in $\delta c^2/c^2$, relative to an underlying

[4] However, the resulting rotation profile was quite far from matching the observed rotation rate, as noted above for the rather similar models of Pinsonneault *et al.* (1989).

smooth variation; they noted that the resulting width, $w = 0.019R_\odot$, (Elliott & Gough 1999) was substantially smaller than the width inferred from the rotational transition (cf. Section 3.4).

Although mixing in a limited region below the convection zone has a substantial effect on the inferred sound-speed differences in Figure 3.6a, it would have modest influence on the much larger differences in the bulk of the radiative interior in Figure 3.6b, resulting when the revised solar surface abundances are used. Even so, the smoother hydrogen profile inferred from helioseismic estimates of the composition would probably still be valid in this case, remaining therefore as evidence for partial mixing of this region. Also, it is interesting that, as discussed in Section 3.3, Bahcall *et al.* (2005c) found that most of the discrepancy in Figure 3.6b can be removed through an increase in the solar neon abundance; the remaining $\delta c^2/c^2$ is fairly similar to the behaviour found in Figure 3.6a and hence can probably be suppressed through partial mixing.

The light elements lithium, beryllium and boron are destroyed at relatively low temperature (2.5×10^6 K, 3.5×10^6 K and 5×10^6 K, respectively) over a period corresponding to the present age of the Sun. As already mentioned, the observed solar surface lithium abundance, relative to silicon, is reduced by a factor of around 150 relative to the corresponding meteoritic value. This indicates that there has been substantial mixing to the depth where lithium is destroyed over the evolution of the Sun. On the other hand, recent analyses of the solar beryllium abundance (Balachandran & Bell 1998; Asplund 2004) indicate that there has been no significant depletion of beryllium, thus limiting the extent to which such mixing has taken place. The indication that there has similarly been no boron depletion (Cunha & Smith 1999) is consistent with this constraint. It is evident that the present solar composition, including also the inferred hydrogen profile, is the result of the combined effects of settling, mixing and nuclear reactions integrated over the lifetime of the Sun. Even so, these abundance determinations clearly provide strong constraints on the motion in the solar interior resulting, for example, from rotational instabilities (e.g. Brun *et al.* 2002).

3.6 Outlook

Since discovering the tachocline, helioseismology has pinned down with reasonable precision its location and thickness in the radial direction, with some latitudinal resolution (Table 3.1). Helioseismology has also established the location of the base of the convection zone (more precisely the base of the essentially adiabatically stratified envelope) to lie at radius $r_{cz} = 0.713R_\odot$ assuming that the structure is spherically symmetric. Thus we can also make qualitative statements that the layer of rotational shear at low latitudes is essentially wholly confined beneath the convection zone, while at higher latitudes it straddles the base of the convection zone although it is still mostly located beneath the convection zone (Figure 3.10).

The tachocline is thinner than the intrinsic resolution of present-day inversions, and so its parameters have been determined by fitting prescribed functional forms, chosen thus far without physical motivation, either to the results of inversions or via forward modelling to the frequency splittings data directly. Through a steady but incremental reduction of the error bars on the frequencies and splittings, through observation over longer periods of time and through better resolution of future observations from the Solar Dynamics Observatory satellite, we can expect the precision with which the dynamical and structural properties of the tachocline region are determined to improve. It must also be admitted that optimal use has not yet been made of the observations in hand. In particular, the observations are typically analysed in chunks of two to three months' data: for low-frequency modes with lifetimes longer than this, some improvement in the determination of their frequencies and splittings can be expected from making coherent peak-fitting of their spectra from longer observational time series.

Resolution might also be improved by extending the mode-sets currently used in the analyses reported here. Lower-frequency modes, although they have poorer intrinsic resolving power because of their larger vertical wavelengths when compared to higher-frequency modes, also live longer and have smaller linewidths and hence their frequencies can be determined more precisely. Higher-frequency modes, on the other hand, though more poorly determined, have higher resolving power. Extending the set of modes used to both lower and higher frequencies may therefore yield improvements in the seismic probing of the tachocline region.

Given that the tachocline is not properly resolved radially by inversions, there is little prospect in the foreseeable future that the shape of its radial variation can be determined from inversion. Similarly, the finite resolution of the inversions in the latitudinal direction means that one could not tell if the radial profile of the tachocline varied as a function of latitude on scales smaller than the spatial extent of the averaging kernels, for example if there were such locations where the rotation rate was a discontinuous function of radius. However, if different candidate theoretical models for the tachocline shape were forthcoming, these might be discriminated using forward modelling to the observed frequency splittings.

In the not-too-distant future we may expect time-distance measurements between points on the surface sufficiently far apart that the ray connecting them extends to the tachocline and beneath it. Using, say, the Solar Orbiter satellite and the Earth as two vantage points, such measurements will enable travel times to be measured along paths that may provide some longitudinal resolution of the tachocline. Thus it may become possible to look for variations of low azimuthal order in the dynamical and structural properties of the tachocline region.

Acknowledgments

We are grateful to Rachel Howe and Mario Monteiro for providing Figures 3.11, 3.12 and 3.13. We thank Alexander Kosovichev for his comments as referee. Our thanks also go to Juri Toomre and Michael Knölker for hospitality at JILA and HAO respectively, during the time when much of this chapter was written.

References

Allende Prieto, C., Lambert, D. L. & Asplund, M. (2001). *Astrophys. J.*, **556**, L63.

Anders, E. & Grevesse, N. (1989). *Geochim. Cosmochim. Acta*, **53**, 197.

Antia, H. M. & Basu, S. (2005). *Astrophys. J.*, **620**, L129.

Antia, H. M. & Chitre, S. M. (1998). *Astron. Astrophys.*, **339**, 239.

Antia, H. M, Basu, S. & Chitre, S. M. (1998). *Mon. Not. Roy. Astron. Soc.*, **298**, 543.

Asplund, M. (2004). *Astron. Astrophys.*, **417**, 769.

Asplund, M., Grevesse, N. & Sauval, A. J. (2005a). In *Cosmic Abundances as Records of Stellar Evolution and Nucleosynthesis*, ed. F. N. Bash & T. G. Barnes (*ASP Conf. Ser.*, **336**), p. 25.

Asplund, M., Grevesse, N., Sauval, A. J., Allende Prieto, C. & Blomme, R. (2005b). *Astron. Astrophys.*, **431**, 693.

Asplund, M., Grevesse, N., Sauval, A. J., Allende Prieto, C. & Kiselman, D. (2004). *Astron. Astrophys.*, **417**, 751 (Erratum: *Astron. Astrophys.*, **435**, 339).

Bahcall, J. N. & Pinsonneault, M. H. (1995), with an appendix by G. J. Wasserburg. *Rev. Mod. Phys.*, **67**, 781.

Bahcall, J. N., Basu, S., Pinsonneault, M. & Serenelli, A. M. (2005a). *Astrophys. J.*, **618**, 1049.

Bahcall, J. N., Serenelli, A. M. & Basu, S. (2005b). *Astrophys. J.*, **621**, L85.

Bahcall, J. N., Serenelli, A. M. & Basu, S. (2005c). *Astrophys. J.*, **631**, 1281.

Bahcall, J. N., Serenelli, A. M. & Pinsonneault, M. (2004). *Astrophys. J.*, **614**, 464.

Balachandran, S. C. & Bell, R. A. (1998). *Nature*, **392**, 791.

Basu, S. (1997). *Mon. Not. Roy. Astron. Soc.*, **288**, 572.

Basu, S. (1998). *Mon. Not. Roy. Astron. Soc.*, **298**, 719.

Basu, S. & Antia, H. M. (1994). *Mon. Not. Roy. Astron. Soc.*, **269**, 1137

Basu, S. & Antia, H. M. (1997). *Mon. Not. Roy. Astron. Soc.*, **287**, 189.

Basu, S. & Antia, H. M. (2001). *Mon. Not. Roy. Astron. Soc.*, **324**, 498.

Basu, S. & Antia, H. M. (2003). *Astrophys. J.*, **585**, 553.

Basu, S. & Antia, H. M. (2004). *Astrophys. J.*, **606**, L85.

Basu, S., Antia, H. M. & Narasimha, D. (1994). *Mon. Not. Roy. Astron. Soc.*, **267**, 209.

Basu, S., Chaplin, W. J., Christensen-Dalsgaard, J. *et al.* (1997). *Mon. Not. Roy. Astron. Soc.*, **292**, 243.

Brown, T. M. (1985). *Nature*, **317**, 591.

Brown, T. M. & Morrow, C. A. (1987). *Astrophys. J.*, **314**, L21.

Brown, T. M., Christensen-Dalsgaard, J., Dziembowski, W. A. *et al.* (1989). *Astrophys. J.*, **343**, 526.

Brummell, N. H., Clune, T. L. & Toomre, J. (2002). *Astrophys. J.*, **570**, 825.

Brun, A. S., Antia, H. M., Chitre, S. M. & Zahn, J.-P. (2002). *Astron. Astrophys.*, **391**, 725.

Brun, A. S., Turck-Chièze, S. & Zahn, J.-P. (1999). *Astrophys. J.*, **525**, 1032 (Erratum: *Astrophys. J.*, **536**, 1005).

Chaboyer, B., Demarque, P. & Pinsonneault, M. H. (1995). *Astrophys. J.*, **441**, 865.

Chaplin, W. J., Christensen-Dalsgaard, J., Elsworth, Y. *et al.* (1999). *Mon. Not. Roy. Astron. Soc.*, **308**, 405.

Chaplin, W. J., Elsworth, Y., Howe, R. *et al.* (1996). *Solar Phys.*, **168**, 1.

Charbonneau, P., Christensen-Dalsgaard, J., Henning, R. *et al.* (1999a). *Astrophys. J.*, **527**, 445.

Charbonneau, P., Dikpati, M. & Gilman, P. A. (1999b). *Astrophys. J.*, **526**, 523.

Charbonneau, P., Tomczyk, S., Schou, J. & Thompson, M. J. (1998). *Astrophys. J.*, **496**, 1015.

Christensen-Dalsgaard, J. (2002). *Rev. Mod. Phys.*, **74**, 1073.

Christensen-Dalsgaard, J. & Schou, J. (1988). In *Seismology of the Sun & Sun-like Stars*, ed. V. Domingo & E. J. Rolfe (ESA SP-286), p. 149.

Christensen-Dalsgaard, J., Corbard, T., Dikpati, M., Gilman, P. A. & Thompson, M. J. (2005). In *Large Scale Structures and their Role in Solar Activity*, ed. K. S. Sankarasubramanian, M. J. Penn & A. A. Pevtsov, *ASP Conf. Ser.*, **346**, p. 115.

Christensen-Dalsgaard, J., Däppen, W., Ajukov, S. V. *et al.* (1996). *Science*, **272**, 1286.

Christensen-Dalsgaard, J., Duvall, T. L., Gough, D. O., Harvey, J. W. & Rhodes, E. J. (1985). *Nature*, **315**, 378.

Christensen-Dalsgaard, J., Gough, D. O. & Thompson, M. J. (1991). *Astrophys. J.*, **378**, 413.

Christensen-Dalsgaard, J., Gough, D. O. & Thompson, M. J. (1992). *Astron. Astrophys.*, **264**, 518.

Christensen-Dalsgaard, J., Monteiro, M. J. P. F. G. & Thompson, M. J. (1995). *Mon. Not. Roy. Astron. Soc.*, **276**, 283.

Christensen-Dalsgaard, J., Proffitt, C. R. & Thompson, M. J. (1993). *Astrophys. J.*, **403**, L75.

Corbard, T. & Thompson, M. J. (2002). *Solar Phys.*, **205**, 211.

Corbard, T., Berthomieu, G., Provost, J. & Morel, P. (1998). *Astron. Astrophys.*, **330**, 1149.

Corbard, T., Jiménez-Reyes, S. J., Tomczyk, S., Dikpati, M. & Gilman, P. A. (2001). In *Helio- and Asteroseismology at the Dawn of the Millennium*, ed. A. Wilson (ESA SP-464), p. 265.

Couvidat, S., García, R. A., Turck-Chièze, S. *et al.* (2003). *Astrophys. J.*, **597**, L77.

Cunha, K. & Smith V. V. (1999). *Astrophys. J.*, **512**, 1006.

Däppen, W., Gough, D. O., Kosovichev, A. G. & Thompson, M. J. (1991). In *Challenges to Theories of the Structure of Moderate-Mass Stars*, ed. D. O. Gough & J. Toomre, *Lecture Notes in Physics* **388**. Heidelberg: Springer, p. 111.

Dicke, R. H. (1964). *Nature*, **202**, 432.

Dikpati, M. & Gilman, P. A. (2001a). *Astrophys. J.*, **552**, 348.

Dikpati, M. & Gilman, P. A. (2001b). *Astrophys. J.*, **551**, 536.

Drake, J. J. & Testa, P. (2005). *Nature*, **436**, 525.

Duvall, T. L., Dziembowski, W. A., Goode, P. R. *et al.* (1984). *Nature*, **310**, 22.

Duvall, T. L., Harvey, J. W. & Pomerantz, M. A. (1986). *Nature*, **321**, 500.

Dziembowski, W. A., Goode, P. R. & Libbrecht, K. G. (1989). *Astrophys. J.*, **337**, L53.

Dziembowski, W. A., Pamyatnykh, A. A. & Sienkiewicz, R. (1990). *Mon. Not. Roy. Astron. Soc.*, **244**, 542.

Eff-Darwich, A. & Korzennik, S. G. (2003). In *Local and Global Helioseismology: the Present and Future*, ed. H. Sawaya-Lacoste (ESA SP-517), p. 267.

Eff-Darwich, A., Korzennik, S. G. & Jiménez-Reyes, S. J. (2002). *Astrophys. J.*, **573**, 857.

Elliott, J. R. & Gough, D. O. (1999). *Astrophys. J.*, **516**, 475.

Elliott, J. R., Gough, D. O. & Sekii, T. (1998). In *Structure and Dynamics of the Interior of the Sun and Sun-like Stars*, ed. S. G. Korzennik & A. Wilson (ESA SP-418), p. 763.
Elsworth, Y., Howe, R., Isaak, G. R. *et al.* (1995). *Nature*, **376**, 669.
Fossat, E. (1991). *Solar Phys.*, **133**, 1.
Fossat, E., Salabert, D., Cacciani, A. *et al.* (2003). In *Local and Global Helioseismology: The Present and Future*, ed. A. Wilson (ESA SP-517), p. 139.
Gabriel, A. H., Grec, G., Charra, J. S. *et al.* (1995). *Solar Phys.*, **162**, 61.
García, R. A., Corbard, T., Chaplin, W. J., *et al.* (2004). *Solar Phys.*, **220**, 269.
Giles, P. M., Duvall, T. L., Scherrer, P. H. & Bogart, R. S. (1997). *Nature*, **390**, 52.
Giles, P. M., Duvall, T. L. & Scherrer, P. H. (1998). In *Structure and Dynamics of the Interior of the Sun and Sun-like Stars*, ed. S. G. Korzennik & A. Wilson (ESA SP-418), p. 775.
Gilman, P. A. (1986). In *Physics of the Sun*, Vol. 1, ed. P. A. Sturrock, T. E. Holzer, D. Mihalas & R. K. Ulrich. Dordrecht: Reidel, p. 95.
Gilman, P. A. & Fox, P. A. (1997). *Astrophys. J.*, **484**, 439.
Gilman, P. A. & Howe, R. (2003). In *Local and Global Helioseismology: The Present and Future*, ed. A. Wilson (ESA SP-517), p. 283.
Gilman, P. A. & Miller, J. (1986). *Astrophys. J. Suppl.*, **61**, 585.
Gilman, P. A., Morrow, C. A. & DeLuca, E. E. (1989). *Astrophys. J.*, **338**, 528.
Glatzmaier, G. A. (1985). *Astrophys. J.*, **291**, 300.
Goode, P. R., Dziembowski, W. A., Korzennik, S. G. & Rhodes, E. J. (1991). *Astrophys. J.*, **367**, 649.
Gough, D. O. (1985). *Solar Phys.*, **100**, 65.
Gough, D. O. (1990). In *Progress of Seismology of the Sun and Stars*, ed. Y. Osaki & H. Shibahashi, *Lecture Notes in Physics*, **367**. Berlin: Springer, p. 283.
Gough, D. O. (1993). In *Astrophysical Fluid Dynamics, Les Houches Session XLVII*, ed. J.-P. Zahn & J. Zinn-Justin. Amsterdam: Elsevier, p. 399.
Gough, D. O. & Kosovichev, A. G. (1988). In *Seismology of the Sun & Sun-like Stars*, ed. V. Domingo & E. J. Rolfe (ESA SP-286), p. 195.
Gough, D. O. & Kosovichev, A. G. (1990). In *Inside the Sun*, ed. G. Berthomieu & M. Cribier. Dordrecht: Kluwer, p. 327.
Gough, D. O. & Thompson, M. J. (1988). In *Advances in Helio- and Asteroseismology*, ed. J. Christensen-Dalsgaard & S. Frandsen. Dordrecht: Reidel, p. 155.
Gough, D. O. & Thompson, M. J. (1991). In *Solar Interior and Atmosphere*, ed. A. N. Cox, W. C. Livingston & M. Matthews. Tucson: University of Arizona Press, p. 519.
Grevesse, N. & Noels, A. (1993). In *Origin and Evolution of the Elements*, ed. N. Prantzos, E. Vangioni-Flam & M. Cassé. Cambridge: Cambridge University Press, p. 15.
Haber, D. A., Hindman, B. W., Toomre, J. *et al.* (2002). *Astrophys. J.*, **570**, 855.
Harvey, J. W., Hill, F., Hubbard, R. P. *et al.* (1996). *Science*, **272**, 1284.
Hill, H. A. & Stebbins, R. T. (1975). *Astrophys. J.*, **200**, 471.
Howard, R. & LaBonte, B. J. (1980). *Astrophys. J.*, **239**, L33.
Howe, R. & Thompson, M. J. (1996). *Mon. Not. Roy. Astron. Soc.*, **281**, 1385.
Howe, R., Christensen-Dalsgaard, J., Hill, F. *et al.* (2000a). *Astrophys. J.*, **533**, L163.
Howe, R., Christensen-Dalsgaard, J., Hill, F. *et al.* (2000b). *Science*, **287**, 2456.
Howe, R., Christensen-Dalsgaard, J., Komm, R., Schou, J. & Thompson, M. J. (2005). *Astrophys. J.*, **634**, 1405.
Iglesias, C. A., Rogers, F. J. & Wilson, B. G. (1992). *Astrophys. J.*, **397**, 717.

Julien, K., Werne, J., Legg, S. & McWilliams, J. (1997). In *Solar Convection and Oscillations and their Relationship*, ed. F. P. Pijpers, J. Christensen-Dalsgaard & C. S. Rosenthal. Dordrecht: Kluwer, p. 231.

Kosovichev, A. G. (1988). *Pis'ma Astron. Zh.*, **14**, 344 (English translation: *Sov. Astron. Lett.*, **14**, 145).

Kosovichev, A. G. (1996). *Astrophys. J.*, **469**, L61.

Kosovichev, A. G. (1997). In *Robotic Exploration Close to the Sun: Scientific Basis*, ed. S. R. Habbal, *AIP Conf. Proc.*, **385**. Woodbury, NY: American Institute of Physics, p. 159.

Kosovichev, A. G. (1999). *J. Comp. Appl. Math.*, **109**, 1.

Kosovichev, A. G. & Fedorova, A. V. (1991). *Astron. Zh.*, **68**, 1015 (English translation: *Sov. Astron.*, **35**, 507).

Kosovichev, A. G. & Schou, J. (1997). *Astrophys. J.*, **482**, L207.

Lazrek, M., Fossat, E., Grec, G., Renaud, C. & Schmider, F. X. (2004). In *Helio- and Asteroseismology: Towards a Golden future*, ed. D. Danesy (ESA SP-559), p. 528.

Libbrecht, K. G., Woodard, M. F. & Kaufman, J. M. (1990). *Astrophys. J. Suppl.*, **74**, 1129.

Mestel, L. (1999). *Stellar Magnetism*. Oxford: Clarendon Press.

Michaud, G. & Proffitt, C. R. (1993). In *Inside the Stars*, ed. A. Baglin & W. W. Weiss (*ASP Conf. Ser.*, **40**), p. 246.

Miesch, M. S. (2005). *Living Rev. Solar Phys.*, **2**, 1 (www.livingreviews.org/lrsp-2005-1).

Monteiro, M. J. P. F. G. & Thompson, M. J. (1998). In *Structure and Dynamics of the Interior of the Sun and Sun-like Stars*, ed. S. G. Korzennik & A. Wilson (ESA SP-418), p. 819.

Monteiro, M. J. P. F. G., Christensen-Dalsgaard, J. & Thompson, M. J. (1994). *Astron. Astrophys.*, **283**, 247.

Nordlund, Å., Stein, R. F. & Brandenburg, A. (1996). *Bull. Astr. Soc. India*, **24**, 261.

Pedlosky, J. (1987). *Geophysical Fluid Dynamics*, 2nd edn. New York: Springer-Verlag.

Pijpers, F. P. & Thompson, M. J. (1992). *Astron. Astrophys.*, **262**, L33.

Pinsonneault, M. H., Kawaler, S. D., Sofia, S. & Demarque, P. (1989). *Astrophys. J.*, **338**, 424.

Rabello-Soares, M. C., Basu, S. & Christensen-Dalsgaard, J. (1999). *Mon. Not. Roy. Astron. Soc.*, **309**, 35.

Rempel, M. (2004). *Astrophys. J.*, **607**, 1046.

Rempel, M. & Dikpati, D. (2003). *Astrophys. J.*, **584**, 524.

Ritzwoller, M. H. & Lavely, E. M. (1991). *Astrophys. J.*, **369**, 557.

Rogers, F. J., Swenson, F. J. & Iglesias, C. A. (1996). *Astrophys. J.*, **456**, 902.

Roxburgh, I. W. & Vorontsov, S. V. (1994). *Mon. Not. Roy. Astron. Soc.*, **268**, 880.

Scherrer, P. H., Bogart, R. S., Bush, R. I. *et al.* and the MDI engineering team (1995). *Solar Phys.*, **162**, 129.

Schmitt, J. H. M. M., Rosner, R. & Bohn, H. U. (1984). *Astrophys. J.*, **282**, 316.

Schou, J. (1999). *Astrophys. J.*, **523**, L181.

Schou, J., Antia, H. M., Basu, S. *et al.* (1998). *Astrophys. J.*, **505**, 390.

Schou, J., Christensen-Dalsgaard, J. & Thompson, M. J. (1994). *Astrophys. J.*, **433**, 389.

Schou, J., Howe, R., Basu, S. *et al.* (2002). *Astrophys. J.*, **567**, 1234.

Shibahashi, H. & Takata, M. (1996). *Publ. Astron. Soc. Japan*, **48**, 377.

Skumanich, A. (1972). *Astrophys. J.*, **171**, 565.

Spiegel, E. A. & Zahn, J.-P. (1992). *Astron. Astrophys.*, **265**, 106.

Takata, M. & Shibahashi, H. (1998). *Astrophys. J.*, **504**, 1035.

Takata, M. & Shibahashi, H. (2003). *Publ. Astron. Soc. Japan*, **55**, 1015.

Thompson, M. J. (1990). *Solar Phys.*, **125**, 1.

Thompson, M. J., Christensen-Dalsgaard, J., Miesch, M. S. & Toomre, J. (2003). *Ann. Rev. Astron. Astrophys.*, **41**, 599.
Tomczyk, S., Streander, K., Card, G. *et al.* (1995). *Solar Phys.*, **159**, 1.
Turck-Chièze, S., Couvidat, S., Piau, L. *et al.* (2004). *Phys. Rev. Lett.*, **93**, 211 102.
van Ballegooijen, A. A. (1982). *Astron. Astrophys.*, **113**, 99.
Vorontsov, S. V. (1988). In *Advances in Helio- and Asteroseismology*, ed. J. Christensen-Dalsgaard & S. Frandsen. Dordrecht: Reidel, p. 151.
Vorontsov, S. V., Christensen-Dalsgaard, J., Schou, J., Strakhov, V. N. & Thompson, M. J. (2002). *Science*, **296**, 101.
Zahn, J.-P. (1991). *Astron. Astrophys.*, **252**, 179.
Zhao, J. & Kosovichev, A. G. (2004). *Astrophys. J.*, **603**, 776.

Part III

Hydrodynamic models

4

Hydrodynamic models of the tachocline

Jean-Paul Zahn

I recall here how latitude-dependent rotation imposed by the solar convection zone on the top of the radiation zone would burrow deep into the interior, owing to thermal diffusion, in any laminar and purely hydrodynamic model. Since helioseismology has shown that this differential rotation remains confined in a thin boundary layer, the tachocline, it means that the radiative spread is inhibited by another physical process; this process may be purely hydrodynamic (non-MHD), which is the scope of this chapter, or it may involve magnetic fields: those are considered by Garaud in Chapter 7 of this book. I will show that the confinement of the tachocline can be achieved through an anisotropic turbulent viscosity, whose cause and plausibility are discussed. Other hydrodynamic mechanisms are examined, such as internal gravity waves, which may also play a role in the tachocline. An alternative possibility is that the tachocline is fully embedded in the layer of penetrative convection, in which case no differential rotation would be applied on to the radiation zone.

4.1 Introduction

In 1990, I was invited with Ed Spiegel to give the principal lectures at the Woods Hole summer school. The theme of that year, 'Stellar Fluid Dynamics', was covered extensively by Ed, and I chose to focus on problems related to the rotation of stars. My last lecture, as it happened, was devoted to 'flow between the Sun's convection and radiation zones and transport of chemicals'. Helioseismology had just shown that the whole solar convection zone was rotating differentially, much as observed at the surface, whereas the radiation zone seemed to rotate uniformly (Brown *et al.* 1989). The transition between these two regimes occurred in a thin layer, too thin to be resolved by the available data, and the situation has not changed much since (see Chapter 3 by Christensen-Dalsgaard & Thompson in this book). It thus appeared that the turbulent convection zone was applying a differential rotation on the stably

The Solar Tachocline, D. W. Hughes, R. Rosner and N. O. Weiss.
Published by Cambridge University Press.

stratified radiation zone, and that this would certainly drive a meridional flow in the boundary layer.

My lecture captured some properties of what we later called the tachocline, but it did not address the fundamental question, which we are still striving to answer today:

'Why is the tachocline so thin?'

Ed had a long experience in related matters – he had even coined a name for a similar boundary layer he encountered when he was engaged in the hot debate concerning the solar spin-down (Spiegel 1972). We joined our efforts and shared the pleasure of exploring this new problem: the result was our paper on 'The Solar Tachocline' (Spiegel & Zahn 1992), which I shall discuss next. By tachocline we meant that shallow layer, revealed through helioseismology (Brown *et al.* 1989), which connects the two regimes of differential rotation above and quasi-uniform rotation below. Douglas Gough kindly agreed to read our manuscript, and he suggested that the original name (tachycline) be changed into what it is now. The tachocline is somewhat similar to the ocean thermocline, a layer where it is the temperature that changes rapidly with depth.

4.2 Setting the stage

Experts in helioseismology still debate whether the tachocline is located entirely in the convection zone, or in the radiation zone, or whether it straddles the boundary between the two. One reason is that the location of the tachocline can be determined only within 1% or 2%, owing to the limited resolution of the inversion of the rotation profile. The derivation of the temperature profile is much more precise, but there one faces another problem: because of convective penetration and overshoot, there is no sharp transition between the superadiabatic convective region and the subadiabatic radiative interior below, which would leave a clear seismic signature.

Everybody agrees that one of the main weaknesses of stellar physics remains our poor description of thermal convection. The widely used mixing-length treatment permits us to construct models that represent fairly well the gross properties of stars, but it fails when one attempts to apply it to more subtle features, such as convective penetration.

The situation is rapidly changing, however. Significant progress has been achieved through numerical simulations of increasingly 'turbulent' convection in a stratified medium. These have shown that compressible convection is highly intermittent, displaying strong, long-lived, downwards-directed flows, which contrast with the slower upward motions (e.g. Hurlburt *et al.* 1986; Cattaneo *et al.* 1991; Nordlund *et al.* 1992; Brummell *et al.* 1996). These coherent structures are called plumes, by analogy with those observed in the Earth's atmosphere. They originate

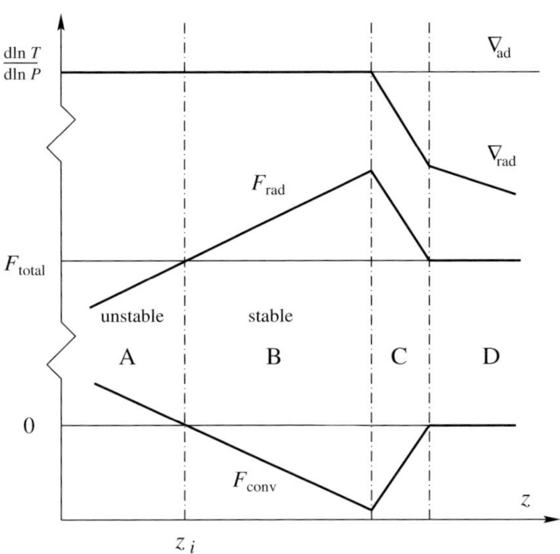

Figure 4.1. Schematic structure of the base of a convective envelope as a function of depth z: A is the superadiabatic convection zone, B is the subadiabatic penetration region, C is the radiative adjustment layer, and D is the radiative interior (after Zahn 1991; courtesy *A&A)*.

in the upper boundary layer, where they are initiated by the strong temperature and density fluctuations, and where they have been detected in the Sun through time–distance tomography (D'Silva *et al.* 1996; Duvall *et al.* 1997).

When these plumes reach the bottom of the unstable region, they still possess a finite velocity, which enables them to penetrate into the stable, subadiabatic interior, where they establish a nearly adiabatic stratification by releasing their excess of heat when they come to rest. A first attempt to estimate the extent of penetration of such plumes was made by Schmitt *et al.* (1984). They found empirically that the penetration depth varies as $f^{1/2}V^{3/2}$, where V is the vertical velocity of the plumes when they enter the stable domain and f their filling factor. This scaling can be easily explained (Zahn 1991).

The stratification at the base of the convective envelope is sketched in Figure 4.1. The letter A designates the unstable region, where the temperature gradient is maintained close to adiabatic by the convective motions. One thus assumes that the radiative leaks are negligible compared to the advective transport of heat, which is the case when the Péclet number is much larger than unity:

$$Pec = \frac{v\ell}{K} \gg 1, \quad K = \frac{\chi}{\rho c_p}, \tag{4.1}$$

where K is the thermal diffusivity and χ is the radiative conductivity; v and ℓ are the velocity and size characterizing the convective motions, i.e. here the central velocity V and the width b of the plumes. At the base of the solar convection zone, $Pec \approx 10^6$.

Owing to the steady increase with depth of the radiative conductivity χ, the radiative flux $F_{\text{rad}} = \chi(\mathrm{d}T/\mathrm{d}z)_{\text{ad}}$ rises until it equals the total flux F_{total} at the depth $z = z_i$, where also the radiative gradient equals the adiabatic gradient: $(\mathrm{d}T/\mathrm{d}z)_{\text{rad}} = (\mathrm{d}T/\mathrm{d}z)_{\text{ad}}$. If there were no convective penetration, this would be the edge of the convection zone, as predicted by the Schwarzschild criterion, and the temperature gradient would thereafter decrease as $\mathrm{d}T/\mathrm{d}z = F_{\text{total}}/\chi$. But the plumes penetrate into the stable region B and they render it nearly adiabatic over some distance d_{pen}, while being decelerated by the buoyancy force. When the Péclet number has dropped below unity, the temperature gradient settles from adiabatic to radiative in a thermal boundary layer (C).

Since the penetration depth is rather small, one may simplify the problem by ignoring the variation with depth of most quantities (density, width of the plume, etc.), and by neglecting the kinetic energy flux and the turbulent entrainment (or detrainment). One keeps of course the variation with z of the conductivity; in the vicinity of z_i, the radiative flux is approximated by

$$F_{\text{rad}} = \chi \left(\frac{\mathrm{d}T}{\mathrm{d}z}\right)_{\text{ad}} = F_{\text{total}} \left[1 + \left(\frac{\mathrm{d}\chi}{\mathrm{d}z}\right)_i (z - z_i)\right]. \tag{4.2}$$

Therefore the convective flux varies as

$$F_{\text{conv}} = -F_{\text{total}} \left(\frac{\mathrm{d}\chi}{\mathrm{d}z}\right)_i (z - z_i)\,. \tag{4.3}$$

This enthalpy flux may be expressed in terms of the vertical velocity V and of the temperature contrast ΔT in the plumes:

$$F_{\text{conv}} = -f \rho c_p V \Delta T, \tag{4.4}$$

where we have introduced the filling factor f of the plumes, defined as the fractional area covered by them.

To estimate the penetration depth, we follow the plumes from $z = z_i$, where their velocity is V_i, until they stop at $z = z_i + d_{\text{pen}}$. Their deceleration is described to lowest order by

$$\frac{1}{2}\frac{\mathrm{d}V^2}{\mathrm{d}z} = g\frac{\Delta\rho}{\rho} = -g\frac{\Delta T}{T}. \tag{4.5}$$

After elimination of ΔT with (4.4), the integration of (4.5) yields the following expression for the penetration depth $d_{\rm pen}$:

$$d_{\rm pen}^2 = \frac{3}{5} H_p H_\chi \left\{ f \frac{\rho V_i^3}{F_{\rm total}} \right\} = \frac{3}{5} H_p H_\chi Q, \tag{4.6}$$

where H_p is the scale-height of the pressure and H_χ that of the radiative conductivity. This is the scaling which was obtained empirically by Schmitt *et al.* (1984).

The term Q in curly brackets is determined by the dynamics in the convection zone; probably it does not depend much on its size, provided it is large enough, compared to the scale-height. In the mixing-length treatment, $Q = (1/10)\, \Lambda/H_p$ in the bulk of the convection zone. This value, with $\Lambda/H_p \approx 1.5$ and $H_\chi \approx H_p/2$, would yield a penetration depth of the order of $1/5$ of the pressure scale-height at the base of the solar convection zone.

However, the vertical velocity must vary from plume to plume, and so also does the extent of their penetration. Therefore the horizontal average of the temperature gradient is probably a smooth function of depth, changing gradually from adiabatic to radiative, a property which is not captured by the crude model above, which predicts a jump in the temperature gradient that depends on the amount of penetration.

What we have called convective *penetration* is that part of the excursion of convective motions into the stable region where they enforce an almost adiabatic stratification ($\nabla_{\rm ad} - \nabla \ll 1$). It corresponds to region B in Figure 4.1, in which the Péclet number is substantially larger than unity (we recall that at the base of the solar convection zone, $Pec \approx 10^6$). In comparison, the thermal adjustment layer C is extremely thin: about 1 km in the Sun; it is this layer that we suggest should be called *overshoot*.

This is not just a semantic point. For lack of spatial resolution, present day simulations are unable to achieve a Péclet number which would realistically describe convective penetration. Even in the very-high-resolution calculations performed by Brummell *et al.* (2002) heat leaks too fast from the plumes: this was pointed out by Rempel (2004), who recently developed a more sophisticated version of Zahn's semi-analytical model. The two-dimensional simulations by Rogers & Glatzmaier (2005) seem to suffer from the same shortcoming. Thus we will have to wait until the numerical simulations reach sufficient resolution to represent this convective penetration in a realistic way.

4.3 The tachocline: governing equations and simplifying assumptions

From now on, I shall ignore the possible complexity of that transition layer, and I shall assume that some amount of differential rotation is applied on to the top of the radiation zone, in other words that at least part of the tachocline is stably stratified, with negligible transport of heat through penetrative convection. There, the dynamics is governed by the continuity equation

$$\frac{\partial \rho}{\partial t} + \nabla \cdot (\rho \mathbf{V}) = 0, \tag{4.7}$$

the momentum equation

$$\rho \left[\frac{\partial \mathbf{V}}{\partial t} + (\mathbf{V} \cdot \nabla)\mathbf{V} + 2\Omega \times \mathbf{V} \right] = -\nabla P + \rho \mathbf{g} + \nabla \cdot \overline{\overline{\mathrm{T}}}, \tag{4.8}$$

and by the heat equation

$$\rho T \frac{\partial S}{\partial t} + \rho T \mathbf{V} \cdot \nabla S = \nabla \cdot (\chi \nabla T), \tag{4.9}$$

where we have deliberately ignored the possible effect of a magnetic field. The usual notations have been taken for density ρ, pressure P, temperature T, specific entropy S and gravity $\mathbf{g}$; χ is the radiative conductivity and $\overline{\overline{\mathrm{T}}}$ is the viscous stress tensor. We neglect the time variation of the angular velocity Ω, which is that of the reference frame, and thus we do not address here the spin-down problem. Using spherical coordinates (r, θ, ϕ) centred on the star, we look only for axisymmetric solutions; then the velocity field contains both a meridional and an azimuthal component: $\mathbf{V} = (u, v, r \sin\theta\, \widehat{\Omega})$, where $\widehat{\Omega}(r, \theta)$ is the differential rotation.

We linearize the governing equations by assuming that the perturbations of pressure, etc. are small; thus $P(r, \theta, t) \rightarrow P(r, t) + \widehat{P}(r, \theta, t)$, and likewise for temperature and entropy. We use the perfect gas equation of state $\widehat{P}/P = \widehat{\rho}/\rho + \widehat{T}/T$, and assume that the velocity $|\mathbf{V}|$ is small compared to the rotational velocity $r\Omega$. *A fortiori* $|\mathbf{V}|$ is then small compared to the sound speed, and we may use the anelastic approximation; we thus neglect the time derivative of ρ in Equation (4.7), which allows us to write the mass flux in terms of a stream function $\widehat{\Psi}(r, \theta, t)$:

$$r^2 \rho u = \frac{\partial \widehat{\Psi}}{\partial \mu}, \qquad r(1-\mu^2)^{\frac{1}{2}} \rho v = \frac{\partial \widehat{\Psi}}{\partial r}, \tag{4.10}$$

with $\mu = \cos\theta$. Furthermore, we assume that the centrifugal force is small compared to gravity, so that we may neglect the oblateness of the level surfaces.

Our final assumption is that the layer we are dealing with is very thin compared to its radius, and thus that the vertical variation scale of all perturbation functions $(\widehat{P}, \widehat{T}, \widehat{\Psi})$ is much shorter than both the radius r and the scale-height of the structure functions (P, ρ, T) which describe the unperturbed model.

In this thin-layer limit, the governing equations take the following form:

$$-\frac{1}{\rho}\frac{\partial \widehat{P}}{\partial r}+g\frac{\widehat{T}}{T}=0, \tag{4.11}$$

$$-2\Omega\mu r\widehat{\Omega}=\frac{1}{\rho r}\frac{\partial \widehat{P}}{\partial \mu}, \tag{4.12}$$

$$\frac{\partial \widehat{\Omega}}{\partial t}+\frac{2\Omega\mu}{(1-\mu^2)}\frac{1}{\rho r^2}\frac{\partial \widehat{\Psi}}{\partial r}=\nu\frac{\partial^2 \widehat{\Omega}}{\partial r^2}, \tag{4.13}$$

$$\frac{\partial \widehat{T}}{\partial t}+\frac{N^2}{g}\frac{T}{\rho r^2}\frac{\partial \widehat{\Psi}}{\partial \mu}=K\frac{\partial^2 \widehat{T}}{\partial r^2}. \tag{4.14}$$

We have introduced here the viscosity ν, the thermal diffusivity $K=\chi/\rho C_p$ and the square of the buoyancy frequency, $N^2=(g/H_p)(\nabla_{\rm ad}-\nabla)$, where H_p is the pressure scale-height and where $\nabla=\partial \ln T/\partial \ln P$ designates the logarithmic temperature gradient. Note that we have filtered out the fastest times, which characterize the hydrostatic and baroclinic (or geostrophic) adjustments, and thus that we have eliminated the gravity and inertial waves.

To integrate this system one has to impose four boundary conditions. One specifies the latitude-dependent angular velocity $\widehat{\Omega}(\theta)$ which is applied by the convection zone at $r=r_{\rm cz}$. Another imposes the continuity of the vertical gradient of $\widehat{\Omega}(r,\theta)$, which ensures the continuity of the temperature fluctuation $\widehat{T}$; since according to helioseismology the differential rotation varies little with depth in the convection zone, we may assume that $\partial_r\widehat{\Omega}(r,\theta)=0$ at $r=r_{\rm cz}$. Finally the differential rotation must vanish deep enough in the star: hence $\widehat{\Omega}$, $\partial_r\widehat{\Omega}\to 0$ as $r\to 0$. The top boundary is permeable to the meridional flow; therefore no Ekman layer is required to match there the interior solution.

4.4 Laminar penetration

The differential system above admits solutions which separate in radius and colatitude, if the perturbation functions are expanded as

$$\widehat{P}(r,\mu,t)=\sum_{i>0}\widetilde{P}_i(r,t)f_i(\mu), \tag{4.15}$$

and likewise for $\widehat{T}$, while

$$\widehat{\Psi}=\sum_{i>0}\widetilde{\Psi}_i(r,t)\int_0^{\mu}f_i(\mu')\mathrm{d}\mu', \tag{4.16}$$

and

$$\mu\widehat{\Omega} = \sum_{i>0} \widetilde{\Omega}_i(r,t)\frac{\mathrm{d}f_i(\mu)}{\mathrm{d}\mu}. \tag{4.17}$$

One readily finds that the horizontal functions $f_i(\mu)$ obey the second-order differential equation

$$\frac{\mathrm{d}}{\mathrm{d}\mu}\left[\frac{(1-\mu^2)}{\mu^2}\frac{\mathrm{d}f_i}{\mathrm{d}\mu}\right] + \lambda_i^2 f_i = 0. \tag{4.18}$$

The solutions obtained by imposing that they be regular at the poles $\mu = \pm 1$ constitute a set of orthogonal eigenfunctions with l nodes in μ, much like the Legendre functions.

The mode amplitudes are solutions of the following differential system:

$$-\frac{1}{\rho}\frac{\partial \widetilde{P}_i}{\partial r} + g\frac{\widetilde{T}_i}{T} = 0, \tag{4.19}$$

$$-2\rho r^2 \Omega \widetilde{\Omega}_i = \widetilde{P}_i, \tag{4.20}$$

$$\frac{\partial \widetilde{\Omega}_i}{\partial t} - \frac{2\Omega}{\lambda_i^2 \rho r_{\mathrm{cz}}^2}\frac{\partial \widetilde{\Psi}_i}{\partial r} = \nu\frac{\partial^2 \widetilde{\Omega}_i}{\partial r^2}, \tag{4.21}$$

$$\frac{\partial \widetilde{T}_i}{\partial t} + \frac{N^2}{g}\frac{T}{\rho r_{\mathrm{cz}}^2}\widetilde{\Psi}_i = K\frac{\partial^2 \widetilde{T}}{\partial r^2}. \tag{4.22}$$

Let us examine the solutions of this system, starting with an unperturbed interior and applying the differential rotation $\widehat{\Omega} = \sum_i \widetilde{\Omega}_i(r_{\mathrm{cz}})\mu^{-1}\mathrm{d}f_i/\mathrm{d}\mu$ at $t = 0$ on the top of the radiation zone. After a quick dynamical adjustment, geostrophic balance is achieved: from Equations (4.19) and (4.20) one has

$$\frac{\widetilde{T}_i}{T} = -2\left(\frac{r_{\mathrm{cz}}^2\Omega^2}{g}\right)\frac{\partial}{\partial r}\left(\frac{\widetilde{\Omega}_i}{\Omega}\right), \tag{4.23}$$

a relation which tightly couples the temperature perturbation with the differential rotation. Likewise, before the effects of diffusion are felt, the vertical advection of heat is linked to the horizontal advection of angular momentum, as can be seen by eliminating $\widetilde{\Psi}_i$ between (4.21) and (4.22), and integrating in time:

$$\frac{\widetilde{\Omega}_i}{\Omega} = -\frac{2}{\lambda_i^2}\frac{g}{N^2}\frac{\partial}{\partial r}\left(\frac{\widetilde{T}_i}{T}\right). \tag{4.24}$$

These two equations, (4.23) and (4.24), are readily combined and solved to yield a boundary layer whose scale-height is the Rossby height

$$h_{\mathrm{Ro}} = \frac{r_{\mathrm{cz}}}{\lambda_i}\left(\frac{2\Omega}{N}\right). \tag{4.25}$$

This layer is well-known in stratified spin-down theory and before that in the theory of stratospheric disturbances (Rossby 1938; Holton 1965; Sakurai 1966; Walin 1969; Clark *et al.* 1971; Spiegel 1972): over its range rotation overpowers stratification to locally establish a Taylor–Proudman regime. With the parameters characterizing the present Sun $h_{\mathrm{Ro}} \approx 1000\,\mathrm{km}$, a value which was larger in the past, when the Sun was a fast rotator. However, this layer plays little role in the present problem: it is extremely short-lived because it does not fulfil the boundary conditions, and promptly thermal and viscous diffusion begin to operate.

By some straightforward eliminations, the full system (4.19)–(4.22) can be cast into the following evolution equation for the modal amplitudes of the differential rotation:

$$\left[\frac{\partial}{\partial t} - \nu\frac{\partial^2}{\partial r^2}\right]\widetilde{\Omega}_i - \left(\frac{2\Omega}{N}\right)^2 \left(\frac{r_{\mathrm{cz}}}{\lambda_i}\right)^2 \left[\frac{\partial}{\partial t} - K\frac{\partial^2}{\partial r^2}\right]\frac{\partial^2\widetilde{\Omega}_i}{\partial r^2} = 0. \tag{4.26}$$

It describes how a differential rotation which is imposed at the top of the radiation zone ($r = r_{\mathrm{cz}}$) propagates inwards. After a rapid thermal relaxation this equation reduces to

$$\frac{\partial\widetilde{\Omega}_i}{\partial t} + \left(\frac{2\Omega}{N}\right)^2 \left(\frac{r_{\mathrm{cz}}}{\lambda_i}\right)^2 K\frac{\partial^4\widetilde{\Omega}_i}{\partial r^4} - \nu\frac{\partial^2\widetilde{\Omega}_i}{\partial r^2} = 0. \tag{4.27}$$

In the Sun, owing to the smallness of the Prandtl number ($\nu/K \approx 10^{-6}$), the spread of the tachocline is governed mainly by thermal diffusion – or rather by 'hyper-diffusion'.[1] To understand this process, we must keep in mind that the differential rotation and the temperature perturbation are coupled through the geostrophic balance (4.23). If a latitude-dependent rotation is applied at the top of the radiation zone, the temperature fluctuation associated with it tends to spread inwards through radiative diffusion, as allowed by Equation (4.22). In contrast, since the viscosity is very low, the differential rotation can be modified only by advection, hence through a meridional circulation. This circulation works against the diffusion of heat, and this is why the whole process is turned into a hyper-diffusion, which at long time operates much slower than diffusion: the spread evolves as $t^{1/4}$.

Taking into account that the Sun was rotating faster in the past, by assuming for instance that the rotation rate followed the law $\Omega(t) \propto t^{-1/2}$ (Skumanich 1972), one can estimate that the tachocline would at present stretch down to $r = 0.3R_{\odot}$.[2]

Since such a rotation profile is clearly ruled out by acoustic sounding (Kosovichev 1996; Basu 1997; Antia *et al.* 1998; Charbonneau *et al.* 1999a,b; Corbard *et al.*

[1] This is not the case in most numerical simulations so far, where viscous diffusion dominates.

[2] Gilman & Miesch (2004) reach a different conclusion: according to them, an upper limit to the extent of penetration is given by the buoyancy-diffusion layer, whose thickness is derived from their Equation (5) by equating the first and third terms. Actually, the dominant terms are the second and third, and it can be shown that they do not allow for a stationary solution that satisfies the boundary conditions imposed on the angular velocity. This and related points are discussed in detail by M. E. McIntyre in Chapter 8 of this volume.

1999), we must conclude that another physical process – at least – interferes with the radiative diffusion of the tachocline.

Thus a model of the tachocline can claim to be consistent only if it identifies that process, and if it proves its efficacy in preventing the spread of the tachocline.

4.5 A hydrodynamic remedy against the spread of the tachocline: anisotropic turbulence

The mechanism that was invoked in Spiegel & Zahn (1992) to account for the thinness of the tachocline was a mild, but strongly anisotropic, turbulent viscosity, such as one may expect from shear instabilities generated by the latitudinal differential rotation in this stratified medium (its plausibility will be discussed later on). If the degree of anisotropy is high enough, we may neglect the vertical diffusion of angular momentum compared to its horizontal diffusion, and the equation describing the transport of angular momentum becomes, instead of Equation (4.13):

$$\rho r^2(1-\mu^2)\frac{\partial\widehat{\Omega}}{\partial t} + 2\Omega\mu\frac{\partial\widehat{\Psi}}{\partial r} = \rho\frac{\partial}{\partial\mu}\left[\nu_{\rm h}(1-\mu^2)^2\frac{\partial\widehat{\Omega}}{\partial\mu}\right], \tag{4.28}$$

where $\nu_{\rm h}$ is the horizontal component of the turbulent viscosity. The problem no longer separates into (r, t) and μ, as it did in the former case, except in the stationary regime, where one can again expand the perturbations as in Equations (4.15), (4.16) and (4.17), but where the horizontal functions F_i now are the solutions of the following fourth-order differential equation:

$$\frac{\mathrm{d}}{\mathrm{d}\mu}\left\{\frac{1}{\mu}\frac{\mathrm{d}}{\mathrm{d}\mu}\left[(1-\mu^2)^2\frac{\mathrm{d}}{\mathrm{d}\mu}\left(\frac{1}{\mu}\frac{\mathrm{d}F_i}{\mathrm{d}\mu}\right)\right]\right\} - (\mu_i)^4 F_i = 0. \tag{4.29}$$

With regularity conditions applied at the poles $\mu = \pm 1$, these eigenfunctions form an orthogonal set, and $(\mu_i)^4 \geq 0$.

The modal amplitude of the differential rotation here obeys the fourth-order differential equation:

$$\left(\frac{2\Omega}{N}\right)^2 \frac{K}{\nu_{\rm h}}\left(\frac{r_{\rm cz}}{\mu_i}\right)^4 \frac{\partial^4\widetilde{\Omega}_i}{\partial r^4} + \widetilde{\Omega}_i = 0. \tag{4.30}$$

It is the lowest-order mode ($i = 4$) that penetrates the deepest, and its first node may be taken to define the thickness of the tachocline:

$$h \approx r_{\rm cz}\left(\frac{\Omega}{N}\right)^{1/2}\left(\frac{K}{\nu_{\rm h}}\right)^{1/4}. \tag{4.31}$$

This result was established in the thin-layer limit; it was later confirmed through complete two-dimensional calculations performed by Elliott (1997) – see Figure 4.2.

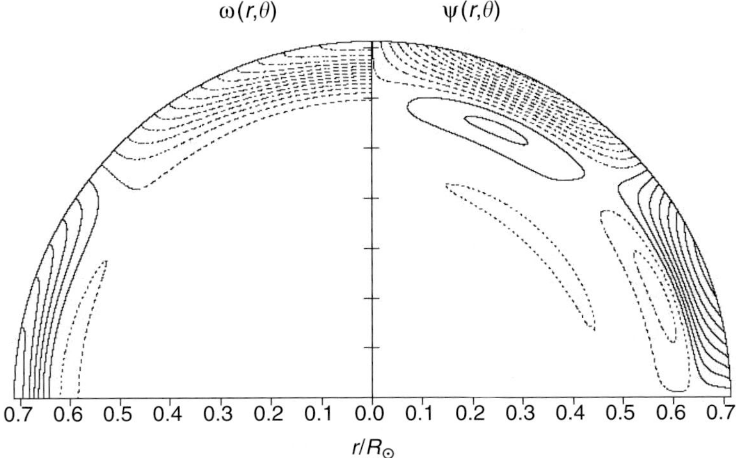

Figure 4.2. Stationary tachocline model, obtained through anisotropic turbulent viscosity, with $\nu_\mathrm{h}/\nu_\mathrm{v} = 1000$ and $\nu_\mathrm{h} = 5 \times 10^4\,\mathrm{cm}^2\,\mathrm{s}^{-1}$ (Elliott 1997, courtesy *A&A*). The left panel displays the angular velocity, the right panel a sample of meridional streamlines.

We thus conclude that anisotropic turbulence is capable of halting the spread of the tachocline, by smoothing out the differential rotation in latitude. If this mechanism operates in the present Sun, the horizontal component of the turbulent viscosity must exceed $\nu_\mathrm{h} \approx 10^5\,\mathrm{cm}^2\,\mathrm{s}^{-1}$ to ensure a tachocline thickness of less than 5%, which appears to be the upper limit derived from helioseismology (Corbard *et al.* 1999).

One may be tempted also to invoke a stable composition gradient as another remedy against the spread; such a gradient certainly exists below the convection zone, owing to the gravitational settling of helium (Noerdlinger & Arigo 1980; Stringfellow *et al.* 1983). At closer inspection, however, this gradient would contribute to the spread by adding a diffusivity

$$D_\mu = K \left(\frac{g}{r_\mathrm{zc} N^2} \right) \frac{\mathrm{d} \ln \mu}{\mathrm{d} \ln r}, \tag{4.32}$$

a somewhat paradoxical result which has been derived in Spiegel & Zahn (1992).

4.6 Which turbulence?

In our 1992 paper, Spiegel and I declared from the onset that we didn't want to 'go into the details of the linear and nonlinear instability mechanisms that may produce the turbulence, except to remark that it is to be expected on account of the large Reynolds number of the horizontal shear'. This statement was admittedly somewhat cavalier, and our assumption, that of an anisotropic turbulence acting such as to restore uniform rotation, has since come under criticism – quite deservedly.

One objection raised against our model was that the tachocline may not be turbulent at all. The linear stability of a latitude-dependent rotation profile has been widely discussed, starting with Watson (1981), who derived a criterion similar to Rayleigh's famous inflection point theorem, for horizontal, two-dimensional perturbations, in spherical geometry. He applied it to a rotation law of the form $\Omega = s_0 - s_2\mu^2$, and concluded that it would be unstable provided the difference of angular velocity between equator and pole exceeds 29%. But later it was found that the stability threshold is very sensitive to the presence of a quartic term: $\Omega = s_0 - s_2\mu^2 - s_4\mu^4$ (Dziembowski & Kosovichev 1987; Charbonneau *et al.* 1999a,b). Thus, depending on the results of helioseismic inversions, the solar profile turned out to be either stable (Charbonneau *et al.* 1999a,b) or marginally unstable (Garaud 2001). The investigation was extended by Dikpati & Gilman (2001) to three-dimensional perturbations, in the shallow-water approximation; they confirmed the two-dimensional results in the presence of strong stratification, and discovered a new type of unstable mode in the case of weak stratification. Very recently Arlt *et al.* (2005) apparently settled the case: considering unrestricted three-dimensional perturbations, they showed that the pole-equator difference of angular velocity can be as large as 52% before instability sets in.

But such turbulence could well be caused by nonlinear processes. Indeed, rotating shear flows that are linearly stable are observed to become turbulent in the laboratory, above a critical Reynolds number which is largely exceeded in the Sun (see Richard & Zahn 1999; Dubrulle *et al.* 2005). The question is thus less that of which instability causes the turbulence, rather than what are the properties of developed turbulence in stratified rotating shearing flows.

A crucial assumption made in the Spiegel–Zahn model is that the turbulence is anisotropic, and that it acts to suppress the latitudinal shear. The postulated anisotropy seems plausible, given the strong stratification in the solar radiation zone, and the fact that no restoring force opposes the horizontal displacements; such turbulence was conjectured already in Zahn (1975). But the second property is far from guaranteed, although it is displayed in laboratory flows, as we shall recall below. Michael McIntyre (1994, 2003b) discussed the question in detail, based on his experience of the Earth's stratosphere and on theories of potential vorticity inversion: he argues that two-dimensional layered turbulence tends to transport potential vorticity, and that its effect would tend to be anti-frictional, as observed in the Earth's atmosphere, rather than eddy-viscosity like, as assumed in Equation (4.28) (see his Chapter 8 in this book).

How can one reconcile the behaviour of the laboratory experiment with that of the Earth's atmosphere? Presumably the determining factor is what produces the

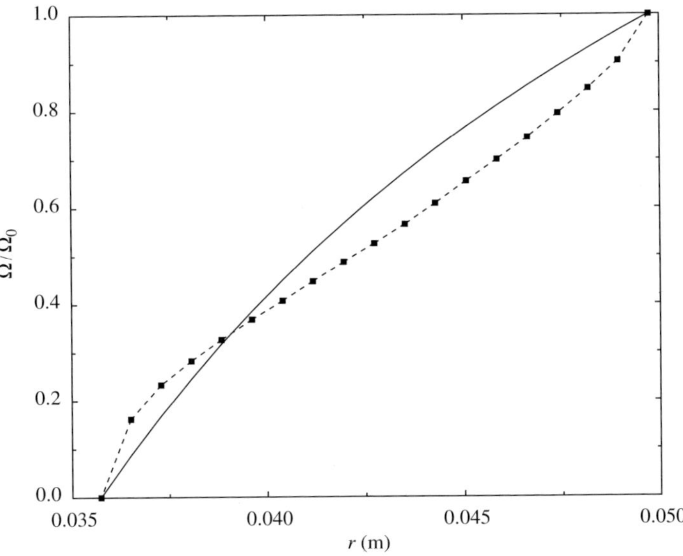

Figure 4.3. Angular velocity profiles in a Couette–Taylor experiment, with the axial distance as abscissa (in metres). The inner cylinder is at rest, the Reynolds number characterizing the angular velocity Ω_0 of the outer cylinder is 7×10^4. The laminar (linearly stable) profile is drawn as a continuous line, the dotted line joins the experimental points. The effect of turbulence is to reduce the angular velocity gradient, thus to suppress the cause of turbulence, which here is the shear (Richard 2001).

turbulence. If it is the latitudinal shear, in other words the barotropic instability, turbulence will act to suppress its cause, namely the differential rotation. This property is observed in the classical Couette–Taylor experiment (cf. Richard & Zahn 1999). When the outer cylinder is rotating and the inner is at rest, the angular velocity profile is linearly stable; then turbulence is caused by nonlinear shear instabilities, which are still not well understood yet, but which clearly tend to reduce the shear, and hence to flatten the angular velocity profile (see Figure 4.3). On the contrary, when the inner cylinder is rotating and the outer is at rest, turbulence results from the Rayleigh instability, which occurs when angular momentum decreases outwards; turbulence acts then to smooth out the angular momentum profile (at least at a moderate Reynolds number – fully developed turbulence tends again to suppress the shear). In both cases, the turbulent stresses are tuned to suppress the cause of turbulence.

However, it is quite possible that the turbulence in the tachocline is not due to a local instability, but that it is linked to the much more vigorous turbulence in the convection zone. This aspect has been explored by Mark Miesch (2003), with global-scale, three-dimensional simulations of stably stratified turbulence driven by penetrative convection, with imposed differential rotation. They are described by him in Chapter 5 of this book, and it suffices here to give the salient results. The

Reynolds stresses are such that they carry momentum poleward and outward, thus implying diffusive latitudinal transport and anti-diffusive vertical transport, much like in the Spiegel–Zahn model. Surprisingly, this is achieved with mildly anisotropic turbulence, and it is not the differential rotation which causes the turbulence.

4.7 Ventilation time and mixing

Once the stationary regime is established, the ventilation time of the tachocline through its meridional flow is given by

$$\frac{1}{t_{\rm vent}} \approx \frac{u}{h} \approx \frac{K}{r^2}\left(\frac{\Omega}{N}\right)^2 \left(\frac{r_{\rm cz}}{h}\right)^4 \frac{\delta\Omega}{\Omega}, \tag{4.33}$$

where $\delta\Omega$ measures the differential rotation in latitude. This expression makes no reference to the process which opposes the spread of the tachocline, but it does so implicitly, through the value of the tachocline thickness h. In the Sun, taking $h/r_{\rm cz} = 0.05$, $\Omega/N = 10^{-4}$ and $\delta\Omega/\Omega = 0.1$, we estimate the ventilation time to be $t_{\rm vent} \approx 2.5 \times 10^6$ yr.

This time is very short compared to the nuclear evolution time, and it conveys the impression that the tachocline is very well mixed. In fact, this property was used by Elliott (1997) to calculate the helium profile below the solar convection zone, which in his model resulted from the competition between gravitational settling and advection by the tachocline circulation. In their discussion of lithium depletion, Elliott & Gough (1999) again assumed that the tachocline is completely mixed.

However, this is not necessarily the case. In the presence of anisotropic turbulence the advection of chemicals is severely eroded by the horizontal diffusion, and it is turned into a weak vertical diffusion. This was shown by Chaboyer & Zahn (1992), who derived the following expression for the resulting effective diffusivity, when the vertical circulation velocity is expanded in spherical functions $u = \sum U_n(r)P_n(\cos\theta)$:

$$D_{\rm eff} = \frac{r^2}{D_{\rm h}} \sum_n \frac{U_n^2(r)}{n(n+1)(2n+1)} \tag{4.34}$$

($D_{\rm h} \approx \nu_{\rm h}$ is the horizontal turbulent diffusivity). Since the dominant term in the tachocline is the octupolar component $n = 4$, the time characterizing this effective diffusion is substantially longer than the ventilation time:

$$\frac{t_{\rm diff}}{t_{\rm vent}} = 180 \left(\frac{\delta\Omega}{\Omega}\right)^{-1}. \tag{4.35}$$

Even so, in spite of its reduced efficiency due to the strong horizontal transport, mixing in the tachocline contributes to shaping the helium distribution and to depleting lithium. This was demonstrated by Brun *et al.* (1999, 2002); the result is displayed

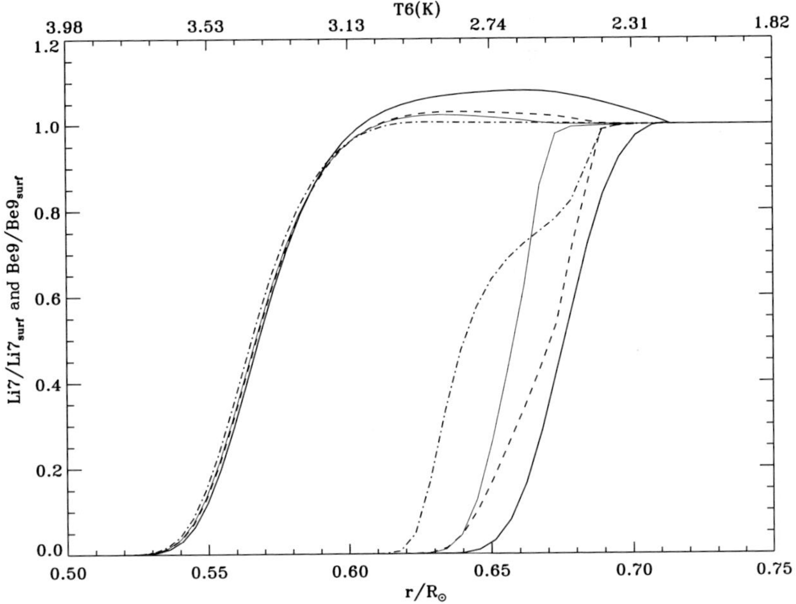

Figure 4.4. Radial profiles of the ^{7}Li and ^{9}Be composition, normalized to their initial values, for a model of the present Sun including only microscopic diffusion (solid line) and for two models where macroscopic mixing has been added, according to Spiegel & Zahn's model (Brun *et al.* 1999, courtesy *ApJ*).

in Figure 4.4, where it is shown that although lithium is depleted by such local mixing, beryllium is preserved.

4.8 Internal gravity waves

Internal gravity waves are certainly emitted at the base of the convection zone and, since they transport angular momentum, they may well play a role in the dynamics of the tachocline. Owing to radiative damping, these waves deposit in the tachocline some of their momentum flux, whose sign depends on the vertical shear, and therefore on latitude. The effect of such waves has been described by Fritts *et al.* (1998), assuming that the mean shear was maintained by an extra mechanism, which they did not mention or identify. In other words, uniform rotation was *imposed* at some arbitrary depth. Under those conditions, the body force exerted by the waves induces a stationary meridional circulation, whose ventilation time is of order 10^5 yr, thus shorter than that driven by thermal hyper-diffusion, which has been discussed above. This suggests that internal gravity waves may indeed contribute to the dynamics of the tachocline, although that model cannot claim to be self-consistent, as it stands, since it ignores the feedback of the waves on the differential rotation and since it lacks the ingredient which prevents the spread of the tachocline.

As Gough & McIntyre (1998) reminded us, geophysicists know well that broadband internal waves can act as another anti-frictional process, tending to increase the ambient shear, because prograde waves (in the sense of mean relative velocity) are more damped than retrograde waves, since they are Doppler-shifted to lower frequency. Once the flow becomes shear-unstable, turbulence sets in to suppress the shear, and a new shear layer builds up, which is of opposite direction, and so on, leading to periodic oscillations. This phenomenon is observed in the Earth's atmosphere, where it is called the quasi-biennial oscillation (QBO; see McIntyre 2003b).

Such an oscillation presumably operates below the solar convection zone, and thus in the tachocline. It has been described by Kim & MacGregor (2001), who considered a pair of waves of the same frequency (one prograde and one retrograde). They found that the behaviour of the shear layer depends on the value of the (turbulent) viscosity: at high enough viscosity, the shear is stationary; as the viscosity is lowered an oscillation appears, first with a single period, next quasi-periodic, etc., until the regime is chaotic.

Similar conclusions were reached by Talon *et al.* (2002), who included a whole (albeit sparse) spectrum of waves. The chosen parameters (frequencies, turbulent viscosity) were meant to represent the solar conditions. The rotation profile was then found to oscillate periodically around a mean value, in a very thin shear layer, owing to the damping of the high-order, short-period waves. This shear layer is transparent to the low-order, long-period waves, and these are able to extract or to deposit angular momentum in the core of the Sun, depending on whether the mean angular velocity increases or decreases just below the convection zone, and hence in the oscillating shear layer. More recently, Charbonnel & Talon (2005; see also Talon & Charbonnel 2005) succeeded in demonstrating that these low-period internal waves are efficient enough to enforce uniform rotation in the solar radiative interior, when they are coupled with thermally driven meridional circulation and shear induced turbulence.

Let us stress, however, that both teams ignored the differential rotation in latitude, and considered only the radial dependence of the angular velocity; therefore these works are hardly relevant to the dynamics of the tachocline as such. In particular, they give no clue on whether internal gravity waves are able or not to confine the tachocline.

4.9 Conclusion

The editors of this book assigned me the task of discussing the purely hydrodynamic (non-MHD) models of the tachocline. Although I proposed the first of such models with Ed Spiegel, this does not mean that we necessarily believe that the ultimate model of the tachocline will be purely hydrodynamical. But it ought to be self-consistent, and therefore it must explain how the radiative spread of the tachocline is prevented. What we have shown is that, in the strict frame of hydrodynamics, this can be achieved through anisotropic turbulence, which could be due either to a local shear instability or to the convection zone above.

We know well that the tachocline could also be confined by magnetic fields, either by a fossil field anchored in the deep interior, as described by Gough & McIntyre (1998), or by the cyclic dynamo field imposed from above, as was suggested recently by Forgács-Dajka & Petrovay (2002). These possibilities are discussed by Pascale Garaud in Chapter 7 of this volume.

In our model, as in most models so far, the convection zone plays a passive role: it is just invoked to provide the upper boundary conditions for the tachocline, i.e. to enforce the latitude dependent rotation. However, as McIntyre (2003a,b) points out, the tachocline circulation could well be driven by turbulent stresses in the convection zone. He goes even further in claiming that, 'in order to predict the Sun's differential rotation in the convection zone, even qualitatively, a convection zone model must be fully coupled with a tachocline model'. This has been undertaken by Miesch (2003, Miesch *et al.* 2006; see also his Chapter 5 in this book), and the picture he draws is that of a tachocline fully embedded in the layer of penetrative convection.

Obviously, the time has come where we have to go beyond the semi-analytical models used in the first exploration of the problem. Its solution probably depends on the high-resolution three-dimensional simulations which are now being performed by several teams.

Thus I wouldn't be too surprised if our 1992 model were quoted as merely a curiosity, in the next meeting devoted to the solar tachocline.

Acknowledgments

I wish to thank David Hughes, Bob Rosner and Nigel Weiss for organizing this very stimulating workshop, and the Isaac Newton Institute in Cambridge for its hospitality. Sacha Brun, Michael McIntyre and Juri Toomre kindly read the manuscript, and their remarks were most helpful.

References

Antia, H. M., Basu, S. & Chitre, S. M. (1998). *Mon. Not. Roy. Astron. Soc.*, **298**, 543.
Arlt, R., Sule, A. & Rüdiger, G. (2005). *Astron. Astrophys.*, **411**, 1171.
Basu, S. (1997). *Mon. Not. Roy. Astron. Soc.*, **288**, 572.
Bretherton, F. P. & Spiegel, E. A. (1968). *Astrophys. J.*, **153**, 277.
Brown, T. M., Christensen-Dalsgaard, J., Dziembowski, W. A. *et al.* (1989). *Astrophys. J.*, **343**, 526.

Brummell, N. H., Clune, T. L. & Toomre, J. (2002). *Astrophys. J.*, **570**, 825.
Brummell, N. H., Hurlburt, N. E. & Toomre, J. (1996). *Astrophys. J.*, **473**, 494.
Brun, A. S., Antia, H. M., Chitre, S. M. & Zahn, J.-P. (2002). *Astron. Astrophys.*, **391**, 725.
Brun, A. S., Turck-Chièze, S. & Zahn, J.-P. (1999). *Astrophys. J.*, **525**, 1032.
Cattaneo, F., Brummell, N. H., Toomre, J., Malagoli, A. & Hurlburt, N. E. (1991). *Astrophys. J.*, **370**, 282.
Chaboyer, B. & Zahn, J.-P. (1992). *Astron. Astrophys.*, **253**, 173.
Charbonneau, P., Christensen-Dalsgaard, J., Henning, R. *et al.* (1999a). *Astrophys. J.*, **527**, 445.
Charbonneau, P., Dikpati, M. & Gilman, P. A. (1999b). *Astrophys. J.*, **526**, 523.
Charbonnel, C. & Talon, S. (2005). *Science*, **309**, 2189.
Clark, A., Clark, P. A., Thomas, J. H. & Lee, N.-H. (1971). *J. Fluid Mech.*, **45**, 131.
Corbard, T., Blanc-Férau, L., Berthomieu, G. & Provost, J. (1999). *Astron. Astrophys.*, **344**, 696.
Dikpati, M. & Gilman, P. A. (2001). *Astrophys. J.*, **551**, 536.
D'Silva, S., Duvall, T. L., Jefferies, S. M. & Harvey, J. W. (1996). *Astrophys. J.*, **471**, 1030.
Dubrulle, B., Dauchot, O., Daviaud, F. *et al.* (2005). *Phys. Fluids*, **17**, 5103.
Duvall, T. L., Kosovichev, A. G., Scherrer, P. H. *et al.* (1997). *Solar Phys.*, **170**, 63.
Dziembowski, W. A. & Kosovichev, A. G. (1987). *Acta Astron.*, **37**, 341.
Elliott, J. R. (1997). *Astron. Astrophys.*, **327**, 1222.
Elliott, J. R. & Gough, D. O. (1999). *Astrophys. J.*, **516**, 475.
Forgács-Dajka, E. (2004). *Astron. Astrophys.*, **413**, 1143.
Forgács-Dajka, E. & Petrovay, K. (2000). *Sol. Phys.*, **203**, 195.
Forgács-Dajka, E. & Petrovay, K. (2002). *Astron. Astrophys.*, **389**, 629.
Fritts, D. C., Vadas, S. L. & Andreassen, O. (1998). *Astron. Astrophys.*, **333**, 343.
Garaud, P. (2001). *Mon. Not. Roy. Astron. Soc.*, **324**, 68.
Gilman, P. A. & Miesch, M. S. (2004). *Astrophys. J.*, **611**, 568.
Gough, D. O. & McIntyre, M. E. (1998). *Nature*, **394**, 755.
Holton, J. R. (1965). *J. Atmosph. Sci.*, **22**, 402.
Hurlburt, N. E., Toomre, J. & Massaguer, J. M. (1986). *Astrophys. J.*, **311**, 563.
Kim, E.-J. & MacGregor, K. B. (2001). *Astrophys. J.*, **556**, L117.
Kosovichev, A. G. (1996). *Astrophys. J.*, **469**, L61.
McIntyre, M. E. (1994). In *The Solar Engine and its Influence on the Terrestrial Atmosphere and Climate*, ed. E. Nesme-Ribes. Heidelberg: Springer-Verlag, p. 293.
McIntyre, M. E. (2003a). In *Stellar Astrophysical Fluid Dynamics*, ed. M. J. Thompson & J. Christensen-Dalsgaard. Cambridge: Cambridge University Press, p. 111.
McIntyre, M. E. (2003b). In *Perspectives in Fluid Dynamics: A Collective Introduction to Current Research*, ed. G. K. Batchelor, H. K. Moffatt & M. G. Worster. Cambridge: Cambridge University Press, p. 557.
Miesch, M. S. (2003). *Astrophys. J.*, **586**, 663.
Miesch, M. S., Brun, A. S. & Toomre, J. (2006). *Astrophys. J.*, **641**, 618.
Noerdlinger, P. D. & Arigo, R. J. (1980). *Astrophys. J.*, **237**, L15.
Nordlund, Å., Brandenburg, A., Jennings, R. *et al.* (1992). *Astrophys. J.*, **392**, 647.
Rempel, M. (2004). *Astrophys. J.*, **607**, 1046.
Richard, D. (2001). Thèse de Doctorat (Université Paris 7).
Richard, D. & Zahn, J.-P. (1999). *Astron. Astrophys.*, **347**, 734.
Rogers, T. M. & Glatzmaier, G. A. (2005). *Astrophys. J.*, **620**, 432.
Rossby, C.-G. (1938). *Beitr. Phys. Freien Atmos.*, **24**, 53.
Sakurai, T. (1966). *Publ. Astron. Soc. Japan*, **18**, 174.
Schmitt, J. H. M. M., Rosner, R. & Bohn, H. U. (1984). *Astrophys. J.*, **282**, 316.

Skumanich, A. (1972). *Astrophys. J.*, **171**, 563.
Spiegel, E. A. (1972). In *Physics of the Solar System*, ed. S. I. Rasool (NASA SP-300), p. 61.
Spiegel, E. A. & Zahn, J.-P. (1992). *Astron. Astrophys.*, **265**, 106.
Stringfellow, G. S., Bodenheimer, P., Noerdlinger, P. D. & Arigo, R. J. (1983). *Astrophys. J.*, **264**, 228.
Talon, S. & Charbonnel, C. (2005). *Astron. Astrophys.*, **440**, 981.
Talon, S., Kumar, P. & Zahn, J.-P. (2002). *Astrophys. J.*, **574**, L175.
Walin, G. (1969). *J. Fluid Mech.*, **36**, 289.
Watson, M. (1981). *Geophys. Astrophys. Fluid Dyn.*, **16**, 285.
Zahn, J.-P. (1975). *Mém. Soc. Roy. Sci. Liège 6 série*, **8**, 31.
Zahn, J.-P. (1991). *Astron. Astrophys.*, **252**, 179.

5

Turbulence in the tachocline

Mark S. Miesch

Helioseismic inversions suggest that the tachocline straddles the base of the convection zone, incorporating the overshoot region and extending into the stably stratified radiative interior. Thus, the upper tachocline is dominated by penetrative convection while the lower tachocline is a stably stratified shear flow under the influence of rotation and magnetism. We review the nature of the turbulence that is likely to exist in these two disparate regions, focusing on the interaction between turbulence and differential rotation. It is argued that turbulent angular momentum transport is likely to be poleward throughout the tachocline, tending to suppress the latitudinal differential rotation maintained by turbulent stresses in the overlying convective envelope. Meanwhile, vertical angular momentum transport in the lower tachocline may be anti-diffusive, tending to amplify the vertical shear. The turbulent alignment of convective plumes may also drive an equatorward meridional circulation in the upper tachocline where it overlaps with the overshoot region.

5.1 Introduction

The solar tachocline lies near the base of the solar convection zone. This is a well-known result of course, but it is essential to establish precisely what *near* means in this context. Helioseismic structure inversions reveal a stiff transition between the nearly adiabatic stratification of the convection zone and the strongly subadiabatic stratification of the radiative interior, mediated by only a narrow region of convective overshoot. As others have argued in this volume, tachocline dynamics is very sensitive to where the rotational shear occurs relative to this structural transition.

Rotational inversions are subject to artificial smoothing arising from the finite width of the inversion kernels so they may overestimate the extent of the tachocline. Still, recent estimates do suggest that the tachocline overlaps with the overshoot

The Solar Tachocline, D. W. Hughes, R. Rosner and N. O. Weiss.
Published by Cambridge University Press.

region and may extend further into the convection zone at high latitudes. As an illustration we consider the recent work of Charbonneau *et al.* (1999). By using several inversion techniques, they estimated the location r_t and width Δ_t (defined in terms of an error function parameterization) to be $r_t/R_\odot = 0.693 \pm 0.002$ and $\Delta_t/R_\odot = 0.039 \pm 0.013$ at the equator and $r_t/R_\odot = 0.717 \pm 0.003$ and $\Delta_t/R_\odot = 0.042 \pm 0.013$ at a latitude of 60°. Thus, the tachocline appears to be prolate in shape and somewhat wider at high latitudes. Similar results were reported by Basu & Antia (2001) who furthermore investigated the base of the convection zone r_b by using structure inversions. According to their estimates, $r_b = 0.7134 \pm 0.0002$ with no detectable latitudinal variation and no significant evidence for overshoot. This is consistent with earlier estimates of r_b by Christensen-Dalsgaard *et al.* (1991). For further elaboration on these and other helioseismic results, see Chapter 3 by Christensen-Dalsgaard & Thompson in this volume.

Thus the tachocline apparently spans a diverse range of physical conditions. The upper portion overlaps with the overshoot region and possibly with the convection zone. Motions are highly turbulent and the dynamical timescales are weeks to months. Here be dinosaurs. By contrast, vertical motions in the lower tachocline are suppressed by the strongly subadiabatic stratification. However, intermittent or sustained turbulence may still exist, driven by breaking gravity waves, magnetic buoyancy, and shear instabilities (see Chapter 10 by Gilman & Cally and Chapter 11 by Hughes).

In this chapter we'll discuss the distinct nature of turbulence in the upper and lower tachocline. Since rotational inversions currently provide the most reliable observational insight into tachocline dynamics, we'll focus throughout on the interaction between turbulence and differential rotation.

5.2 The upper tachocline: penetrative convection

5.2.1 Turbulent plumes

In the past few decades, numerical simulations and laboratory experiments have demonstrated conclusively that plumes are the dominant coherent structures in turbulent convection (e.g. Cattaneo *et al.* 1991; Siggia 1994; Brummell *et al.* 1996; Julien *et al.* 1996; Stein & Nordlund 1998; Porter & Woodward 2000). Intermittent, vortical plumes originate in the boundary layers and penetrate well into the interior of the domain, transporting heat, mass and momentum. Density stratification induces an asymmetry such that downflow lanes form an interconnected network near the top of the domain which fragments into isolated plumes deeper in the convection zone (Cattaneo *et al.* 1991; Brummell *et al.* 1996; Stein & Nordlund 1998; Miesch *et al.* 2000; Brun & Toomre 2002). Individual plumes may remain

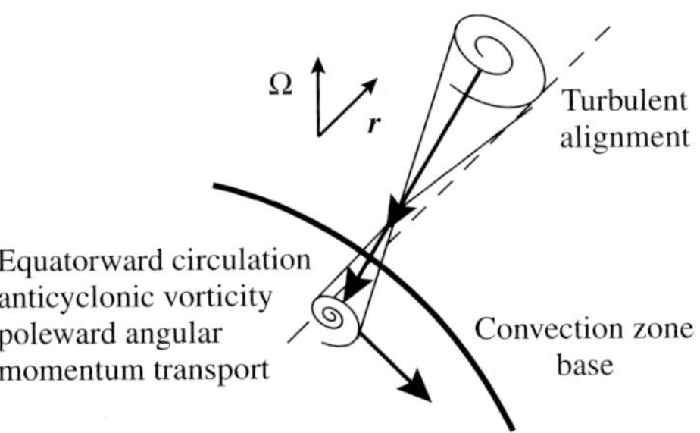

Figure 5.1. Schematic diagram of the behaviour of turbulent plumes in rotating spherical shells. The process of turbulent alignment (see text) causes downflow plumes to tilt toward the rotation axis relative to the vertical (indicated by the dashed line). These plumes are fed by converging fluid in the upper convection zone which acquires cyclonic vorticity due to the Coriolis force. As this fluid moves downward and enters the overshoot region, it diverges and acquires anticyclonic vorticity. Buoyancy removes the plume's vertical momentum, causing it to veer equatorward. The equatorward advection of anticyclonic vorticity produces poleward angular momentum transport via the Reynolds stress, $\mathbf{F}^{RS}$ (Equation (5.2)). Advection of retrograde differential rotation by the equatorward meridional circulation, $\mathbf{F}^{DR}$ (Equation (5.4)), enhances this poleward angular momentum transport.

coherent over multiple density scale heights and, if a stable layer is incorporated below the convection zone, they dominate the convective overshoot (Brummell *et al.* 2002).

Downflow plumes in turbulent compressible convection are profoundly influenced by rotation (Julien *et al.* 1996; Brummell *et al.* 1996). Unless the plume is located at one of the poles, its downward velocity will have a component that is perpendicular to the rotation axis. The Coriolis force deflects this velocity component while motion parallel to the rotation axis proceeds unimpeded. The net effect is to tilt downward plumes away from the vertical toward the rotation axis in a process known as turbulent alignment (Brummell *et al.* 1996). Turbulent alignment can decrease the vertical transport efficiency of plumes and can inhibit convective penetration (Brummell *et al.* 2002).

In a spherical shell, the turbulent alignment of plumes has important implications with regard to the maintenance of mean flows, as illustrated in Figure 5.1. In the upper convection zone, the Coriolis force induces cyclonic vorticity as fluid converges into the plume, tending to conserve angular momentum locally. If the plume remains coherent throughout the convection zone it will continue to converge due to the density stratification although this may be offset to a large extent by turbulent entrainment. When the plume reaches the overshoot region it will be decelerated by buoyancy and will diverge as fluid circulates back up to the convection zone. As it diverges, the Coriolis force will induce anticyclonic vorticity. Although the plume will no longer possess vertical momentum it will still have latitudinal momentum as a consequence of turbulent alignment, driving an equatorward circulation.

5.2.2 Angular momentum transport induced by overshooting plumes

What effect do turbulent plumes have on the differential rotation in the tachocline? To address this question we consider the angular momentum flux due to advection of the differential rotation by meridional circulation $\mathbf{F}^{\rm MC}$ and due to the Reynolds stress $\mathbf{F}^{\rm RS}$ (Brun *et al.* 2004; Miesch 2005):

$$\mathbf{F}^{\rm MC} = \langle \overline{\rho} \mathbf{v}_{\rm M} \rangle \mathcal{L}, \tag{5.1}$$

$$\mathbf{F}^{\rm RS} = \overline{\rho} r \sin\theta \left(\left\langle v'_r v'_\phi \right\rangle \hat{\boldsymbol{r}} + \left\langle v'_\theta v'_\phi \right\rangle \hat{\boldsymbol{\theta}} \right), \tag{5.2}$$

where $\overline{\rho}$ is the density of the spherically symmetric background state under the anelastic approximation, $\mathbf{v}_{\rm M} = v_r \hat{\boldsymbol{r}} + v_\theta \hat{\boldsymbol{\theta}}$ is the velocity in the meridional plane, and $\mathcal{L}$ is the specific angular momentum, defined as follows:

$$\mathcal{L} \equiv r \sin\theta \left(\Omega_0 r \sin\theta + \langle v_\phi \rangle \right) \equiv (r \sin\theta)^2 \, \Omega. \tag{5.3}$$

These expressions correspond to a spherical polar coordinate system (r, θ, ϕ) rotating at an angular velocity of Ω_0, with unit vectors $\hat{\boldsymbol{r}}$, $\hat{\boldsymbol{\theta}}$, $\hat{\boldsymbol{\phi}}$ and velocity components v_r, v_θ, v_ϕ. Angular brackets denote longitudinal averages and primes indicate that the longitudinal average has been removed.

Under the anelastic approximation, the uniform rotation component of $\mathbf{F}^{\rm MC}$ cannot produce any net angular momentum transport across surfaces aligned with the rotation axis (Elliott *et al.* 2000; Miesch 2005). Thus, any global redistribution of angular momentum between the equator and poles by meridional circulation must be caused solely by the differential rotation component of $\mathbf{F}^{\rm MC}$:

$$\mathbf{F}^{\rm DR} = r \sin\theta \, \langle \overline{\rho} \mathbf{v}_{\rm M} \rangle \langle v_\phi \rangle = \mathbf{F}^{\rm MC} - (r \sin\theta)^2 \, \langle \overline{\rho} \mathbf{v}_{\rm M} \rangle \, \Omega_0. \tag{5.4}$$

In the convection zone, the net angular momentum transport must be equatorward in order to maintain a solar-like rotation profile, with a relatively fast equator and slow poles. According to Equation (5.4), a persistent equatorward circulation in the overshoot region (positive v_θ in the northern hemisphere) would oppose this, producing a poleward angular momentum transport at high latitudes where $\langle v_\phi \rangle$ is negative. An equatorward circulation in the tachocline has been proposed in the context of flux-transport solar dynamo models where it plays an essential role in setting the pace and the nature of the solar activity cycle (e.g. Dikpati & Charbonneau 1999). If present, an equatorward circulation in the overshoot region may also contribute to tachocline confinement by suppressing latitudinal shear.

We now turn to the Reynolds stress. If the vertical momentum of a plume in the overshoot region is suppressed by buoyancy, then the vertical component of $\mathbf{F}^{\mathrm{RS}}$ may be neglected and the Reynolds stress divergence may be approximated as

$$\nabla \cdot \mathbf{F}^{\mathrm{RS}} \sim -\overline{\rho} r \sin\theta \left\langle v_\theta' \omega_r' \right\rangle , \tag{5.5}$$

where ω_r is the radial component of the fluid vorticity relative to the rotating frame. This implies the following maintenance equation for the angular velocity:

$$\frac{\partial \Omega}{\partial t} \sim -\frac{1}{r \sin\theta} \left\langle v_\theta' \omega_r' \right\rangle + \cdots . \tag{5.6}$$

Thus, the equatorward advection of anticyclonic radial vorticity in plumes implies a convergence of angular momentum which would tend to accelerate the local rotation rate. Other contributions to the right-hand-side of Equation (5.6) may arise from the meridional circulation flux $\mathbf{F}^{\mathrm{MC}}$, the Lorentz force, and viscous diffusion (although the latter is negligible in the Sun).

The turbulent alignment process is most efficient at high and mid-latitudes where the rotation vector is nearly vertical. At low latitudes, convection can minimize the inhibiting effects of the Coriolis force in a different way, by forming quasi-two-dimensional rolls or downflow lanes oriented in a north–south direction with motions perpendicular to the rotation axis (Miesch *et al.* 2000; Brummell *et al.* 2002; Brun & Toomre 2002; Miesch 2005). Thus, turbulent alignment may be expected to produce a convergence of angular momentum at high latitudes in the overshoot region which would spin up the poles relative to lower latitudes.

In summary, the angular momentum transport by the Reynolds stress $\mathbf{F}^{\mathrm{RS}}$ and the meridional circulation $\mathbf{F}^{\mathrm{DR}}$ may both be poleward in the overshoot region owing to the turbulent alignment of downflow plumes. This of course assumes that plumes dominate the turbulent transport, which is justified if they are the principal coherent structures in an otherwise chaotic flow. However, there is no guarantee that the surrounding flow is truly chaotic. Systematic velocity correlations in the broad and relatively laminar upflow lanes may oppose the Reynolds stress exerted by plumes, particularly at high Reynolds numbers where the filling factor of plumes is small.

5.2.3 Convection simulations

There is reason to believe that the processes illustrated in Figure 5.1 are indeed playing a significant role in global-scale numerical simulations of penetrative convection. Figures 5.2 and 5.3 show results from a numerical simulation of penetrative

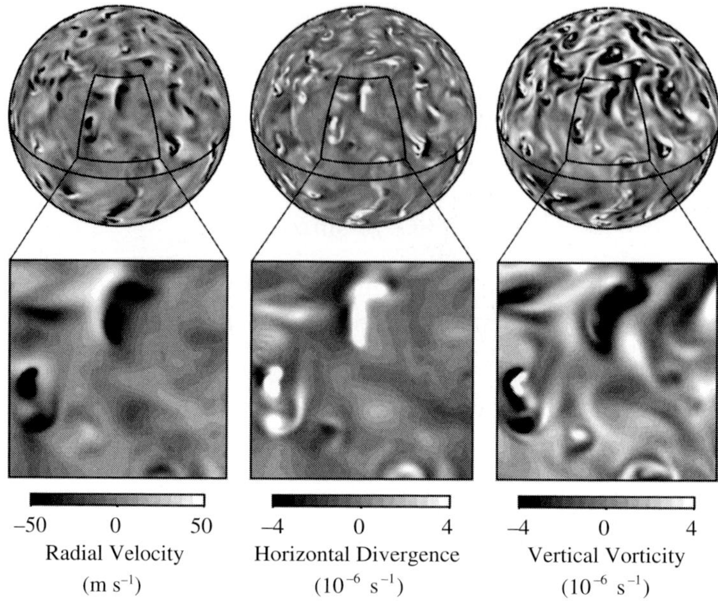

Figure 5.2. Snapshots of the radial velocity (left), horizontal divergence (middle) and vertical vorticity (right) are shown at a level within the overshoot region in a simulation of penetrative convection. Upper frames show orthographic projections tilted 30° toward the observer and lower frames show zoomed-in perspectives of selected areas as indicated. The equator is indicated by a solid line. Bright tones denote outward velocities, horizontal divergence, and outward vorticity as indicated by the grey scales.

convection in a rotating spherical shell carried out using the anelastic spherical harmonic (ASH) code described by Clune *et al.* (1999) (see also Brun *et al.* 2004). The simulation shown is a variation of Case TUR, described in detail by Miesch *et al.* (2000), with a higher horizontal resolution and lower dissipation. The resolution in this case is $N_\theta \times N_\phi \times N_r = 512 \times 1024 \times 98$ and the viscosity and thermal diffusivity at the top of the shell are $\nu_{\rm top} = 2.5 \times 10^{12}$ cm^2 s^{-1} and $\kappa_{\rm top} = 10^{12}$ cm^2 s^{-1} respectively. The ASH code uses cgs units and the background state is obtained from a solar structure model. The viscosity ν and thermal diffusivity κ vary with depth as $\overline{\rho}^{\,-1/2}$ where $\overline{\rho}$ is the background density. Solar values are used for the luminosity and mean angular velocity Ω_0. The computational domain is from $0.62R_\odot$ to $0.96R_\odot$. Above $\sim 0.98R_\odot$ the anelastic approximation breaks down and the density and pressure scale heights decrease dramatically, driving relatively small-scale motions which cannot presently be resolved in a global model. For further details see Miesch *et al.* (2000).

Downflow plumes in the overshoot region are visible as dark patches in the radial velocity images shown in the left column of Figure 5.2. These downflow

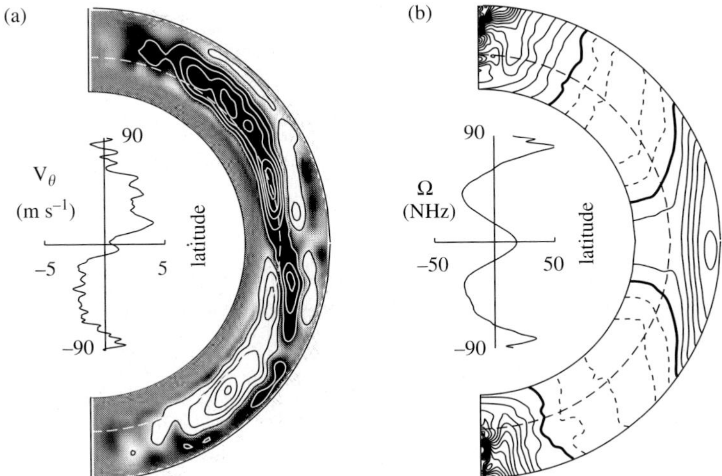

Figure 5.3. The meridional circulation (a) and angular velocity (b) are shown for the same simulation as in Figure 5.2, averaged over longitude and time (143 days). In (a), dark tones and bright contours denote counter-clockwise circulation while bright tones and dark contours denote clockwise circulation. Angular velocity contours in (b) range from −20 to 190 nHz relative to the rotating coordinate frame, in intervals of 10 nHz. Solid and dotted contours denote positive and negative angular velocities respectively and the bold solid line marks the zero contour. Dashed lines in (a) and (b) denote the base of the convection zone. The insets in (a) and (b) illustrate the mean meridional velocity $\langle v_\theta \rangle$ and angular velocity Ω as a function of latitude in the overshoot region (at the same level as shown in Figure 5.2). Positive values indicate southward and prograde flow, respectively.

sites correlate well with regions of horizontal divergence and anticyclonic vorticity as is apparent from the centre and right columns of Figure 5.2. This is consistent with the picture put forth in Figure 5.1 in which downflow plumes diverge and spin down as they encounter the stably-stratified overshoot region. The turbulent alignment of these plumes helps drive a persistent equatorward circulation in the overshoot region of a few metres per second, as demonstrated in Figure 5.3a. Time-averaged statistics verify that the $\langle v'_\theta v'_r \rangle$ correlation in regions of strong downflow is indeed negative (positive) in the northern (southern) hemisphere as would be expected from turbulent alignment (Miesch *et al.* 2000).

The local $v'_\theta \omega'_r$ correlation in each individual plume acts to accelerate higher latitudes through the Reynolds stress (Equation (5.6)) while the collective contribution from ensembles of plumes drives an equatorward circulation which enhances the poleward angular momentum transport (Equation (5.4)). Both processes contribute to the relatively rapid polar rotation evident in Figure 5.3b.

The equatorward circulation near the base of the convection zone extends to latitudes of about ±70°, which coincides well with the boundary of the prograde polar vortex in each hemisphere. Above this the circulation is weakly poleward on average. Thus the latitudinal angular momentum transport by meridional circulation, proportional to $\langle v_\theta \rangle \langle v_\phi \rangle$ (see Equation (5.4)) is generally poleward in the overshoot region above latitudes of about 20°. The Reynolds stress contribution $\mathbf{F}^{\rm RS}$ fluctuates substantially but consistently produces a convergence of angular momentum in downward plumes.

Thus, to some extent, the rapid polar rotation can be attributed to the turbulent nature of the fluid motions; laminar convection does not generally possess plumes and therefore does not exhibit turbulent alignment (Brummell *et al.* 1996). Relatively laminar simulations of penetrative convection in spherical shells do not exhibit a persistent equatorward circulation in the overshoot region comparable to that seen in Figure 5.3a (Miesch *et al.* 2000). However, laminar convection simulations can still exhibit a polar vortex resulting from a Coriolis-induced Reynolds stress in low-azimuthal-wavenumber convection cells which are the preferred linear modes at high latitudes, inside the cylinder which is aligned with the rotation axis and tangent to the base of the convection zone (Busse & Cuong 1977; Gilman 1979). Both turbulent and laminar processes likely contribute to the polar vortex seen in Figure 5.3b and in other spherical shell simulations.

Many spherical convection simulations do not exhibit a polar vortex. In the laminar case, this typically occurs for relatively deep convection zones such that the columnar convection cells which are the preferred modes outside the tangent cylinder can extend to high latitudes and transport angular momentum away from the poles (Gilman 1979). When applied to the Sun, simulations are constrained by solar structure models and helioseismic inversions so the convection zone depth is not a free parameter. Here the common thread among simulations which do not exhibit a polar vortex is the absence of large-scale meridional circulation cells which extend from low to high latitudes (Brun & Toomre 2002). Such global circulations tend to conserve angular momentum, spinning up the polar regions as they converge on the rotation axis.

A localized vortex within 10–15° of the pole cannot be ruled out by helioseismic inversions or surface Doppler measurements. However, at least in the upper convection zone, rotational inversions suggest the opposite; the poles appear to be rotating relatively slowly (Schou *et al.* 1998; Thompson *et al.* 2003). A polar vortex would be more consistent with helioseismic measurements if it were confined to the tachocline where it may be maintained by penetrative convection (Section 5.2.2) or by stably-stratified turbulence (Section 5.3). If present, a polar vortex would have important implications for the structure of the tachocline and chemical mixing between the convective envelope and the radiative interior, as discussed by McIntyre in Chapter 8.

5.2.4 Solar implications

To date, non-penetrative simulations of convection in rotating spherical shells have yielded more solar-like differential rotation profiles than those which incorporate penetration into an underlying stable layer (see Miesch 2005 for references). Penetrative simulations tend to be plagued by fast polar rotation throughout the convection zone, which is not consistent with helioseismic inversions. This may be attributed at least in part to the processes discussed here in connection with Figure 5.1. Global simulations cannot currently resolve the stiff transition from superadiabatic to subadiabatic stratification at the base of the solar convection zone (see Section 5.1). This transition is softened by artificially decreasing the subadiabatic stratification and thus increasing the extent of the overshoot region.

Poleward angular momentum transport in the upper tachocline induced by the turbulent alignment of plumes probably also occurs in the Sun, but we appear to be overestimating it in numerical simulations which have an artificially wide overshoot region. The relatively gentle deceleration of plumes in such simulations as they encounter the stable zone permits a relatively laminar spreading and spin-down. In non-penetrative simulations the deceleration of plumes is more abrupt and more violent. Vorticity of all orientations is generated as plumes strike the impermeable lower boundary, reducing the net $\langle v'_\theta \omega'_r \rangle$ correlation and therefore reducing the Reynolds stress. Turbulent alignment still occurs but plumes are less efficient at driving an equatorward circulation near the base of the convection zone, possibly because of the turbulent entrainment of fluid as they splash against the boundary.

The Sun is probably somewhere between these two extremes. The absence of a polar vortex suggests that it might correspond better to non-penetrative simulations than to current simulations which possess an artificially wide overshoot region. However, other dynamics may also contribute which is not currently captured by global simulations. For example, the much higher Péclet numbers[1] in the Sun relative to simulations may permit downflow plumes to carry low-entropy fluid equatorward more efficiently. This may establish a latitudinal entropy gradient (warm poles, cool equator) and a corresponding thermal wind differential rotation (e.g. Miesch 2005; see also Chapter 8 by McIntyre in this volume). In any case, it is clear that the complex dynamics occurring near the base of the convection zone must play an important role in maintaining the global rotation profile in the solar interior. This poses yet another challenge to numerical modelling efforts: accurately reproducing the solar internal rotation profile may require a more realistic depiction of convective penetration.

[1] The Péclet number is defined as UL/κ, where U and L are characteristic velocity and length scales for the flow and κ is the thermal diffusivity.

As promised in Section 5.1, we have focused on how penetrative convection may drive mean flows in the upper tachocline. There are many other aspects of penetrative convection which we have left untouched for the sake of brevity. In particular, there have been innumerable investigations of how the penetration depth varies with the stiffness of the subadiabatic–superadiabatic transition, with the Reynolds and Péclet numbers, with the rotation rate, and with latitude. There is also some current debate on whether overshooting convection in the Sun is efficient enough to establish a nearly adiabatic penetration region as described by Zahn (1991). Furthermore, penetrative convection excites gravity waves which have important implications for helioseismology as well as for the exchange of chemical tracers and angular momentum between the convective envelope and the radiative interior. For a review of these issues and many more references, see the online article by Miesch (2005). Many of these issues are also discussed by Brummell *et al.* (2002).

Thus far in our discussion we have neglected magnetic fields; the simulation shown in Figures 5.2 and 5.3 is non-magnetic. In comparable magnetic simulations with self-sustained dynamo action the field never becomes strong enough to substantially alter the structure of the convection (Brun *et al.* 2004). Thus, the arguments put forth in this section should still be valid in the presence of magnetism. This is in contrast to the lower tachocline where the presence or absence of magnetism has a substantial influence on the dynamics (Section 5.3.4).

Although magnetism should not greatly alter the convective structure, it introduces new dynamical phenomena of great relevance to solar dynamo theory. Numerical simulations of penetrative MHD convection in spherical shells promise to shed new light on the turbulent pumping of fields into the overshoot region, the amplification of these fields due to differential rotation in the tachocline, and the subsequent rise of toroidal flux due to magnetic buoyancy. Such simulations are now underway.

5.3 The lower tachocline: stably-stratified turbulence and shear

5.3.1 Overview

The lower tachocline is a stably-stratified shear flow. Global motions are confined to a thin shell by buoyancy and are sufficiently slow that the Coriolis force plays a dominant role (small Froude and Rossby numbers). In these respects, the lower tachocline has more in common dynamically with the Earth's atmosphere and oceans than it does with the highly turbulent solar envelope. However, magnetism still sets tachocline dynamics apart from geophysical applications.

The stabilizing effect of buoyancy on vertical shear is quantified by the Richardson number $Ri = (N/\mathrm{d}\overline{u}/\mathrm{d}z)^2$, where N is the Brunt–Väisälä frequency and $d\overline{u}/dz$ is the mean strain rate. In the solar tachocline $Ri \gg 1$, so the vertical shear is thought to be hydrodynamically stable (Schatzman *et al.* 2000). However, the latitudinal shear is thought to be unstable, particularly in the presence of magnetic fields (see Chapter 10 by Gilman & Cally). Furthermore, gravity waves generated by penetrative convection propagate into the lower tachocline and dissipate by nonlinear breaking or radiative diffusion, redistributing momentum in the process. Other instabilities may also occur, driven by magnetism, buoyancy and shear (see, for example, Chapter 11 by Hughes). In light of the small molecular viscosity, any global-scale motions that exist in the lower tachocline will be characterized by very high Reynolds numbers ($>10^{10}$), implying turbulent flow.

How might turbulence and waves in the lower tachocline interact with differential rotation? Specifically, will they tend to suppress the mean radial and latitudinal shear via down-gradient angular momentum transport as in turbulence models which employ eddy diffusion? Or, alternatively, might turbulent transport be counter-gradient (anti-diffusive), tending to drive mean flows rather than dissipate them?

Down-gradient transport might be expected under two particular circumstances. First; if the turbulence is driven by hydrodynamic shear instabilities then it will extract energy from the mean flow, tending to reduce its amplitude. Second; if the turbulence is approximately homogeneous and isotropic and occurs on small scales relative to the mean flow (scale separation) then turbulent mixing will be local and diffusive in nature. In this case, an effective eddy viscosity may be defined under the framework of mean-field theory.

The alternative, counter-gradient transport, might be expected if the flow field is dominated by waves. Waves induce long-range momentum transport between regions of excitation and dissipation which is inherently non-diffusive. This has been argued at length by McIntyre (1994, 1998, 2003; see also Miesch 2005).

5.3.2 Angular momentum transport in stably-stratified shear flows

Numerical simulations of stably-stratified turbulence in the presence of vertical shear exhibit a transition from diffusive to anti-diffusive transport as the stratification is increased. This is demonstrated in Figure 5.4, which is reproduced from the work of Galmiche *et al.* (2002). Three realizations of freely-evolving turbulent shear flow are shown. In all cases, the initial conditions consist of a non-divergent, random velocity field which is isotropic and homogeneous and which possesses a Gaussian energy spectrum peaked at intermediate wavenumbers ($k_p \sim 30/L$, where L is the linear extent of the cubical domain). This random velocity field is superposed on a background horizontal flow with vertical shear $\overline{u}(z)$ and the system is then allowed to evolve freely with no additional forcing.

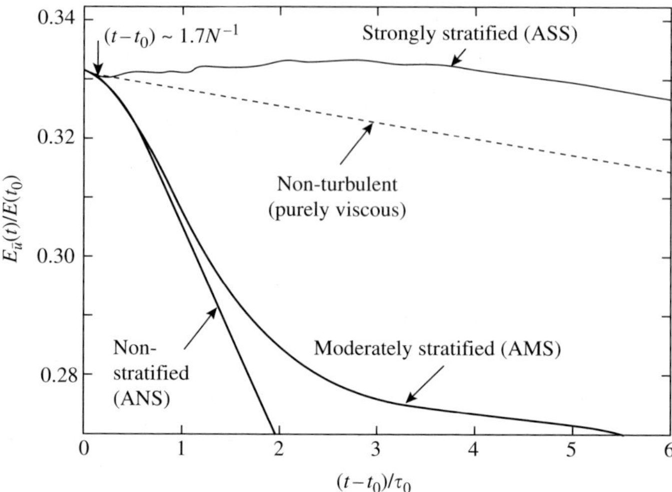

Figure 5.4. The kinetic energy in the mean zonal flow component is shown as a function of time for three realizations of decaying stably-stratified turbulence with vertical shear. Each realization is an ensemble average of six numerical simulations. Results are shown for neutral (ANS), moderate (AMS), and strong stratification (ASS) as indicated. The straight line indicates viscous decay of the mean shear in the absence of turbulence. All curves are normalized with respect to the kinetic energy at the initial time, t_0, and time is normalized with respect to the advective time scale of the turbulent flow component, τ_0. (From Galmiche *et al.* 2002.)

The three realizations differ only in the strength of the background stratification. This is quantified by the Froude number $Fr = (N\tau_0)^{-1}$, where τ_0 is the advective or turnover timescale of the turbulence. If $Fr \ll 1$ then buoyancy dominates over nonlinear advection in the vertical momentum equation. In the neutrally-stratified case ANS ($Fr = \infty$) the mean shear decays within a few turnover timescales, much more rapidly than the viscous decay in the absence of turbulence. Thus, turbulent mixing can be characterized by means of an effective eddy viscosity which is much larger than the molecular viscosity. Moderate stratification inhibits vertical mixing, resulting in a slower decay rate in case AMS ($Fr = 1.2$).

In the strongly-stratified case ASS ($Fr = 0.12$) the behaviour is qualitatively different. The mean flow initially begins to decay at a rate comparable to cases ANS and AMS but the momentum flux soon reverses, becoming counter-gradient after a time $t - t_0 \sim 1.7N^{-1}$. The mean zonal kinetic energy begins to grow in an oscillatory manner, implying an acceleration of the background shear flow $\overline{u}(z)$. Most of this energy remains in the fundamental mode $k_z = 1$ so the transfer of energy from the turbulence to the mean flow may be characterized by means of a negative eddy viscosity. At long times ($t - t_0 > 4\tau_0$), the turbulent angular momentum flux decreases and the mean flow decays at the viscous rate.

Other researchers have similarly reported an oscillatory but persistently counter-gradient momentum flux in numerical simulations of freely-evolving stably-stratified turbulence with vertical shear (Holt *et al.* 1992; Jacobitz 2002). The turbulent transport associated with the fluctuating density field is also oscillatory and counter-gradient for small Froude numbers, implying a negative eddy diffusivity. This counter-gradient momentum and density transport arises from the linear distortion of the initial turbulent flow field by buoyancy and shear and can be understood within the context of Rapid Distortion Theory (Galmiche & Hunt 2002; Hanazaki & Hunt 2004). As the Froude number is progressively decreased, the flow possesses a more wave-like character and the transport can no longer be described by means of turbulent diffusion.

The stably-stratified simulations we have discussed thus far and their interpretation in terms of Rapid Distortion Theory are concerned with freely-evolving turbulence. By contrast, turbulence in the solar tachocline may be continually maintained by penetrative convection and shear instabilities. Furthermore, the solar tachocline is a rotating system and possesses latitudinal as well as vertical shear. How do the results discussed above apply to more complex systems?

Horizontal shear was investigated by Jacobitz (2002) in freely evolving numerical simulations similar to those illustrated in Figure 5.4. He found that the horizontal momentum transport remained diffusive (down-gradient) even for strong stratification but the efficiency of the transport (i.e. the magnitude of the effective eddy viscosity) decreased with decreasing *Fr*. Further correspondence with the solar tachocline was achieved by Miesch (2003), who considered stably-stratified turbulence in rotating spherical shells. These simulations included random high-wavenumber forcing which was homogeneous and isotropic in horizontal planes but concentrated near the top of the shell in order to crudely approximate the influence of sustained penetrative convection. They also included an imposed differential rotation possessing both latitudinal and vertical shear which was maintained against viscous diffusion by an additional drag force. Some results are illustrated in Figure 5.5.

The latitudinal angular momentum transport due to the Reynolds stress in Miesch's (2003) simulations was found to be down-gradient, in the same sense as the viscous diffusion (Figure 5.5a). Both contributions tended to suppress the differential rotation maintained by external forcing. Since the imposed differential rotation was chosen to have a solar-like profile, this corresponds to poleward angular momentum transport. By contrast, the vertical transport of angular momentum by Reynolds stresses was found to be outward, opposing the inward diffusive flux (Figure 5.5b). Thus the vertical transport was counter-gradient, tending to enhance the vertical shear as in the freely-evolving, non-rotating, Cartesian simulations illustrated in Figure 5.4.

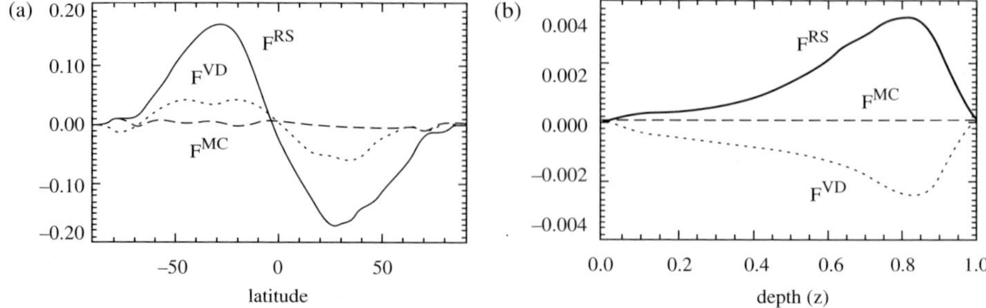

Figure 5.5. The angular momentum flux is shown for a simulation of stably-stratified turbulence in a rotating spherical shell with imposed shear. Here (a) is the latitudinal angular momentum flux integrated over depth and (b) is the vertical angular momentum flux integrated over latitude. All curves are integrated over longitude and time. The temporal integration spans ten advective timescales after the shear was introduced. Solid, dotted, and dashed lines indicate contributions from the Reynolds stress, viscous diffusion, and meridional circulation respectively. All quantities are non-dimensional, normalized with respect to the velocity scale of the imposed forcing, the background density, and the radius and thickness of the shell. (From Miesch 2003.)

5.3.3 *Self-organization in quasi-two-dimensional turbulence*

Stable stratification inhibits vertical motions and rotation induces vertical coherence, both effects tending to make the dynamics quasi-two-dimensional. Although it is dangerous to take the analogy too far, much insight into rotating, stably-stratified turbulence can be obtained by considering strictly two-dimensional turbulence on horizontal surfaces. It is well known that two-dimensional turbulence exhibits self-organization: smaller vortices merge into progressively larger vortices, giving rise to an inverse cascade of kinetic energy from small to large scales (e.g. Lesieur 1997). This may be regarded as another example of counter-gradient momentum transport, implying a negative eddy viscosity. Meanwhile, enstrophy (vorticity squared) undergoes a forward cascade from large to small scales where it is dissipated by viscous diffusion.

As the inverse cascade proceeds to larger scales, the characteristic turnover timescales of eddies increase and they become more influenced by rotation. Eventually, if the rotation is rapid enough, the Rossby phase speed will exceed the advection velocity[2] and eddies will disperse as Rossby wave packets before they merge. This suppression of the inverse cascade by Rossby wave dispersion is inherently anisotropic; nonlinear energy transfer toward large latitudinal scales is inhibited but the cascade can proceed toward large longitudinal scales (Rhines 1975;

[2] The phase speed of two-dimensional Rossby–Haurwitz waves in spherical shells scales as $\Omega/(\ell(\ell+1))$, where ℓ is the spherical harmonic degree (see e.g. Miesch 2005).

Vallis & Maltrud 1993). The net result is banded zonal flows: alternating eastward and westward jets of fluid with a latitudinal extent determined by the effective Rossby number $Ro = (2\Omega\tau_0)^{-1}$.

The emergence of banded zonal flows from rotating turbulence in two-dimensional and quasi-two-dimensional (shallow-water, two-layer) spherical systems has been observed in both freely-evolving and randomly-forced numerical simulations (Williams 1978; Cho & Polvani 1996; Huang & Robinson 1998; Peltier & Stuhne 2002; Kitamura & Matsuda 2004). Related dynamics have also been studied under the β-plane approximation (Rhines 1975; Vallis & Maltrud 1993; Panetta 1993). Perhaps the most familiar and well-studied observational manifestation of persistent banded zonal flows in rotating turbulence occurs in the atmospheres of the Jovian planets (Ingersoll 1990; Vasavada & Showman 2005).

Similar dynamics may occur in the solar tachocline but, more likely, it may be suppressed by shear and magnetism. Differential rotation is maintained in the tachocline by large-scale stresses from the convective envelope and radiative interior. Deformation of eddies by an imposed shear such as this fundamentally alters the nonlinear transfer of energy and enstrophy among spectral modes, favouring non-local interactions between the mean flow and disturbance modes possessing the same azimuthal wavenumber. This was investigated by Shepherd (1987) who considered two-dimensional turbulence on a β-plane in the presence of an imposed zonal flow with latitudinal shear. His analytical and numerical results indicate that shear-induced transfer operating on the large-scale Rossby wave field tends to accelerate the mean flow, implying counter-gradient momentum transport. However, turbulent isotropization on smaller scales disrupts this process and tends to extract energy from the mean flow. The sense and magnitude of the net energy exchange between the mean flow and the fluctuations is sensitive to the details of the problem such as the Rossby number, the forcing mechanism, and the amplitude and profile of the imposed shear.

5.3.4 *The profound influence of magnetism*

In the presence of magnetism the enstrophy is no longer an ideal invariant and an inverse cascade of kinetic energy no longer occurs, not even in two-dimensional flows (Biskamp 1993). Instead, there is an inverse cascade of a new ideal invariant, the squared magnetic vector potential. These results apply to homogeneous, isotropic, two-dimensional, MHD turbulence. More complex circumstances may yet exhibit counter-gradient momentum transport. Kim & Dubrulle (2002) have made analytic estimates of the transport coefficients in two-dimensional MHD turbulence in the presence of a background shear and toroidal field, under the assumptions of scale separation and quasi-linearity (first-order smoothing). They find that the effective eddy viscosity is generally positive in the presence of magnetic fields (down-gradient transport) but it can become negative if the mean field is strong relative to the fluctuations and if highly anisotropic forcing (elongated perpendicular to the mean flow) is applied to the magnetic potential. In the purely hydrodynamic case their analysis indicates that the eddy viscosity is always negative, regardless of the forcing.

In an MHD system, both the Reynolds and Maxwell stresses contribute to the net momentum transport and both may be incorporated into the definition of the 'eddy' or 'turbulent' viscosity. If left to evolve freely in the presence of a mean background field, the fluctuating velocity and magnetic fields in MHD turbulence tend to align such that $\mathbf{v}' = \pm\mathbf{B}'$ in what is known as the Alfvén effect (Biskamp 1993). If this occurs, Reynolds and Maxwell stresses will tend to cancel and the turbulent angular momentum transport $\langle v_i'v_j' - B_i'B_j'\rangle$ will decrease substantially. In Kim & Dubrulle's (2002) analysis the Alfvén effect leads to a suppression of turbulent transport as the strength of the background field B_0 is increased, yielding an effective eddy viscosity which scales as B_0^{-2} (see also Kim *et al.* 2001).

This raises the more general issue of transport barriers in magnetized plasmas which is addressed by Diamond *et al.* in Chapter 9. Briefly, shear and magnetic fields can suppress turbulent transport perpendicular to the flow or field direction by distorting turbulent eddies and by dissipating or deflecting gravity and Rossby waves. In a solar context, this implies that the vertical shear and strong toroidal fields thought to be present in the tachocline may act to decouple the radiative interior and convective envelope. The positive nonlinear feedback may help maintain the tachocline; radial shear and the toroidal fields it generates suppresses vertical transport which may further enhance the shear and produce stronger toroidal fields. Alternatively, poloidal field components and motions induced by magnetic buoyancy may tend to increase the vertical coherence relative to non-magnetized stratified flows, which would suppress vertical shear. In short, vertical angular momentum transport in the lower tachocline is complex and remains enigmatic. Much more work is needed to clarify the dominant processes.

By contrast, latitudinal angular momentum transport in the tachocline is better understood. Magnetic fields are likely to induce down-gradient angular momentum transport by destabilizing the latitudinal differential rotation. There are two classes of MHD shear instabilities which are probably operating in the lower tachocline. The first is the magneto-rotational instability (MRI), which operates wherever the angular velocity decreases outward, perpendicular to the rotation axis. The vertical shear in the lower tachocline is stabilized in this respect by the subadiabatic stratification but MRI can still proceed on horizontal surfaces and would tend to suppress positive latitudinal angular velocity gradients on a shear timescale (Balbus & Hawley 1994). However, the latitudinal angular velocity gradient in the tachocline is negative so it is not subject to MRI in a global sense, although local regions may still become unstable. For a more thorough discussion of MRI see Chapter 12 by Ogilvie.

The second type of MHD shear instability which is probably occurring in the lower tachocline is a global one. As first demonstrated by Gilman & Fox (1997), the latitudinal differential rotation in the solar tachocline is unstable in the presence of a toroidal field with a variety of possible amplitudes and profiles. The most unstable modes are generally those with azimuthal wavenumber $m = 1$ and have the character of tipping instabilities whereby rings of toroidal field evolve such that their central axis tilts away from the rotation axis. The resulting Maxwell stresses induce poleward angular momentum transport but not in a uniform manner. Rather, angular momentum converges near latitudes which represent singular points in the linear perturbation equations or which coincide with imposed toroidal field bands, producing localized zonal jets. Many subsequent papers have further clarified these mechanisms. The reader is referred to Chapter 10 by Gilman & Cally for an up-to-date review. These global MHD shear instabilities as well as the MRI may also operate in the upper tachocline where their growth rates are generally larger owing to the weaker subadiabatic stratification.

5.4 Summary

If the tachocline overlaps with the convection zone as helioseismic inversions suggest (Section 5.1), then the upper and lower portions must be dramatically different in terms of their dynamics. Turbulent penetrative convection dominates the upper tachocline, efficiently exchanging mass, momentum, and magnetic flux with the overlying convective envelope. By contrast, vertical transport is much less efficient in the lower tachocline where the strong subadiabatic stratification and rotational influence makes the dynamics quasi-two-dimensional.

In Section 5.2 it was argued that the turbulent alignment of convective plumes in the solar convection zone will induce an equatorward circulation and a poleward angular momentum transport in the overshoot region. The simulations and theoretical arguments reviewed in Section 5.3 in the context of stably-stratified turbulence and MHD shear instabilities furthermore suggest that the latitudinal angular momentum in the lower tachocline is probably poleward (down-gradient) as well, although inverse cascades and wave-induced momentum and energy fluxes may tend to drive mean flows such as zonal jets.

Poleward angular momentum transport by anisotropic turbulence in the tachocline was first proposed by Spiegel & Zahn (1992) as a mechanism for suppressing the downward spread of the convection zone differential rotation into the radiative interior via a thermally-driven meridional circulation. This turbulence was attributed to a nonlinear hydrodynamic instability of the latitudinal shear, but other mechanisms which produce poleward transport have similar implications with regard to tachocline confinement. Turbulent alignment, MHD shear instabilities, and stably-stratified turbulence driven by penetrative convection may all contribute to suppress radiative spreading and thus keep the tachocline thin. This may result in an angular momentum cycle such as that proposed by Gilman *et al.* (1989) whereby equatorward angular momentum transport in the bulk of the convection zone is balanced by poleward transport in the tachocline and possibly in the near-surface shear layer.

The strong stable stratification in the lower tachocline may further contribute to tachocline confinement by inducing counter-gradient angular momentum transport in the vertical direction which would maintain or even enhance the vertical shear. Moreover, interactions between gravity waves and differential rotation in the lower tachocline may drive oscillatory zonal flows analogous to the Quasi-Biennial Oscillation (QBO) in the Earth's stratosphere (Kim & MacGregor 2001; Talon *et al.* 2002). Such interactions may lie at the root of tachocline oscillations detected in helioseismic inversions by Howe *et al.* (2000).

Below the tachocline, wave-induced transport dominates over turbulent transport and the angular momentum redistribution is likely to be non-local and non-diffusive. As argued by Gough & McIntyre (1998), gravity waves alone cannot maintain the nearly uniform rotation of the radiative interior inferred from helioseismology. Other mechanisms are necessary, the most plausible being large-scale torques from a fossil magnetic field as reviewed in Chapter 7 of this book by Garaud. Thus, although turbulent transport in the tachocline may help prevent the downward spread of the differential rotation in the convection zone, it is not the final word on tachocline confinement.

Despite recent progress, much uncertainty remains regarding vertical transport in the lower tachocline and the coupling between the radiative interior and the convective envelope. Further numerical and theoretical modelling is needed to sort out the dominant forcing mechanisms, the subtleties of wave-induced transport, and the role of magnetic fields.

5.4.1 Acknowledgments

The simulations presented in Figures 5.2 and 5.3 were done in collaboration with Juri Toomre, A. Sacha Brun, and Matthew Browning. I also thank Nigel Weiss, Bob Rosner, and David Hughes for organizing a stimulating and productive programme at the Newton Institute and Mausumi Dikpati for helpful comments on the manuscript. This work was partly supported by NASA through award numbers W-10,177 and W-10,175.

References

Balbus, S. A. & Hawley, J. F. (1994). *Mon. Not. Roy. Astron. Soc.*, **266**, 769.
Basu, S. & Antia, H. M. (2001). *Mon. Not. Roy. Astron. Soc.*, **324**, 498.
Biskamp, D. (1993). *Nonlinear Magnetohydrodynamics.* Cambridge: Cambridge University Press.
Brummell, N. H., Hurlburt, N. E. & Toomre, J. (1996). *Astrophys. J.*, **473**, 494.
Brummell, N. H., Clune, T. L. & Toomre, J. (2002). *Astrophys. J.*, **570**, 825.
Brun, A. S. & Toomre, J. (2002). *Astrophys. J.*, **570**, 865.
Brun, A. S., Miesch, M. S. & Toomre, J. (2004). *Astrophys. J.*, **614**, 1073.
Busse, F. H. & Cuong, P. G. (1977). *Geophys. Astrophys. Fluid Dyn.*, **8**, 17.
Cattaneo, F., Brummell, N. H., Toomre, J., Malagoli, A. & Hurlburt, N. E. (1991). *Astrophys. J.*, **370**, 282.
Charbonneau, P., Christensen-Dalsgaard, J., Henning, R. *et al.* (1999). *Astrophys. J.*, **527**, 445.
Cho, J. Y.-K. & Polvani, L. M. (1996). *Phys. Fluids*, **8**, 1531.
Clune, T. L., Elliott, J. R., Miesch, M. S., Toomre, J. & Glatzmaier, G. A. (1999). *Parallel Computing,* **25**, 361.
Dikpati, M. & Charbonneau, P. (1999). *Astrophys. J.*, **518**, 508.
Elliott, J. R., Miesch, M. S. & Toomre, J. (2000). *Astrophys. J.*, **533**, 546.
Galmiche, M. & Hunt, J. C. R. (2002). *J. Fluid Mech.*, **455**, 243.
Galmiche, M., Thual, O. & Bonneton, P. (2002). *J. Fluid Mech.*, **455**, 213.
Gilman, P. A. (1979). *Astrophys. J.*, **231**, 284.
Gilman, P. A. & Fox, P. A. (1997). *Astrophys. J.*, **484**, 439.
Gilman, P. A., Morrow, C. A. & DeLuca, E. E. (1989). *Astrophys. J.*, **338**, 528.
Gough, D. O. & McIntyre, M. E. (1998). *Nature*, **394**, 755.
Hanazaki, H. & Hunt, J. C. R. (2004). *J. Fluid Mech.*, **507**, 1.
Holt, S. E., Koseff, J. R. & Ferziger, J. H. (1992). *J. Fluid Mech.*, **237**, 499.
Howe, R., Christensen-Dalsgaard, J., Hill, F. *et al.* (2000). *Science*, **287**, 2456.
Huang, H.-P. & Robinson, W. A. (1998). *J. Atmos. Sci.,* **55**, 611.
Ingersoll, A. P. (1990). *Science*, **248**, 308.
Jacobitz, F. G. (2002). *J. Turb.*, **3**, 55.
Julien, K., Legg, S., McWilliams, J. & Werne, J. (1996). *J. Fluid Mech.*, **322**, 243.
Kim, E.-J. & MacGregor, K. B. (2001). *Astrophys. J.*, **556**, L117.
Kim, E.-J., Hahm, T. S. & Diamond, P. H. (2001). *Phys. Plasmas*, **8**, 3576.
Kim, E.-J. & Dubrulle, B. (2002). *Physica D*, **165**, 213.
Kitamura, Y. & Matsuda, Y. (2004). *Fluid Dyn. Res.*, **34**, 33.
Lesieur, M. (1997). *Turbulence in Fluids*, 3rd edition. Dordrecht: Kluwer.
McIntyre, M. E. (1994). In *The Solar Engine and its Influence on the Terrestrial Atmosphere and Climate*, ed. E. Nesme-Ribes. Heidelberg: Springer-Verlag, p. 293.
McIntyre, M. E. (1998). *Prog. Theor. Phys. Suppl.*, **130**, 137 (Corrigendum: *Prog. Theor. Phys.* **101**, 189 (1999)).
McIntyre, M. E. (2003). In *Stellar Astrophysical Fluid Dynamics*, ed. M. J. Thompson & J. Christensen-Dalsgaard. Cambridge: Cambridge University Press, p. 111.
Miesch, M. S. (2003). *Astrophys. J.*, **586**, 663.
Miesch, M. S. (2005). *Living Rev. Solar Phys.*, **2**, 1 (www.livingreviews.org/lrsp-2005-1).
Miesch, M. S., Elliott, J. R., Toomre, J. *et al.* (2000). *Astrophys. J.*, **532**, 593.
Panetta, R. L. (1993). *J. Atmos. Sci.*, **50**, 2073.

Peltier, W. R. & Stuhne, G. R. (2002). In *Meteorology at the Millennium*, ed. R. P. Pearce. San Diego: Academic Press, p. 43.
Porter, D. H. & Woodward, P. R. (2000). *Astrophys. J. Suppl.*, **321**, 323.
Rhines, P. B. (1975). *J. Fluid Mech.*, **69**, 417.
Schatzman, E., Zahn, J.-P. & Morel, P. (2000). *Astron. Astrophys.*, **364**, 876.
Schou, J. *et al.* (1998). *Astrophys. J.*, **505**, 390.
Shepherd, T. G. (1987). *J. Fluid Mech.*, **467**, 509.
Siggia, E. D. (1994). *Ann. Rev. Fluid Mech.*, **26**, 137.
Spiegel, E. A. & Zahn, J.-P. (1992). *Astron. Astrophys.*, **265**, 106.
Stein, R. F. & Nordlund, Å. (1998). *Astrophys. J.*, **499**, 914.
Talon, S., Kumar, P. & Zahn, J.-P. (2002). *Astrophys. J.*, **574**, L175.
Thompson, M. J., Christensen-Dalsgaard, J., Miesch, M. S. & Toomre, J. (2003). *Ann. Rev. Astron. Astrophys.*, **41**, 599.
Vallis, G. K. & Maltrud, M. E. (1993). *J. Phys. Ocean.*, **23**, 1346.
Vasavada, A. R. & Showman, A.P. (2005). *Rep. Prog. Phys.*, **68**, 1935.
Williams, G. P. (1978). *J. Atmos. Sci.*, **35**, 1399.
Zahn, J.-P. (1991). *Astron. Astrophys.*, **252**, 179.

6

Mean field modelling of differential rotation

Günther Rüdiger & Leonid L. Kitchatinov

Analytical expressions for the Λ-effect and the heat conductivity tensor for rotating turbulent convection are compared with current results of box simulations with the NIRVANA code. With these results the large-scale flow pattern (rotation plus meridional circulation) in the convection zone is computed in good agreement with the observations. The penetration of the meridional flow into the subadiabatic layer beneath the convection zone (with viscosity $\nu_{\rm core}$) appears to vary with $\sqrt{\nu_{\rm core}}$ so that in a non-turbulent tachocline the penetration would be extremely small. New mean field model calculations are also presented for the rotation laws in F stars and M dwarfs and finally the question is discussed whether mean field models may also lead to 'antisolar' rotation, i.e. to rotation laws with a decelerated equator.

6.1 Introduction

In order to explain the internal rotation of solar/stellar convection zones, the theory of the Λ-effect has been developed. It describes the angular momentum transport in rigidly rotating anisotropic fields of free turbulence. The preferred direction is radial, owing to the stratification of both the density and the intensity of the turbulence.

The cross-correlations $\langle u'_r u'_\phi \rangle$ and $\langle u'_\theta u'_\phi \rangle$ of the one-point-correlation tensor $Q_{ij} = \langle u'_i(\boldsymbol{x},t) u'_j(\boldsymbol{x},t) \rangle$ provide the radial and latitudinal turbulent transport of angular momentum. For those terms the general formulation

$$Q_{ij} = \cdots + \lambda_{ijk}\omega_k \tag{6.1}$$

has been introduced, or, in more detail,

$$Q_{r\phi} = \nu_{\rm T} V \omega \sin\theta, \qquad Q_{\theta\phi} = \nu_{\rm T} H \omega \cos\theta, \tag{6.2}$$

with $\nu_{\rm T}$ as the eddy viscosity. The dimensionless functions V and H are normalized expressions for the vertical and horizontal cross-correlations. A quasi-linear theory

The Solar Tachocline, D. W. Hughes, R. Rosner and N. O. Weiss.
Published by Cambridge University Press.

for a special turbulence model has been given by Kitchatinov & Rüdiger (1993) to reveal the dependence on the basic angular velocity Ω and the colatitude θ. The main result of such analytical calculations for the case of *rapid rotation* can be summarized in the form

$$Q_{r\phi} \simeq -\hat{H}\cos^2\theta\sin\theta, \qquad Q_{\theta\phi} \simeq \hat{H}\sin^2\theta\cos\theta, \tag{6.3}$$

with positive $\hat{H}$. The vanishing of $Q_{\theta\phi}$ at the equator is a simple consequence of the prevailing symmetries, but the vanishing of $Q_{r\phi}$ at the equator of a rapid rotator is a surprising and non-trivial result. Also, the signs of the cross-correlations are non-trivial results of the calculations. The resulting angular momentum transport is always inwards ($V < 0$) and equatorwards ($H > 0$). By a direct inspection of the results of helioseismology one finds an increase of angular velocity with depth in the uppermost layers of the solar convection zone, which means that the angular momentum is transported inwards, $V < 0$ (Section 6.4). Also, almost all of the presented numerical simulations lead to $V < 0$ for both slow and fast rotation (see below).

6.2 The Λ-effect

Quasi-linear analytical derivations of the angular momentum transport by rotating turbulence in stratified fluids result in the expressions

$$\begin{aligned} V &= V^{(0)}\left(\Omega^*\right) - H^{(1)}\left(\Omega^*\right)\cos^2\theta, \\ H &= H^{(1)}\left(\Omega^*\right)\sin^2\theta, \end{aligned} \tag{6.4}$$

for the normalized fluxes. The coefficients $V^{(0)}$ and $H^{(1)}$ depend on the Coriolis number $\Omega^* = 2\tau_{\rm corr}\Omega$ (see Figure 6.1 derived from an actual model of the solar convection zone). The usual concept to determine the correlation tensor of density-stratified rotating turbulence is to *prescribe* the turbulence field without rotation and then to derive the influence of rotation on the original turbulence. As the first step in this procedure an anelastic flow ($\nabla\cdot(\rho\boldsymbol{u}^{(0)}) = 0$) in a non-rotating fluid must be considered. The spectral tensor of the momentum density for such a non-uniform original turbulence is

$$\begin{aligned} \hat{M}_{ij} = {}& \frac{\hat{E}(k,\omega,\boldsymbol{\kappa})}{16\pi k^2}\left(\delta_{ij} - \left(1+\frac{\kappa^2}{4k^2}\right)\frac{k_ik_j}{k^2} + \frac{1}{2k^2}(\kappa_ik_j - \kappa_jk_i) + \frac{\kappa_i\kappa_j}{4k^2}\right) \\ &+ \frac{\hat{E}_1(k,\omega,\boldsymbol{\kappa})}{16\pi k^4}\left(\frac{(\boldsymbol{k}\cdot\boldsymbol{\kappa})}{k^2}(\kappa_ik_j + \kappa_jk_i) - \frac{(\boldsymbol{k}\cdot\boldsymbol{\kappa})^2}{k^2}\delta_{ij} - \kappa_i\kappa_j\right. \\ &\left. + \frac{1}{2}\left(\kappa^2 + \frac{(\boldsymbol{k}\cdot\boldsymbol{\kappa})^2}{k^2}\right)(\delta_{ij} - k_ik_j/k^2)\right), \end{aligned} \tag{6.5}$$

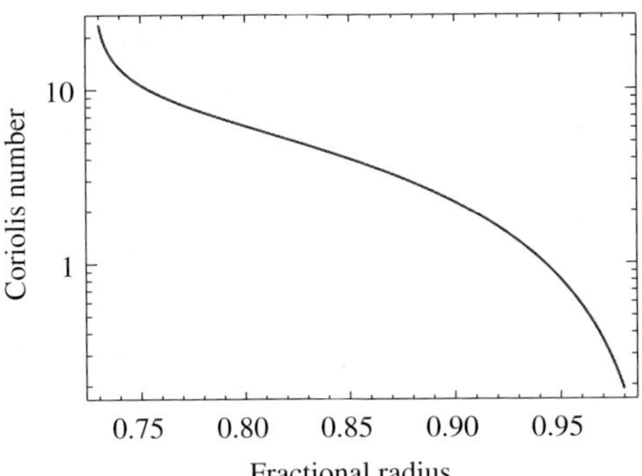

Figure 6.1. Radial profile of the Coriolis number Ω^* for our solar model. The turbulence must be considered as fast (slowly) rotating at the bottom (top) of the convection zone.

where $\boldsymbol{k}$ and $\boldsymbol{\kappa}$ are the wave-vectors for the small scales of turbulence and for the scale of variation of its mean characteristics (Kitchatinov & Rüdiger 2005).

The Λ-effect is derived in the mixing-length approximation, which will be understood as the dominance of one scale (the mixing-length) in the turbulence spectra. In this case, the spectral functions $\hat{E}$ and $\hat{E}_1$ can formally be written as

$$E = 2\rho^2 \langle u^{(0)2} \rangle \, \delta\left(k - \ell_{\mathrm{corr}}^{-1}\right) \delta(\omega), \qquad E_1 = aE/4, \tag{6.6}$$

with the proportionality coefficient a. The spectral functions can be transformed from wave space to real space, for example

$$E(k, \omega, \boldsymbol{x}) = \int \hat{E}(k, \omega, \boldsymbol{\kappa}) \exp(\mathrm{i}\boldsymbol{x} \cdot \boldsymbol{\kappa}) \mathrm{d}\boldsymbol{\kappa}. \tag{6.7}$$

The anisotropy parameter a is not completely free. It is restricted by the condition that in the one-point-correlation tensor for the original turbulence the radial and horizontal intensities must be positive. This yields

$$-\frac{5H_\rho^2}{2\ell_{\mathrm{corr}}^2} \leq a \leq \frac{15}{2} + \frac{5H_\rho^2}{\ell_{\mathrm{corr}}^2}, \tag{6.8}$$

where H_ρ is the density scale height. It is assumed that the density stratification is the dominant spatial inhomogeneity.

Following the procedure described in Kitchatinov & Rüdiger (1993) we arrive at the expression

$$\begin{aligned} Q_{ij}^{\Lambda} &= \nu_{\mathrm{T}} \Omega_k g_l \Big(V^{(0)}\left(\Omega^*\right) \left(g_i \epsilon_{jkl} + g_j \epsilon_{ikl}\right) \\ &\quad - H^{(1)}\left(\Omega^*\right) \frac{(\boldsymbol{g} \cdot \boldsymbol{\Omega})}{\Omega^2} \left(\Omega_i \epsilon_{jkl} + \Omega_j \epsilon_{ikl}\right) \Big), \end{aligned} \tag{6.9}$$

for the Λ-effect, with $\boldsymbol{g}$ as the radial unit vector. The two parameters are

$$V^{(0)} = \left(\frac{\ell_{\rm corr}}{H_\rho}\right)^2 \left(J_0\left(\Omega^*\right) + aI_0\left(\Omega^*\right)\right),$$

$$H^{(1)} = \left(\frac{\ell_{\rm corr}}{H_\rho}\right)^2 \left(J_1\left(\Omega^*\right) + aI_1\left(\Omega^*\right)\right), \tag{6.10}$$

where the functions J_0 and J_1 are the same as in Kitchatinov & Rüdiger (1993), but

$$I_0\left(\Omega^*\right) = \frac{1}{4\Omega^{*4}}\left(-19 - \frac{5}{1+\Omega^{*2}} + \frac{3\Omega^{*2}+24}{\Omega^*}\arctan\Omega^*\right),$$

$$I_1\left(\Omega^*\right) = \frac{3}{4\Omega^{*4}}\left(-15 + \frac{2\Omega^{*2}}{1+\Omega^{*2}} + \frac{3\Omega^{*2}+15}{\Omega^*}\arctan\Omega^*\right) \tag{6.11}$$

are new. In the slow rotation case ($\Omega^* \ll 1$), only the radial flux of angular momentum survives, i.e.

$$J_0 \simeq \frac{4}{15},\quad I_0 \simeq -\frac{3}{10},\quad J_1 \simeq I_1 \simeq O\left(\Omega^{*2}\right); \tag{6.12}$$

hence $V^{(0)}$ is negative with $a > 8/9$ for slow rotation.[1] For fast rotation ($\Omega^* \gg 1$), J_1 is much larger than all other functions,

$$J_1 \simeq \frac{\pi}{4\Omega^*},\quad J_0 \simeq I_0 \simeq I_1 \simeq O\left(\Omega^{*-3}\right), \tag{6.13}$$

so that $V < 0$ and $H > 0$ results in this case (see Figure 6.2). Any uncertainty in the Λ-effect related to the unknown free a parameter of Equations (6.6) and (6.10) disappears for the most relevant application, i.e. for fast rotation.

Our box simulations confirm the above findings (Figures 6.3 and 6.4). The aspect ratio of the box is 1:4, it is discretized with 100^3 grid points (Rüdiger *et al.* 2005b). For $Ta = 10^6$ the quantity V is always negative and becomes very small at the equator. The maximal values are reached in the middle of the box. The situation for the horizontal Λ-effect is more complicated. As first found by Pulkkinen *et al.* (1993), averaged over the entire box it is positive. In our simulation it dominates close to the equator and is large and positive in the top domain and small and negative in the lower half of the box.

The results of the quasi-linear analytical calculations and those of the numerical simulations of the Λ-effect are rather close together. Though the analytical results are quasi-linear *and* though they are only valid for a simplified turbulence model, the differences from the fully consistent nonlinear box simulations are rather small. The strong concentration of the horizontal angular momentum transport towards the equator, however, is not yet fully understood (see also Käpylä *et al.* 2004; Hupfer *et al.* 2005).

[1] Opposite to the case of $a = 0$ considered by Kitchatinov & Rüdiger (1993).

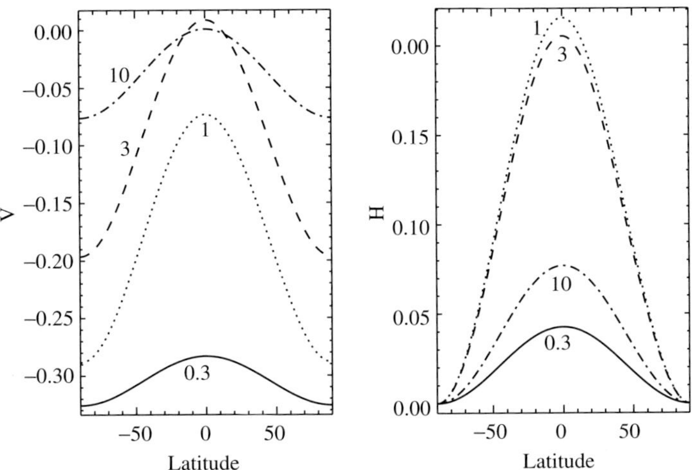

Figure 6.2. Normalized fluxes of angular momentum as functions of latitude for $a = 2$. The lines are marked by the values of Ω^*. Note also the clear Λ-quenching by rapid rotation.

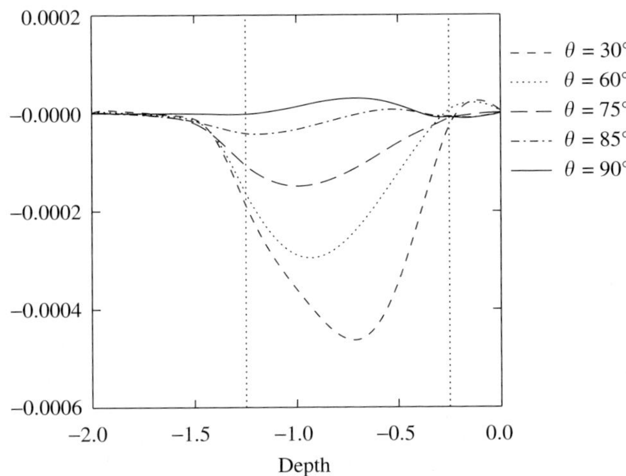

Figure 6.3. Box simulations with NIRVANA for rotating convection. Here $Q_{r\phi}$ is shown as a function of depth, normalized with the square of the sound speed at the surface of the unstable domain. $Ta = 10^6$, $Pr = 0.1$.

6.3 Turbulent heat transport

In rotating turbulent fluids the relation between the turbulent heat flux $\boldsymbol{F} = \rho C_\mathrm{p}\langle \boldsymbol{u}'T'\rangle$ and the superadiabatic temperature gradient $\boldsymbol{\beta} = \boldsymbol{g}/C_\mathrm{p} - \nabla T$ ($\boldsymbol{g}$ gravity) is tensorial, i.e.

$$F_i = \rho C_\mathrm{p} \chi_{ij} \beta_j. \tag{6.14}$$

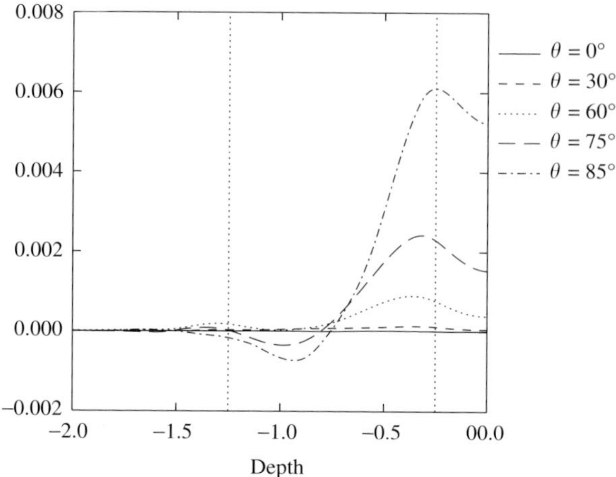

Figure 6.4. As Figure 6.3, but for $Q_{\theta\phi}$.

In the simplest case without rotation it is $\chi_{ij} = \chi_{\rm T}\delta_{ij}$ so that the well-known expression $\boldsymbol{F} = \rho C_{\rm p}\chi_{\rm T}\boldsymbol{\beta}$ results for non-rotating fluids.

There is a close relation between the heat-flux tensor and the one-point correlation tensor. We start from the quasi-linear connection

$$\chi_{ij} = \iint \frac{\chi k^2 \hat{Q}_{ij}(\boldsymbol{k}, \omega)}{\omega^2 + \chi^2 k^4} \mathrm{d}\boldsymbol{k}\mathrm{d}\omega \tag{6.15}$$

between the spectral tensor $\hat{Q}_{ij}$ of the turbulence and the heat-conductivity tensor χ_{ij}. For vanishing microscopic heat conduction ($\chi \to 0$) a Dirac δ-function appears so that

$$\chi_{ij} = \pi \int \hat{Q}_{ij}(k, 0)\mathrm{d}k \equiv \frac{1}{2} \int Q_{ij}(0, \tau)\mathrm{d}\tau \tag{6.16}$$

results. If the τ-integral is approximated by $\tau_{\rm corr}$ then $\chi_{ij} \simeq 0.5\tau_{\rm corr}Q_{ij}$. We have thus to expect that the radial heat flux follows the behaviour of the radial turbulence intensity resulting from the box simulations (Figure 6.5). In the bulk of the convection box $\langle u_r'^2 \rangle$ at the equator dominates $\langle u_r'^2 \rangle$ at the poles. The same is indeed true for the radial heat flux derived with NIRVANA and shown in Figure 6.6 (see Rüdiger *et al.* 2005a).

Figure 6.6 (left-hand plot) shows the depth-profile of the correlation $\langle u_r' T' \rangle$ for various latitudes. Owing to the rotation, the values differ between poles and the equator. The pole–equator difference in the radial heat-flux depends, however, on the radius. Except for the top layer, the eddy heat-flux at the equator exceeds the eddy heat-flux at the poles. In the top layers, where the turbulence is horizontally dominated, the polar heat-flux dominates that at the equator. This is a characteristic but unexpected result.

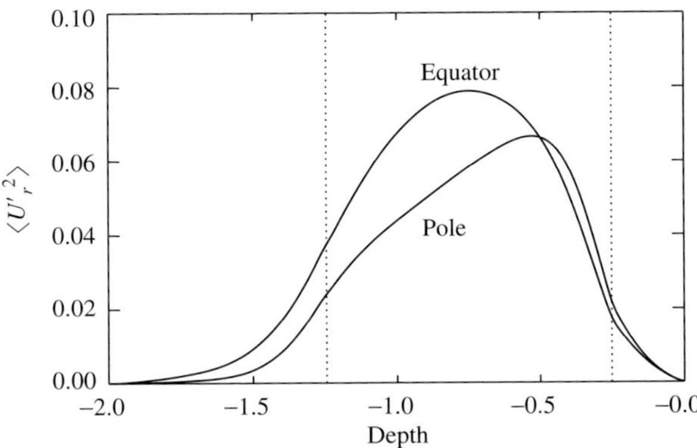

Figure 6.5. The radial turbulence intensity (for the same convection model as used in Figures 6.3 and 6.4). With solar values the maximal r.m.s. velocity is 300 m s^{-1}. Note that, except in the top layer, the turbulence at the equator exceeds the polar values. Note that a similar result also follows from box simulations without any density stratification.

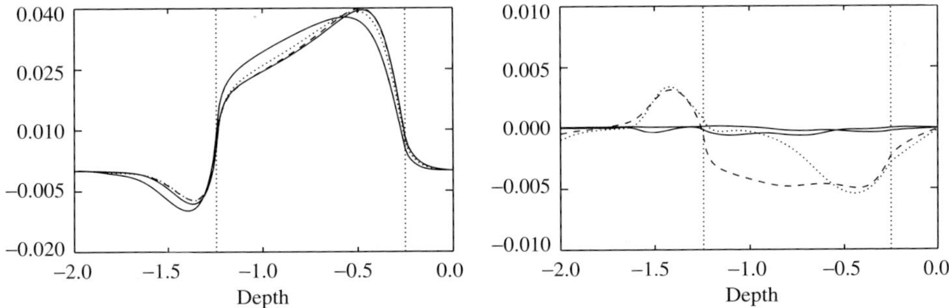

Figure 6.6. The correlations $\langle \boldsymbol{u}'T' \rangle$ vs. depth for different colatitudes in the box simulations after horizontal- and time-averaging; $Ta = 10^6$. Left: $\langle u'_r T' \rangle$. Right: $\langle u'_\theta T' \rangle$. Solid: pole, dashed: $\theta = 30°$, dotted: $\theta = 60°$, triple-dot-dashed: equator.

Rieutord *et al.* (1994, their Figure 8a), Käpylä *et al.* (2004, their Figure 7) and Hupfer *et al.* (2005) found similar results. We are led to the conclusion that a cross-over exists of the pole–equator difference in the radial eddy heat-flux almost at the same depth where the vertically dominated turbulence changes to a horizontally dominated turbulence. As we have demonstrated with Equation (6.16), the behaviour of the radial heat-flux is a direct reflection of the rotation-influenced radial turbulence intensity $\langle u_r'^2 \rangle$. It is shown in Figure 6.5 that in the box (except in the outermost layer) $\langle u_r'^2 \rangle$ at the equator exceeds the value at the poles.

A similar crossover does not exist for the latitudinal eddy heat-flux $\langle u'_\theta T' \rangle$ plotted in Figure 6.6 (right-hand plot). It generally vanishes at the poles and at the equator. Between these extrema the heat flows towards the pole in the convection zone and towards the equator in the lower overshoot region. This is because of the action of the Coriolis force.

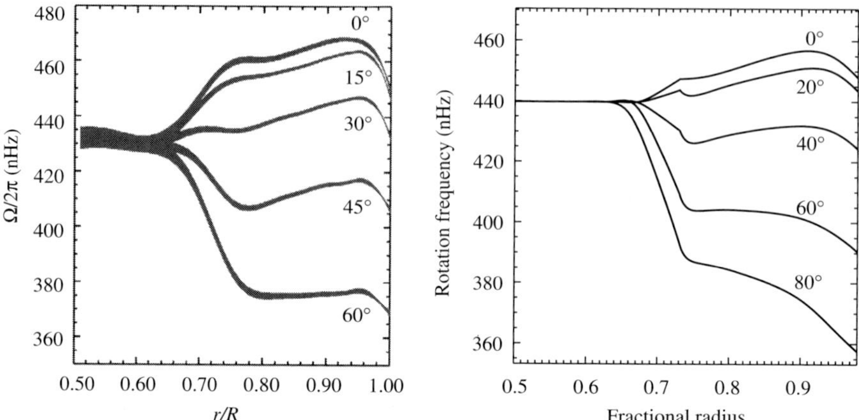

Figure 6.7. Left: The internal solar rotation law as determined by helioseismology. (Courtesy NSF National Solar Observatory, see also Kosovichev *et al.* 1997.) Right: The rotation law from the mean-field model with a rotation period of 25 days and $a = 2$. The curves are marked with the latitude ($= 90° - \theta$).

6.4 Solar models

Differential rotation and meridional flow within the convection zone are the results of the simultaneous solution of the steady axisymmetric equations for the momentum and the mean entropy. The models involve the Λ-effect defined by Equations (6.9)–(6.11), which involve the only free parameter, a, while the entropy equation involves the rotation-induced anisotropic heat-flux tensor (see, however, Rempel 2005). Recent models combine the numerical simulation of the differential rotation in the convection zone with computations of the tachocline resulting from a weak internal magnetic field within the solar radiative core (details given by Kitchatinov & Rüdiger 2005). This combination of mean field hydrodynamics and magnetohydrodynamics leads to the results shown in Figures 6.7 (right-hand plot) and 6.8.

In the simulations the anisotropy parameter is $a = 2$ (enhanced dominance of the radial turbulence intensity), resulting in a considerably improved agreement with helioseismological results of the rotation law in the outer supergranulation layer. Our old models were computed with $a = 0$ because there were no data for slow rotation domains to restrict its value. Now we have both better observations and the results of the numerical box simulations. The rotation laws of Figure 6.7 show a clear subsurface inward increase of the angular velocity, which may be important for the solar dynamo (Brandenburg 2005).

Even more significant may be the meridional ('Kippenhahn') flow as a principal ingredient of the advection-dominated dynamo models (Choudhuri *et al.* 1995; Dikpati & Gilman 2001; Bonanno *et al.* 2002). The resulting meridional flow is

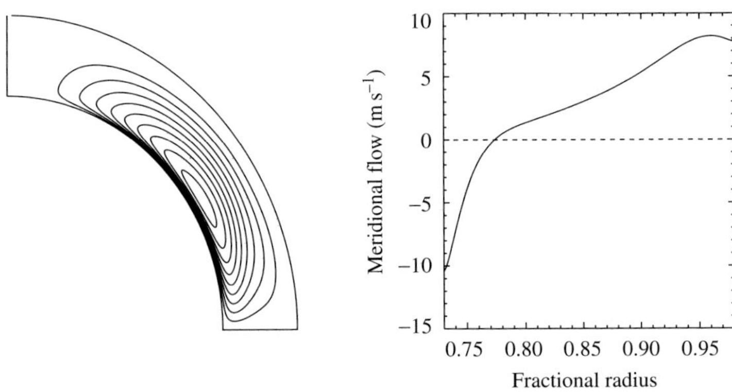

Figure 6.8. Streamlines of the meridional flow (left) and the flow velocity for 45° latitude as a function of depth (right) for $a = 2$. Positive velocity denotes a poleward flow. One cell of anticlockwise circulation occupies the entire convection zone.

shown in Figure 6.8. It consists of a *single circulation cell* with poleward flow on the top – opposite to the result of Kippenhahn (1963). This direction of the surface flow complies with direct Doppler measurements (Komm *et al.* 1993). Helioseismology also indicates that this direction of the flow prevails to a depth of at least 12 Mm (Zhao & Kosovichev 2004). The flow reverses to the return equatorward direction somewhere deeper down. The computed return flow in Figure 6.8 has a velocity of $\lesssim 10\,\mathrm{m\,s^{-1}}$ at the base of the convection zone. This value suffices to transport magnetic fields towards the equator during the 11 year cycle as required by the advection-dominated dynamo models.

Not only the flow amplitude but also the depth of its penetration into the radiative zone beneath the convection zone is important. The extent to which the penetration exists in the Sun is currently debated (Nandy & Choudhuri 2002; Gilman & Miesch 2004). Its value can be computed with the mean-field model if the bottom boundary of the simulation domain is placed inside the region of stable subadiabatic stratification. Our model applies a local mixing-length approximation. The base of the superadiabatically stratified shell, therefore, coincides with the bottom of the convection zone. A finite effective viscosity should be prescribed for the radiative core below the convection zone. The effective viscosity, ν_{core}, can be quite small compared to the turbulent viscosity of the convection zone but must be larger than its microscopic value.

The penetration depth computed with such an approach is shown in Figure 6.9. The dependence on ν_{core} is close to $\sqrt{\nu_{\mathrm{core}}}$, in accordance with the finding of Gilman & Miesch (2004) that the penetration under solar conditions belongs to the Ekman regime. This penetration results from viscous drag imposed by meridional flow at the base of the convection zone on the fluid beneath. The standard Ekman depth is $D_{\mathrm{pen}} \sim \sqrt{\nu_{\mathrm{core}}/\Omega}$. Such a penetration cannot play any role in a dynamo

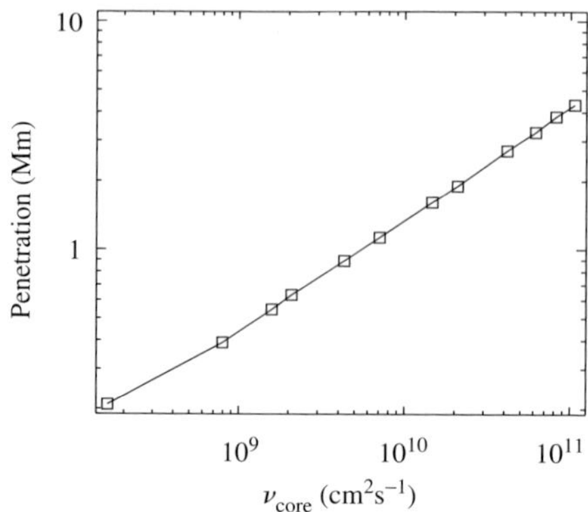

Figure 6.9. Penetration depth D_{pen} of meridional flow into the convection zone at latitude 45° vs. the viscosity ν_{core} of the convectively stable region. The dependence is very close to $D_{pen} \propto \sqrt{\nu_{core}}$. Squares represent the actual computations.

process because the time of magnetic field diffusion across the penetration layer, $\tau_d \sim D_{pen}^2/\eta_{core}$, is very small compared to the advection time, $\tau_{adv} \sim R_\odot/u^m$, since $\tau_d/\tau_{adv} \sim Pm\, u^m/(\Omega R_\odot) \ll 1$, except for the case of unrealistically large magnetic Prandtl numbers, $Pm = \nu_{core}/\eta_{core}$ (Kitchatinov & Rüdiger 2005).

Estimation of penetration by the Ekman depth is further supported by the finding that a variation of thermal conductivity, χ_{core}, for constant ν_{core} does not change D_{pen}. Our computations, however, do not reproduce the Ekman relation $D_{pen} \sim \Omega^{-0.5}$. The slope of the dependence is not constant and is slightly larger than -0.5. The rotation rate dependence is better represented by a relation

$$D_{pen} \propto \left(\Omega^2 + \Omega \frac{\tan\theta}{2} \frac{\partial \Omega}{\partial \theta}\right)^{-0.25}. \tag{6.17}$$

Whether the viscosity ν_{core} is large or small beneath the convection zone depends strongly on the stability of the solar tachocline. If the latter is unstable then the viscosity is large and the penetration is deep. If, however, it is stable then the viscosity is small and the penetration is only very weak. In the hydrodynamical regime our calculations favour the small-viscosity case. If the solar tachocline is considered as a shear flow then for high enough Reynolds numbers of the rotation it is unstable for sufficiently high equator–pole differences of the angular velocity at its upper boundary.

The hydrodynamical stability/instability of the solar tachocline has been probed with the Hollerbach code for a shallow spherical shell subject to differential rotation (Arlt *et al.* 2005). If the shear is formed only by latitudinal gradients of Ω (the Watson case) then the onset of the instability starts at about 30% of rotational shear. If, however, the radial profile of the rotation law (known from helioseismology) also enters the model then a *stronger* latitudinal shear results for the onset

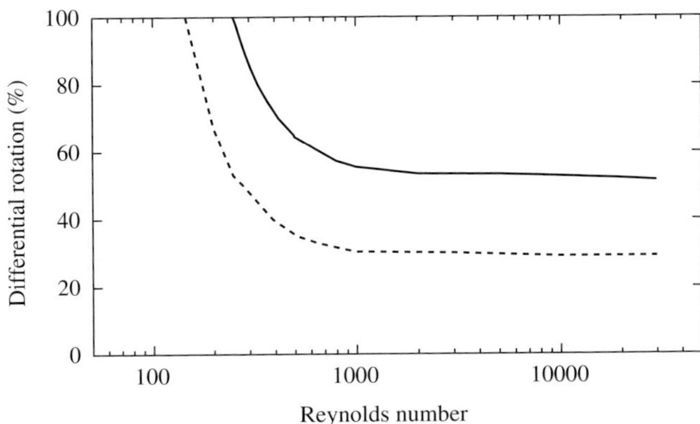

Figure 6.10. Bifurcation map for the hydrodynamic shear instability of the solar tachocline. Dashed line: only latitudinal shear, solid line: rotation law $\Omega = \Omega(r, \theta)$ taken from the helioseismology (Figure 6.7, left-hand plot). Note that the real solar tachocline is *not* unstable in the hydrodynamical regime.

of the instability (Figure 6.10). We, therefore, assume the viscosity as small in the solar tachocline and the penetration of the meridional flow as weak. Of course, a final statement about this problem only depends on the solution of the stability problem in the magnetohydrodynamic regime.

6.5 Stellar models

Hall (1991) found differential rotation for a number of magnetically active stars from the variation of spot rotation periods over the stellar activity cycle. Messina & Guinan (2003) derived surface rotation laws from photometric data of a monitoring programme of stars resembling the Sun in earlier states of its evolution (Sun in time).

Unfortunately, the number of single main-sequence stars to which Doppler imaging has been successfully applied is still small (see Strassmeier 2002). For sufficiently fast rotators, surface differential rotation can be detected, however, from the broadening of spectral lines. Reiners & Schmitt (2003a,b) and Reiners (2006) carried out measurements for F stars with moderate and short rotation periods. They found differential rotation to be much more common for stars with moderate rotation rates than for very rapid rotators.

A model has been developed for the differential rotation of a Main Sequence (MS) star of spectral type F8 (Küker & Rüdiger 2005). The star represents the upper end of the lower MS in the context of differential rotation and stellar activity. With a convection zone depth of 160 Mm and a surface gravity roughly equal to the solar value the main difference from the Sun is the luminosity, which is 1.7 times the solar value.

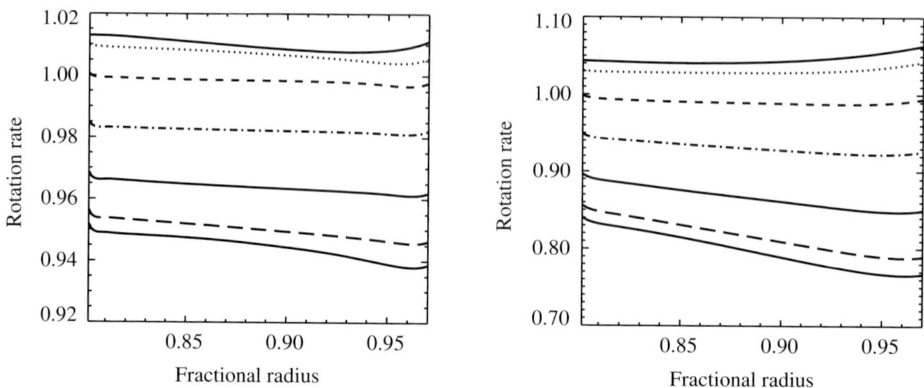

Figure 6.11. Normalized rotation rate as a function of the stellar radius for the latitudes 0°(equator), 15, 30, 45, 60, 75, and 90°(poles) (from top to bottom) for rotation periods of 4 days (left) and 14 days (right).

In Figure 6.11 the rotation rate is plotted vs. the radius. In all cases the equator rotates faster than the poles but the amplitude of the relative shear varies with the rotation rate. The model with the fastest rotation yields the most rigid rotation law. Note also that for fast rotation all lines are rather horizontal, i.e. there is almost no radial shear. While the isocontours are mainly radial for fast rotation the slow rotation case shows a disc-shaped pattern at high latitudes.

Figure 6.12 shows the total horizontal shear, $\delta\Omega$, as a function of the rotation period for various stellar models. For both F and G stars the rotation becomes more rigid for faster rotation. Between the two limiting cases the total horizontal shear has a maximum value. The period at which the maximum is reached is about 1 month in the case of the Sun, and 10 days for the F star. None of the curves for stars later than G shows a sharp peak. There is mostly a broad range of nearly constant shear around the maximum. The distinct maximum of the surface shear for F stars has recently been seen by Reiners (2006). Also the very clear run of the differential rotation with the effective temperature shown by the theory appears in his observational results (his Figure 5).

The model presented here predicts that the equatorial acceleration of lower MS stars should depend more on the luminosity rather than on the rotation rate. A possible empirical trend of the stellar differential rotation with mass and/or the rotation rate has been studied by Barnes *et al.* (2005). They found that the differential rotation decreases rapidly with decreasing mass but varies only slightly with the rotation period for a given spectral type, in general agreement with our theoretical findings.

Also AB Dor and PZ Tel, though rotating much faster than the Sun, show surface differential rotation very similar to the Sun. AB Dor and PZ Tel are PMS stars of

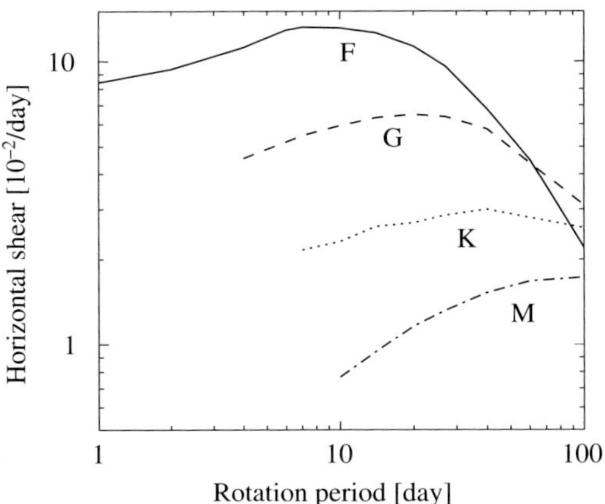

Figure 6.12. Surface shear as a function of the rotation period and the effective temperature for MS stars marked with their spectral class.

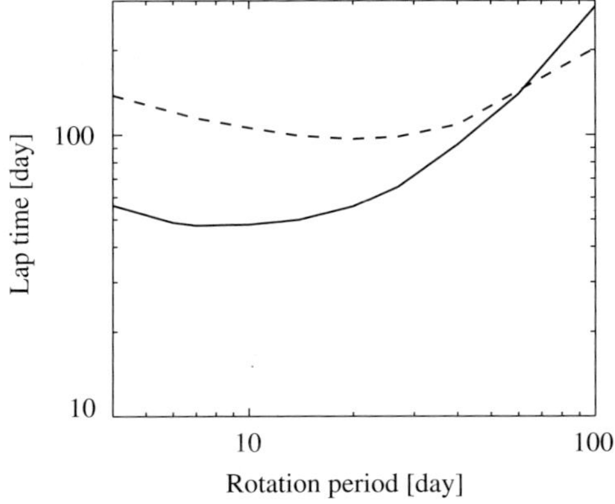

Figure 6.13. Lap time as a function of the rotation period for the F star (solid) and the solar-type star (dashed).

spectral type K0 and G0 with rotation periods of 0.5 day and 1 day, respectively. With $\delta\Omega = 0.056\ \text{day}^{-1}$ (AB Dor) and $\delta\Omega = 0.075\ \text{day}^{-1}$ (PZ Tel) the surface shear values of these stars lie close to the solar value, with the more luminous AB Dor also showing more surface shear.

Reiners & Schmitt (2003a,b) found values between 10 and 30 days for the lap time $2\pi/\delta\Omega$. Figure 6.13 shows the lap time vs. the rotation period for the F and G stars. For both types of stars there is little variation except for very long periods where the lap time strongly increases. The value for the solar-type star is about 100 days at the period of maximum shear. In Figure 6.14 the maximum flow speed at the bottom of the convection zone of the F star is shown. A positive sign means that the flow is toward the equator, negative values indicate poleward flow. For the F8 star the value of the drift decreases from $10\,\text{m}\,\text{s}^{-1}$ for $P_{\text{rot}} = 4$ days to very

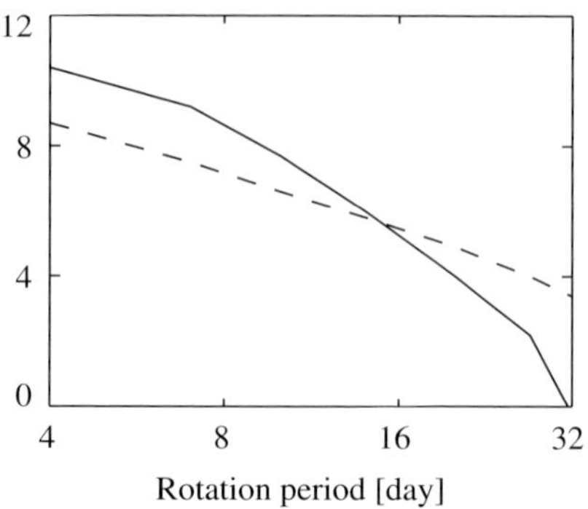

Figure 6.14. Meridional flow speed in metres per second at the bottom of the convection zone as a function of the rotation period. Positive values correspond to gas motion toward the equator. Solid line: F8 star. Dashed line: G star. (From Küker & Rüdiger 2005.)

small values for $P = 30$ days. A possible change of the flow direction at the bottom of the convection zone should have dramatic consequences for the stellar dynamo if indeed the form of the butterfly diagram is dominated by the meridional flow at the bottom of the convection zone.

6.6 Antisolar rotation?

Hydrodynamical models of stellar rotation always lead to solar-type rotation with an accelerated equator. Observations confirm that it is indeed the typical case (Petit *et al.* 2004). However, observational indications of antisolar rotation are numerous enough to demand a consideration of its possible origin.

The clearest possibility for a faster rotation of high latitudes is a rapid meridional flow $u^{\rm m}$. The flow provides a uniform angular momentum along the streamlines when the Reynolds number

$$Re = \frac{u^{\rm m} R_\odot}{\nu_{\rm T}} \tag{6.18}$$

is sufficiently large, thus ensuring antisolar rotation (Rüdiger 1989). A polar vortex results for both directions of the meridional flow. The required Reynolds number depends on the sense of the flow. A faster polar rotation is easier to produce by a flow that is poleward on the top. A moderate $Re \lesssim 100$ can be sufficient in this case.

However, the meridional flow computed with the mean-field models is not fast enough. The Reynolds number for the solar model of Section 6.4 is $\lesssim 10$. An additional driver of meridional circulation is thus required for antisolar rotation. Baroclinic forcing from magnetic-induced large-scale thermal inhomogeneities (or tidal forcing by a close companion) is the possible driver (Kitchatinov & Rüdiger

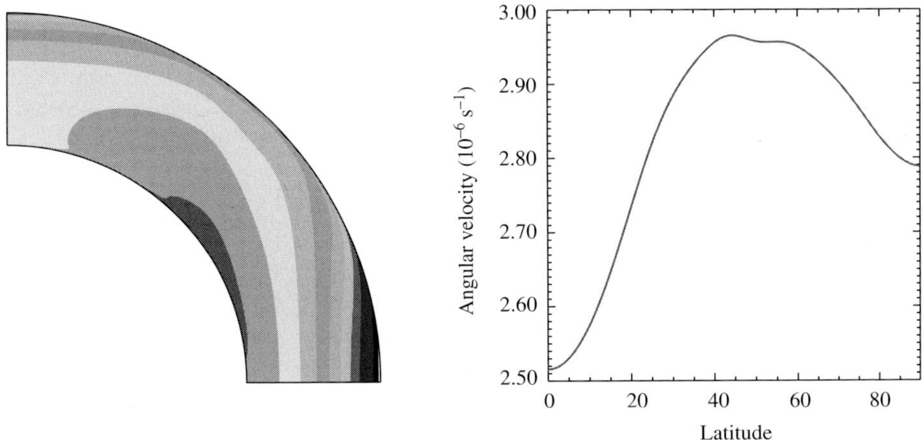

Figure 6.15. A characteristic rotation law resulting under the presence of a poloidal magnetic field with an amplitude of 200 G on a giant star. The angular velocity increases from the equator to the poles by $\sim$10%.

2004). Figure 6.15 shows the resulting rotation law for a magnetic field involved through the boundary condition of a steady radial field penetrating the convection zone at the inner boundary from the radiative core. The bottom field is prescribed by a steady potential at the inner boundary, which can be understood as the penetration of a relic field stored in the radiative core into the convection zone.

This concept is favoured by the observations (Strassmeier 2004). Among nine examples of antisolar rotation he reports six belonging to close binaries and two to giant stars with large dark spots on their surfaces. The remaining star is LQ Hya for which different observations disagree about the sense of its differential rotation (perhaps, because it strongly varies with time, see Donati *et al.* (2003)). Recently Weber *et al.* (2005) suggest that antisolar rotation on giant stars may indeed be accompanied by a fast ($\sim$1000 m s^{-1}) poleward surface flow.

Acknowledgements

L. L. K. is grateful to the Alexander v. Humboldt Foundation for support. Thanks are due also to the Russian Foundation for Basic Research (Project 05-02-16326).

References

Arlt, R., Sule, A. & Rüdiger, G. (2005). *Astron. Astrophys.*, **441**, 1171.

Barnes, J. R., Cameron, A. C., Donati, J.-F. *et al.* (2005). *Mon. Not. Roy. Astron. Soc.*, **357**, L1.

Bonanno, A., Elstner, D., Rüdiger, G. & Belvedere, G. (2002). *Astron. Astrophys.*, **390**, 673.

Brandenburg, A. (2005). *Astrophys. J.*, **625**, 539.

Choudhuri, A. R., Schüssler, M. & Dikpati, M. (1995). *Astron. Astrophys.*, **303**, L29.
Dikpati, M. & Gilman, P. A. (2001). *Astrophys. J.*, **559**, 428.
Donati, J.-F., Cameron, A. C. & Petit, P. (2003). *Mon. Not. Roy. Astron. Soc.*, **345**, 1187.
Gilman, P. A. & Miesch, M. S. (2004). *Astrophys. J.*, **611**, 568.
Hall, D. S. (1991). In *The Sun and Cool Stars: Activity, Magnetism, Dynamos*, ed. I. Tuominen, D. Moss & G. Rüdiger. Berlin: Springer-Verlag, p. 353.
Hupfer, C., Käpylä, P. J. & Stix, M. (2005). *Astron. Nachr.*, **326**, 223.
Käpylä, P. J., Korpi, M. J. & Tuominen, I. (2004). *Astron. Astrophys.*, **422**, 793.
Kippenhahn, R. (1963). *Astrophys. J.*, **137**, 664.
Kitchatinov, L. L. & Rüdiger, G. (1993). *Astron. Astrophys.*, **276**, 96.
Kitchatinov, L. L. & Rüdiger, G. (2004). *Astron. Nachr.*, **325**, 496.
Kitchatinov, L. L. & Rüdiger, G. (2005). *Astron. Nachr.*, **326**, 379.
Komm, R. W., Howard, R. F. & Harvey, J. W. (1993). *Sol. Phys.*, **147**, 207.
Kosovichev, A. G., Schou, J., Scherrer, P. H. *et al.* (1997). *Sol. Phys.*, **170**, 43.
Küker, M. & Rüdiger, G. (2005). *Astron. Astrophys.*, **433**, 1023.
Messina, S. & Guinan, E. F. (2003). *Astron. Astrophys.*, **409**, 1017.
Nandy, D. & Choudhuri, A. R. (2002). *Science*, **296**, 1671.
Petit, P., Donati, J.-F. & Cameron, A. C. (2004). *Astron. Nachr.*, **325**, 221.
Pulkkinen, P. J., Tuominen, I., Brandenburg, A., Nordlund, A. & Stein, R. F. (1993). *Astron. Astrophys.*, **267**, 265.
Reiners, A. (2006). *Astron. Astrophys.*, **446**, 267.
Reiners, A. & Schmitt, J. H. M. M. (2003a). *Astron. Astrophys.*, **398**, 647.
Reiners, A. & Schmitt, J. H. M. M. (2003b). *Astron. Astrophys.*, **412**, 813.
Rempel, M. (2005). *Astrophys. J.*, **622**, 1320.
Rieutord, M., Brandenburg, A., Mangeney, A. & Drossart, P. (1994). *Astron. Astrophys.*, **286**, 471.
Rüdiger, G. (1989). *Differential Rotation and Stellar Convection: Sun and Solar-Type Stars*. New York: Gordon and Breach Science Publishers.
Rüdiger, G., Egorov, P., Kitchatinov, L. L. & Küker, M. (2005a). *Astron. Astrophys.*, **431**, 345.
Rüdiger, G., Egorov, P. & Ziegler, U. (2005b). *Astron. Nachr.*, **326**, 315.
Strassmeier, K. G. (2002). *Astron. Nachr.*, **323**, 309.
Strassmeier, K. G. (2004). In *Stars as Suns*, ed. A.K. Dupree & A.O. Benz. ASP, p. 11.
Weber, M., Strassmeier, K. G. & Washuettl, A. (2005). *Astron. Nachr.*, **326**, 287.
Zhao, J. & Kosovichev, A. G. (2004). *Astrophys. J.*, **603**, 776.

Part IV

Hydromagnetic properties

7

Magnetic confinement of the solar tachocline

Pascale Garaud

Two distinct classes of magnetic confinement models exist for the solar tachocline. The 'slow tachocline' models are associated with a large-scale primordial field embedded in the radiative zone. The 'fast tachocline' models are associated with an overlying dynamo field. I describe the results obtained in each case, their pros and cons, and compare them with existing solar observations. I conclude by discussing new lines of investigation that should be pursued, as well as some means by which these models could be unified or reconciled.

7.1 Introduction

7.1.1 Magnetic fields in the tachocline

Two distinct possible origins for solar magnetic fields in the tachocline region can be identified. The Ohmic decay timescale of a large-scale dipolar field embedded in the radiative interior is much larger than the estimated age of the Sun (Cowling 1945; Garaud 1999), so that a fraction of the magnetic flux initially frozen within the accreting protostellar gas is likely to persist today. In parallel, according to the standard dynamo field theory, small-scale magnetic fields are thought to be constantly generated by fluid motions within the solar interior. Optimal conditions for the generation of large-scale fields require the combination of large-scale azimuthal shear and small-scale helical motion, which are both naturally found in the region of the tachocline (Parker 1993; Ossendrijver 2003; Tobias 2005).

The fundamental differences between primordial and dynamo-generated fields – see the discussion by Tobias & Weiss in Chapter 13 of this book – have naturally led to two distinct classes of tachocline confinement models: a *slow* tachocline, interacting on secular timescales with an underlying large-scale primordial field and slow meridional flows, and a *fast* tachocline, interacting on dynamical timescales with small-scale turbulent flows and an overlapping or overlying dynamo

The Solar Tachocline, D. W. Hughes, R. Rosner and N. O. Weiss.

Table 7.1. *Properties at the base of the convection zone*

Quantity	Value	Quantity	Value
ρ	0.2	η	4.3×10^2
T	2.2×10^6	ν	27
g	5.3×10^4	κ	1.3×10^7
N	9×10^{-4}	r_{cz}	5×10^{10}

Numerical values (in cgs units) of typical values of the density ρ, temperature T, gravity g, the buoyancy frequency N, the molecular magnetic diffusivity η, viscosity ν, thermal diffusivity κ and the radius r_{cz}.

field. This chapter strives to provide a fairly complete overview of the state of this rapidly evolving topic, and presents the two alternative confinement models that were considered at the time of the workshop. Since historically these two types of models have remained clearly separated, I shall take the same path and present them independently in Sections 7.2 and 7.3. Whether the true tachocline genuinely does fall into one category or another was widely debated during the meeting, and is discussed in Section 7.4 (see also Chapter 1 by Gough). A first attempt at constructing a global tachocline model that includes both fast and slow dynamics has been developed since then by McIntyre and is presented in Chapter 8; the reader should bear this new development in mind when reading this chapter.

7.1.2 Characteristic numbers in the tachocline

In order to compare models and observations of the tachocline, I adopt characteristic values for certain quantities in that region as listed in Table 7.1 (see also Table 1.1 in Chapter 1). For consistency, these values are used throughout this review; in some cases, however, they differ from those adopted by various other authors by factors of order unity.

7.2 Primordial field confinement: the slow tachocline

7.2.1 First models

With tremendous insight into today's debate, Mestel (1953) realised early on that even a very weak large-scale primordial field within the solar interior would have a significant impact on the solar angular velocity profile. Indeed, Alfvén waves are possibly the most efficient transporter of angular momentum in a rotating magnetized fluid. They propagate unimpeded along poloidal field lines with a characteristic velocity that depends on the local field amplitude and the local fluid density. Both the field geometry and the density stratification result in spatial inhomogeneities of the Alfvén velocity and the consequent phase mixing and damping of the waves. Angular momentum is then redistributed along (and across) the field lines, leading to a rotation profile satisfying Ferraro's (1937) isorotation law:

$$\mathbf{B} \cdot \nabla \Omega = 0, \tag{7.1}$$

or in other words, with Ω constant along magnetic field lines. It has been argued that field amplitudes as low as 10^{-2} G are capable of enforcing uniform rotation to the entire radiative interior (Mestel 1953; Cowling 1957; Mestel & Weiss 1987).

The first model to study quantitatively the effect of an internal primordial field on the solar radiative zone rotation profile, and in particular its potential role in confining the tachocline, was proposed by Rüdiger & Kitchatinov (1997). Shortly afterward, MacGregor & Charbonneau (1999) complemented their work by studying the effects of different internal field geometries.

Both investigations evaluate the steady-state outcome of the interaction between a primordial field and the latitudinal shear diffusing from the convection zone. Meridional flows are assumed to be negligible, on the grounds that the strong local stratification effectively reduces their amplitude to a few centimetres per second (Gough & McIntyre 1998); given this assumption, the poloidal component of the field decouples from the governing equations and can be chosen arbitrarily. While Rüdiger & Kitchatinov consider only poloidal fields entirely confined within the radiative zone, MacGregor & Charbonneau also study various cases in which at least some field lines overlap the convective zone. The steady-state structure of the toroidal field B_ϕ and angular velocity of the fluid Ω is then obtained by solving the azimuthal component of the momentum equation (here, cast in the form of a conservation equation for angular momentum) and of the induction equation:

$$\nabla \cdot (\rho \nu r^2 \sin^2 \theta \nabla \Omega) + \frac{1}{\mu_0} \mathbf{B}_{\mathrm{p}} \cdot \nabla (r \sin \theta B_\phi) = 0, \tag{7.2}$$

$$r \sin \theta \mathbf{B}_{\mathrm{p}} \cdot \nabla \Omega + \eta \left(\nabla^2 B_\phi - \frac{B_\phi}{r^2 \sin^2 \theta} \right) = 0, \tag{7.3}$$

where the poloidal component of the field, $\mathbf{B}_{\mathrm{p}}$, is fixed. These equations are subject to the boundary conditions at the interface with the convective zone,

$$\Omega(r_{\mathrm{cz}}, \theta) = \Omega_{\mathrm{eq}}(1 - a_2 \cos^2 \theta - a_4 \cos^4 \theta), \tag{7.4}$$

$$B_\phi(r_{\mathrm{cz}}, \theta) = 0, \tag{7.5}$$

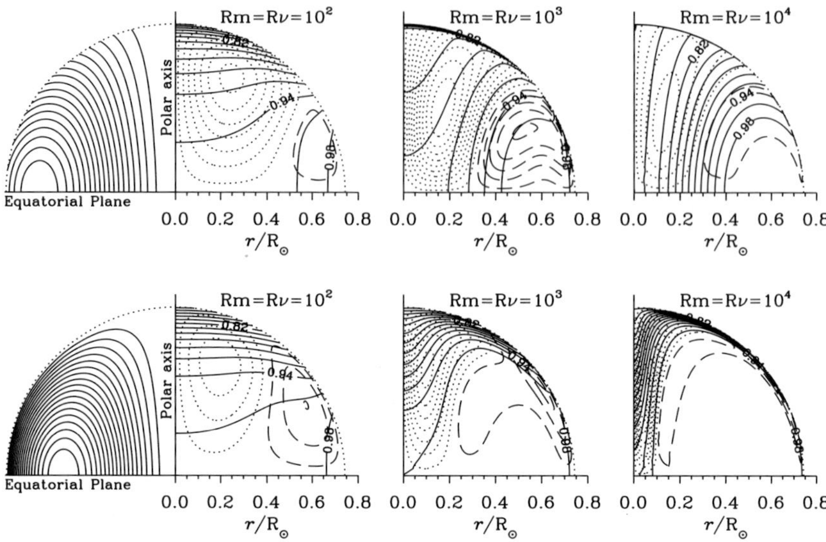

Figure 7.1. Steady-state solutions obtained by MacGregor & Charbonneau (1999) for the open and confined field configurations (top and bottom row respectively), for increasing Reynolds numbers. For this figure, the Reynolds numbers are defined as $R\nu = B_0 r_{\rm cz}/(\nu\sqrt{\mu_0\rho})$ and $Rm = B_0 r_{\rm cz}/(\eta\sqrt{\mu_0\rho})$, so that increasing the Reynolds numbers can be interpreted as increasing B_0 or decreasing ν and η. The left quadrant shows the poloidal field lines, whereas the right quadrants show the rotation profile (solid lines) and the toroidal field amplitude (dashed and dotted lines correspond to positive and negative B_ϕ).

where $\Omega_{\rm eq}$, a_2 and a_4 are derived from helioseismic inversions of the rotation profile in the convective zone; typically, $\Omega_{\rm eq}/2\pi = 460\,{\rm nHz}$, $a_2 = 0.14$ and $a_4 = 0.15$ (Charbonneau *et al.* 1999a). Adequate regularity conditions are applied on the polar axis and at the centre. The boundary condition on the toroidal field is related to the assumption that any toroidal field at the interface with the convection zone is promptly removed through buoyancy instabilities. An alternative boundary condition that is sometimes also used assumes the convection zone to be an excellent insulator (with $\eta \to \infty$), and matches the interior field to a potential field. These two possibilities result in different quantitative estimates for the confining field strength and toroidal field amplitudes, but have otherwise qualitatively similar associated solutions.

The numerical solutions reveal a striking difference in angular velocity profile between the confined field and open field cases (see Figure 7.1). While the former results in a more-or-less uniformly rotating radiative zone, with a thin shear layer effecting the smooth diffusive transition to the differentially rotating convective zone, the latter results in a latitudinally sheared state close to Ferraro isorotation, as field lines connected to the differentially rotating convection zone provide a support for the inward propagation of Alfvén waves. Helioseismic observations appear to set empirical constraints on the geometry of an embedded primordial field.

The angular momentum equation (7.2) illustrates the balance between viscous transport and magnetic transport near the outer boundary. A boundary analysis provides useful quantitative estimates of the tachocline properties in both open and confined geometries: viscous effects are only important in a thin Ekman–Hartmann boundary layer (see the review by Acheson & Hide 1973) of width

$$\delta_{\parallel} = \left(\frac{\mu_0 \rho \nu \eta}{B_0^2 r_{\rm cz}^2}\right)^{1/4} r_{\rm cz} = \left(\frac{E_\nu E_\eta}{\Lambda}\right)^{1/4} r_{\rm cz} \sim 4 \times 10^{-5} B_0^{-1/2} r_{\rm cz}, \tag{7.6}$$

where the field of amplitude B_0 is assumed to be mostly parallel to the outer boundary (as in the case of the confined field) and is measured in gauss. The usual Ekman numbers are defined as

$$E_\nu = \frac{\nu}{r_{\rm cz}^2 \Omega_0}, \quad E_\eta = \frac{\eta}{r_{\rm cz}^2 \Omega_0}, \tag{7.7}$$

and a new parameter Λ is defined as

$$\Lambda = \frac{v_{\rm A}^2}{v_\Omega^2}, \tag{7.8}$$

where $v_\Omega = r_{\rm cz}\Omega_0$ and $v_{\rm A}$ is the Alfvén velocity $v_{\rm A} = B_0/\sqrt{\mu_0 \rho}$. Here, Ω_0 is a mean angular velocity of the system. Equation (7.6) provides the first of many estimates of the relation between the internal field strength and the tachocline thickness. If the poloidal field is given by

$$\mathbf{B}_{\rm p} = \nabla \times \left(\frac{A}{r \sin\theta}\hat{\mathbf{e}}_\phi\right), \quad \text{with } A = B_{\rm in}\frac{r^2}{2}\left(1 - \frac{r}{r_{\rm cz}}\right)^q, \tag{7.9}$$

where the index q controls the field concentration towards the interior, and $B_{\rm in}$ is the amplitude of the field deep in the interior, then a field of amplitude B_0 in a tachocline of thickness Δ corresponds to

$$B_{\rm in} \simeq \frac{q}{2} B_0 \left(\frac{r_{\rm cz}}{\Delta}\right)^{q-1}. \tag{7.10}$$

Combining all of the above estimates suggests that a field strength of 2×10^{-6} G near the edge of the convective zone (which corresponds to an interior field of about 6G for $q = 5$) would confine the tachocline to its observed width of $0.03 r_{\rm cz}$ (Elliott & Gough 1999).

In a very interesting remark, MacGregor & Charbonneau (1999) point out that even in a laminar tachocline, angular momentum transport between the convective and radiative zones would not, in fact, proceed through viscous effects only; as Spiegel & Zahn (1992) (see also Chapter 4 by Zahn) had shown, the tachocline spread is aided by meridional flows, which act approximately as a hyperdiffusion of the kind

$$\frac{\partial \Omega}{\partial t} \sim \frac{r_{\mathrm{cz}}^4}{t_{\mathrm{ES}}} \frac{\partial^4 \Omega}{\partial r^4}, \quad \text{where } t_{\mathrm{ES}} = \frac{1}{4} \frac{N^2}{\Omega_0^2} \frac{r_{\mathrm{cz}}^2}{\kappa} \sim 2 \times 10^{11} \, \mathrm{yr}, \tag{7.11}$$

where N is the local buoyancy frequency in the tachocline. In that case an equivalent boundary layer analysis reveals a different relation between the tachocline thickness and the field strength:

$$\delta = \left(\frac{\mu_0 \rho \eta}{B_0^2 t_{\mathrm{ES}}} \right)^{1/6} r_{\mathrm{cz}} = \left(\frac{E_\eta}{\Lambda \Omega_0 t_{\mathrm{ES}}} \right)^{1/6} r_{\mathrm{cz}} \sim 0.0001 B_0^{-1/3} r_{\mathrm{cz}}. \tag{7.12}$$

The local poloidal field required to confine the tachocline is now of the order of $B_0 \sim 6 \times 10^{-4}$ G, and the resulting toroidal field has a typical amplitude of the order of 10^5 G, which (as MacGregor & Charbonneau point out) is interestingly close to the estimated upper limit for field storage against magnetic buoyancy within the tachocline (Schüssler *et al.* 1994).

7.2.2 Towards a self-consistent model: the governing equations

Despite the great degree of simplification inherent in the model just described, one essential result is of profound generality: Alfvénic angular momentum transport occurs on a very rapid timescale, and does not permit large deviations from isorotation anywhere in the radiative zone. Observed sheared regions (such as the tachocline) must be relatively free of poloidal field. The magnetic confinement problem takes an alternative but equivalent meaning: there must exist some mechanisms that confine the primordial field within the radiative zone in such a way as to be largely disjoint from the convective zone. Very little overlap between the internal field and the convective region is allowed by the upper limits set from observations of the sunspot parity throughout the cycles (Boyer & Levy 1984; Boruta 1996).

The microscopic magnetic diffusivity in the radiative zone does not exceed $\sim$500 cm^2 s^{-1}. Consequently, even apparently slow flows have a large magnetic Reynolds number. Radial motions in the tachocline are heavily suppressed by the strong local stratification, the flow speed for a steady-state system being controlled by the thermal diffusion time. Across the tachocline, the upper limit for the radial flow velocities is $u_r \sim 10^{-4}$ cm s^{-1}, with a corresponding magnetic Reynolds number of a few hundred, which is sufficient to have significant nonlinear interactions with the poloidal field, contrary to the assumptions of the studies described in the previous section.

Gough & McIntyre (1998) realized the importance of meridional flows in the dynamics of the tachocline. They proposed a model in which gyroscopic pumping

Figure 7.2. Schematic representation of the Gough & McIntyre model. The outer convection zone is differentially rotating, and generates meridional flows (black lines) through gyroscopic pumping. These confine the underlying field (thick grey lines) to the radiative interior, while leaving the tachocline virtually magnetic free. See Gough & McIntyre (1998) for full colour figure.

near the convective–radiative interface drives flows whose role is to confine the interior field, thereby completing the missing piece of the tachocline puzzle.

Their model consists of four radially distinct regions (see Figure 7.2). In the convection zone (extended, if necessary, by a few tens of megametres to include the overshoot region, and a corresponding fast tachocline), angular momentum balance is achieved between anisotropic Reynolds or Maxwell stresses, and large-scale advection by meridional flows (zone 1). The flow geometry near the convective-radiative interface is dictated by the steady-state thermal wind and thermal energy balance. The flows burrow into the stably stratified, mostly laminar region directly underneath (zone 2) and interact with the deeply embedded magnetic field within a thin magnetic boundary layer (zone 3). This conveniently results in the simultaneous confinement of the underlying field to the lower part of the radiative zone, and that of the meridional flows within a well-ventilated but mostly magnetic free upper part of the radiative zone. Below the magnetic boundary layer, the confined field imposes uniform rotation to the bulk of the radiative zone (zone 4).

It is perhaps worth pointing out here that the notion of a tachocline has significantly evolved in recent years. Within the well-ventilated, magnetic-free region (zone 2) angular momentum is roughly conserved along the meridional flow lines and the latitudinal shear imposed by the convective zone is not so much suppressed as 'reshuffled'. As a result, given the strict definition of tachocline as 'a strong shear layer beneath the convective region' one could argue that the Gough & McIntyre tachocline is in fact limited to the magnetic diffusion layer. On the other hand, a more modern interpretation of the tachocline as 'the region which operates the dynamical transition between the convection zone and the radiative zone' would then encompass both the magnetic diffusion layer and the magnetic-free region directly above. This distinction will be useful when comparing the various predictions for the tachocline thickness proposed in the literature. Moreover, a third meaning of tachocline confinement now emerges in relation to the tachocline meridional flows. Observations of surface abundances of light elements and helioseismic observation of the sound speed profile in the tachocline suggest that the depth of the mixed layer beneath the convection zone is of the order of a few percent of the solar radius (see Chapter 3 by Christensen-Dalsgaard & Thompson for more detail; Rüdiger & Pipin 2001; Elliott & Gough 1999). Given that the upper limits on the depth of the overshoot region have been recently estimated to be significantly smaller than the tachocline depth (Brummell *et al.* 2002; Rogers & Glatzmaier 2005), these observations can be related with reasonable confidence to the tachocline ventilation depth (zones 2 and 3).

The equations governing laminar fluid motions in the radiative zone consist of the momentum equation, the mass conservation equation, the thermal energy conservation equation, the field advection–diffusion equation, the equation of state and, finally, a solenoidal condition for the field. When considering slow meridional flows in a slowly rotating star like the Sun, one can linearize the equations around a uniformly rotating, spherically symmetric background hydrostatic equilibrium and use the anelastic approximation. The complete set of equations representing the secular laminar dynamics of the radiative interior is then

$$\rho\frac{\partial \mathbf{u}}{\partial t} + 2\rho\boldsymbol{\Omega}_0 \times \mathbf{u} = -\nabla\tilde{p} - \tilde{\rho}\mathbf{g} + \mathbf{j}\times\mathbf{B} + \nabla\cdot\boldsymbol{\pi}, \tag{7.13}$$

$$\nabla\cdot(\rho\mathbf{u}) = 0, \tag{7.14}$$

$$\rho T\frac{\partial s}{\partial t} + \rho T\mathbf{u}\cdot\nabla s = \nabla\cdot(k\nabla T), \tag{7.15}$$

$$\frac{\tilde{p}}{p} = \frac{\tilde{\rho}}{\rho} + \frac{\tilde{T}}{T}, \tag{7.16}$$

$$\frac{\partial \mathbf{B}}{\partial t} = \nabla\times(\mathbf{u}\times\mathbf{B}) + \nabla\times(\eta\nabla\times\mathbf{B}), \tag{7.17}$$

$$\nabla\cdot\mathbf{B} = 0, \tag{7.18}$$

where tildes denote perturbations from hydrostatic equilibrium, s is the entropy, π is the viscous stress tensor, $k = \rho c_{\mathrm{p}}\kappa$ is the thermal conductivity, and all other quantities have their usual meaning. This complete system of equations cannot yet be solved exactly for realistic solar values of the background state quantities. Numerical solutions have difficulties reaching simultaneously the correct thermal, viscous and magnetic diffusivities, while analytical solutions struggle to cope with the complex geometry and the intrinsic nonlinearity of the problem. What follows describes the various attempts at treating the problem that have been proposed so far.

7.2.3 The Gough & McIntyre boundary layer analysis

The insight of Gough & McIntyre's seminal work (1998) is to reduce the above system of equations to a boundary layer analysis, by considering from the outset the thin nature of the tachocline, and retaining in each zone identified only the dominant terms in the dynamical balance.

Thermal-wind balance in the upper region of the tachocline (zone 2). In this region, Gough & McIntyre assume that the amplitude of the confined internal magnetic field is too low to have any significant effect on the flow dynamics. In that case, thermal-wind balance is achieved: the azimuthal component of the vorticity equation reduces to

$$2\Omega_0 r \sin\theta \frac{\partial \tilde{\Omega}}{\partial z} = \frac{g}{rT}\frac{\partial \tilde{T}}{\partial \theta}, \tag{7.19}$$

where the pressure fluctuations in the equation of state have been neglected in accordance with the anelastic approximation. Maintaining the thermal-wind balance against diffusion requires heat and momentum advection by meridional flows; within a thin tachocline this is equivalent to:

$$\frac{N^2 T u_r}{g} = \frac{1}{\rho c_{\rm p} r^2}\frac{\partial}{\partial r}\left(r^2 k \frac{\partial \tilde{T}}{\partial r}\right), \tag{7.20}$$

where k is the thermal conductivity ($\kappa = k/\rho c$).

Additional information on the flow geometry related to the tachocline shear can be deduced qualitatively from Equations (7.19) and (7.20). The observed angular velocity profile in the tachocline, as given by Equation (7.4), corresponds to a significant latitudinal entropy perturbation, positive near the poles and equator, and negative at mid-latitudes. In order to maintain this gradient against diffusion (specifically in the radial direction, since the overlying convective zone is largely isentropic) meridional flows are required, with downwelling near the poles and upwelling in mid-latitudes. This geometry favours the internal field confinement only if the upwelling region is sufficiently narrow.

Advection–diffusion balance in the magnetic diffusion layer (the tachopause, zone 3). In the downwelling regions, the tachocline flow meets the underlying field and confines it to the radiative interior. In a steady state, the system is in equilibrium when the downward advection exactly compensates the outward diffusion of the field. Within a thin diffusion layer, the dominant terms of the advection–diffusion balance are extracted to yield

$$2\Omega_0 u_\theta \cos\theta = \frac{B_0}{\mu_0 \rho r \sin\theta}\frac{\partial}{\partial\theta}(B_\phi \sin\theta), \tag{7.21}$$

$$-B_0 \sin\theta \frac{\partial\tilde{\Omega}}{\partial\theta} = \eta\frac{\partial^2 B_\phi}{\partial r^2}, \tag{7.22}$$

from the angular momentum equation and the azimuthal component of the induction equation, respectively. Here, B_0 is the amplitude of the meridional component of the primordial field in the region of the tachocline. In addition, a rough estimate of the radial flow velocity required to balance the diffusion of the field in the boundary layer of thickness δ_3 is

$$u_r \sim \frac{\eta}{\delta_3}, \tag{7.23}$$

which can be combined with the anelastic mass conservation equation to obtain an estimate of the latitudinal velocity:

$$\frac{1}{r^2}\frac{\partial}{\partial r}(r^2 \rho u_r) + \frac{1}{r\sin\theta}\frac{\partial}{\partial\theta}(\rho\sin\theta u_\theta) = 0. \tag{7.24}$$

Boundary layer scaling. Boundary layer scalings are easily derived using the approximations $\partial/\partial r \sim 1/\delta_2$ in zone 2, $\partial/\partial r \sim 1/\delta_3$ in zone 3 and $\sin\theta \sim \cos\theta \sim 1/\sqrt{2}$ with $\partial/\partial\theta \sim iL$, where L is a latitudinal wavenumber.

Before outlining the results obtained by Gough & McIntyre, it should be noted that in the limit where the magnetic diffusion layer is of similar width to the whole tachocline (in that case, there is no magnetic-free region – zones 2 and 3 are combined) $\delta_2 = \delta_3$ and the combination of Equations (7.19) to (7.24) with $\partial/\partial r = 1/\delta$ yields (as expected) exactly the estimate of the tachocline thickness (7.12) derived by MacGregor & Charbonneau (1999). For this scaling to hold, it is important to verify that the Lorentz force in the vorticity equation can be neglected compared with the thermal-wind balance. This is indeed the case for the field amplitude corresponding to the observed tachocline width.

The Gough & McIntyre model suggests that a different force balance can occur when the magnetic diffusion layer is significantly thinner than the magnetic-free region. In zone 2, a unique expression relating the flow amplitude and the thickness of the region δ_2 can be derived from the thermal-wind balance and the thermal energy equations, namely (7.19) and (7.20):

$$u_r \sim \frac{2}{L}\left(\frac{\kappa}{r_{\rm cz}^2\Omega_0}\right)\left(\frac{r_{\rm cz}}{\delta_2}\right)^3\left(\frac{\Omega_0}{N}\right)^2\left(\frac{\tilde{\Omega}}{\Omega_0}\right) r\Omega_0. \qquad (7.25)$$

Note that if δ_2 is fixed, this equation provides a stringent relation between the imposed shear and the meridional flows permitted within the tachocline.

Two scenarios may then occur depending on the strength of the internal field. The Gough & McIntyre model assumes that the magnetic field amplitude within the tachopause is sufficiently small for the thermal wind relation to hold there as well. Thus, Equations (7.19) and (7.20) complement Equations (7.21) to (7.24) in zone 3 to yield the scaling:

$$\delta_3 \sim \left(\frac{4}{L^4}\frac{v_\Omega^2}{v_{\rm A}^2}\frac{\Omega_0^2}{N^2}\frac{\kappa\eta}{r_{\rm cz}^4\Omega_0^2}\right)^{1/6} r_{\rm cz}. \qquad (7.26)$$

Note that the Gough & McIntyre tachopause is exactly the boundary layer studied by MacGregor & Charbonneau (1999) – see Equation (7.12). Matching the tachopause dynamics with the overlying flow from zone 2, by combining (7.26) with (7.23) and (7.25), yields a unique relation between δ_2 and B_0:

$$\delta_2 \sim \left[\frac{2^8}{L^{10}}\frac{v_\Omega^2}{v_{\rm A}^2}\left(\frac{\kappa}{\eta}\right)^5\frac{\kappa^2}{r_{\rm cz}^4\Omega_0^2}\left(\frac{\Omega_0^2}{N^2}\right)^7\left(\frac{\tilde{\Omega}}{\Omega_0}\right)^6\right]^{1/18} r_{\rm cz}. \qquad (7.27)$$

Comparing the expression for δ_2 to the observed tachocline ventilation depth as measured by Elliott & Gough (1999), Gough & McIntyre deduce that the internal field strength (in the tachocline region) is of the order of 1 G, corresponding to a primordial field strength in the deep interior of the Sun of the order of 10^4 G. As assumed, the thickness of the tachopause is only a few percent of the thickness of the whole tachocline. The tachocline ventilation time is of the order of 3×10^6 yr; while being slow, it provides sufficient mixing of light elements beneath the convective zone to explain the observed abundances of Li and Be. This ventilation timescale is still significantly smaller than the solar spin-down timescale, which accounts for the fact that the interior angular velocity is close to that of the surface layers.

Given this estimate for the field amplitude in the tachopause, it appears that neglecting the Lorentz force in the vorticity equation is only marginally justified. In fact, Gough & McIntyre themselves acknowledge that the thermal-wind relation may not hold in the lower regions of the magnetic boundary layer, where the nonlinear interaction between the field and the flow is maximal. What happens in the alternative case has not yet been evaluated in detail; however, dropping

Equations (7.19) and (7.20) plausibly describes the right balance, and reveals a new boundary layer scaling

$$\delta_3 \sim \left(\frac{2}{L^3} \frac{\Omega_0}{\tilde{\Omega}} \frac{v_\Omega^2}{v_A^2} \frac{\eta^2}{r_{cz}^4 \Omega_0^2} \right)^{1/4} r_{cz}, \tag{7.28}$$

which, when combined with Equation (7.23) from the poloidal advection–diffusion balance, and Equation (7.25) from thermal-wind balance in zone 2, reveals yet another possible relation between the tachocline thickness, the imposed shear and the magnetic field:

$$\delta_2 \sim \left[\frac{2^5}{L^7} \left(\frac{\kappa}{\eta} \right)^4 \frac{\eta^2}{r_{cz}^4 \Omega_0^2} \left(\frac{\Omega_0^2}{N^2} \right)^4 \left(\frac{\tilde{\Omega}}{\Omega_0} \right)^3 \frac{v_\Omega^2}{v_A^2} \right]^{1/12} r_{cz}. \tag{7.29}$$

The main difference between this boundary layer analysis and the one proposed by Gough & McIntyre is the non-thermal nature of the boundary layer[1].

So which (if any) of the above scalings really correspond to the solar tachocline? This question is difficult to answer without a careful quantitative estimate of the force balance in the tachopause, which can only be done through numerical simulations. Moreover, since the Coriolis force and the field geometry vary strongly with latitude, the force balance and the nature of the boundary layer is very likely to differ between the equator, mid-latitudes and the poles.

7.2.4 Numerical solutions of the Gough & McIntyre model

To obtain a more precise view of the geometry of the tachocline dynamics, as well as quantitative predictions for the internal rotation rate, the light element depletion timescale and the amount of overlap between the interior field and the convective zone, one must resort to numerical simulations. Two approaches have recently been considered. Douglas Gough and I have been interested in studying the steady-state tachocline balance, while Brun & Zahn (2006) are looking at its temporal evolution for a given initial poloidal field configuration. While the former is able to bypass the various numerical problems caused by the wide range of timescales inherent in the physics of the system, the latter is ideally suited to the study of potential multiple equilibria, and naturally eliminates from the force balance any processes occurring on a timescale longer than the stellar evolution timescale.

[1] The tachopause in the Gough & McIntyre model is also a thermal boundary layer.

7.2.4.1 Steady-state calculations

Axially symmetric steady-state calculations can be performed by an expansion of all governing equations on a suitably selected basis of orthogonal polynomials in the latitudinal direction, and then by solving the remaining ODEs using a Newton–Raphson relaxation procedure. Note that other methods also exist (expansion in spherical harmonics or finite differences in all directions), but have not been implemented for the steady-state problem.

In 2002, I presented a preliminary numerical study of the nonlinear interaction between the primordial field and the meridional flows, in an idealized setup where the solar tachocline and radiative zone are assumed to be composed of an incompressible, homogeneous and isentropic fluid (Garaud 2002). This assumption largely simplifies the set of governing equations since all thermodynamical quantities decouple from the system; however, it also eliminates the crucial baroclinicity that is thought to drive meridional flows. These must then be artificially replaced by Ekman flows driven by viscous forces on a no-slip impermeable boundary. The latitudinal variation of the Coriolis force implied by the imposed shear from the convection zone (for the Gough & McIntyre model) and in a viscous Ekman layer (in the simplified model) provides gyroscopic pumping with a similar geometry, but of different amplitude. This simplified model clearly could not provide any quantitative estimates of the tachocline dynamics, but the geometrical similarities with the correct model provide an interesting complement to the Gough & McIntyre (1998) boundary layer analysis.

In this simplified model, the equations solved are the following:

$$2\mathbf{\Omega}_0 \times \mathbf{u} = -\nabla p + \mathbf{j} \times \mathbf{B} + \nu \nabla^2 \mathbf{u}, \tag{7.30}$$

$$\nabla \cdot \mathbf{u} = 0, \tag{7.31}$$

$$\nabla \times (\mathbf{u} \times \mathbf{B}) = \eta \nabla \times (\nabla \times \mathbf{B}), \tag{7.32}$$

$$\nabla \cdot \mathbf{B} = 0, \tag{7.33}$$

with a fiducial density $\rho = 1$. No-slip, impermeable boundary conditions are assumed for the meridional flows, and on the upper boundary the rotation profile is given by the convection zone profile (see Equation (7.4)). The lower boundary is a stress-free solid conducting core. The field is matched onto a potential field decaying exponentially outside the radiative zone, and matching on to a point dipole of given amplitude $B_{\rm in}$ located at the centre of the inner core.

The dynamical connection between the interior flow and the top boundary operates through Ekman and Hartmann layers, which have typical scalings of the order of

$$\delta_\nu = E_\nu^{1/2} r_{\rm cz}, \tag{7.34}$$

for a purely viscous Ekman layer, and

$$\delta_{\|} = \left(\frac{E_\nu E_\eta}{\Lambda}\right)^{1/4} r_{\rm cz} \text{ and } \delta_{\perp} = \left(\frac{E_\nu E_\eta}{\Lambda}\right)^{1/2} r_{\rm cz}, \tag{7.35}$$

for Hartmann layers when a magnetic field of amplitude B_0 is respectively parallel and perpendicular to the outer boundary. Ekman numbers of the order of 10^{-5} or less are therefore required to model structures on the scale of the tachocline.

In what follows, it is important to remember that the induction equation is linear in the field amplitude; thus, the ability of the flow to confine the field[2] depends not so much on the field amplitude as on the meridional flow velocity and corresponding magnetic Reynolds number $Rm = u_r\delta/\eta$. Gyroscopic pumping (of the Ekman, or Ekman–Hartmann type) on the outer boundary implies that the latitudinal component of the flow u_θ has amplitude comparable to the azimuthal velocity of the outer boundary u_ϕ, whereas the radial component of the flow is given by $u_r \sim \delta u_\theta / r_{\rm cz}$, where δ is the thickness of the relevant boundary layer. This simple estimate has two important consequences. Since δ is naturally smaller for larger field strengths, the stronger the field, the smaller the effective magnetic Reynolds number. Moreover, for a given field strength $\delta_{\|} \gg \delta_{\perp}$, so that the Ekman–Hartmann flow is much stronger in the confined field case (i.e. parallel to the outer boundary) than for the open field case (i.e. perpendicular to the outer boundary). The system is therefore subject to a strong positive feedback effect: when and where the field lines are confined *because* of an initially large flow amplitude, the resulting field geometry permits large flow amplitudes. The converse is true for the open field case, with weak flows as a cause *and* consequence of the radial field geometry on the boundary. Such dual dynamics with positive feedback in both limits is likely to harbour multiple equilibria. Unfortunately, the numerical algorithm I use is not ideally suited for the search for co-existing steady states; this could however be the subject of an interesting investigation.

The following results are the only steady states found for a given set of parameters. Varying the internal field strength (through Λ) for fixed Ekman numbers reveals three possible dynamical structures. Note that the physical interpretation of the numerical results given here differs from that of the original paper (Garaud 2002), and should be preferred.

For low field strengths ($\Lambda \ll 1$), the internal flow is dominated by Coriolis forces, with a more-or-less cylindrical angular velocity profile (commonly referred to as Taylor–Proudman rotation). Meridional flows are of Ekman type (with $u_r \sim E_\nu^{1/2} r_{\rm cz}\tilde{\Omega}$), penetrate deep into the radiative zone, and confine the field to the interior (except in the polar regions).

[2] Note that there is, in this simulation, an indirect dependence on the field strength through Ekman–Hartmann pumping.

For very high field strengths ($\Lambda \gg 1$), the internal flow is dominated by Lorentz forces, and the angular velocity is in a state of isorotation with the field. In contrast with the previous case, the driven flows are particularly weak ($u_r \sim \delta_\perp \tilde{\Omega}$, so that $Rm \sim E_\nu/\Lambda \ll 1$), and do not have significant effects on the field, which retains a mostly dipolar structure throughout the computational domain. The field lines freely connect with the convective zone, and the shear is propagated inwards accordingly. In this limit, it is in fact possible to linearize the equations around a state of isorotation, which was successfully done by Dormy *et al.* (1998, 2002).

For intermediate field strength, the nonlinear interaction between the internal field and the meridional flows dominates the dynamics of the interior. Two separate regions can be identified. The essentially radial geometry of the flow in the polar regions, as suspected by Gough & McIntyre, provides only weak coupling with the underlying (mostly radial) field. Polar field lines are connected to the convection zone, which results in slowly rotating, strongly sheared polar regions. On the other hand, the downwelling flow near the equator is able to confine the internal field over a broad range of latitudes, which results in a uniform rotation profile below. In this region, a Hartmann layer is observed with flow amplitudes scaling as $u_r \sim \delta_\parallel \tilde{\Omega}$ and corresponding to a magnetic Reynolds number $Rm \sim (E_\nu/E_\eta)^{1/2}\Lambda^{-1} = Pm^{1/2}/\Lambda$ (where $Pm = \nu/\eta$ is the magnetic Prandtl number). The meridional flows themselves are deflected by the underlying field and the resulting radial mixing is strongly suppressed. There is a marginal hint of the type of nested boundary layer structure predicted by Gough & McIntyre (1998), with a largely magnetic-free region overlying a thin diffusion layer. However, this result needs to be confirmed with lower diffusivity simulations.

The intermediate field strength case appears to approach qualitatively the dynamical structure that we may expect to see in the solar tachocline. However, the incompressible and isentropic nature of the fluid is an intrinsic flaw of this preliminary work which needs to be addressed. New results obtained by Douglas Gough and me on the steady-state structure of the Gough & McIntyre tachocline including stratification and thermal diffusion were presented at the workshop. This time, the complete set of Equations (7.13)–(7.18) are solved for a steady-state solution. The boundary conditions are similar to the ones used in the incompressible case for the magnetic field, but the assumption of 'impermeability' of the base of the convection zone to flows was dropped in favour of one which assumes the continuity of Reynolds stresses across the boundary. Several Reynolds stress prescriptions in the convection zone are currently being explored, and the preliminary results presented in Figure 7.3 correspond to a simplistic stress-free assumption (although, as before, the observed rotation profile of the convection zone is still imposed at the top of the computational domain). Finally, we assume that the convection zone acts as a perfect conductor, so our numerical solution is matched to a 'potential solution'

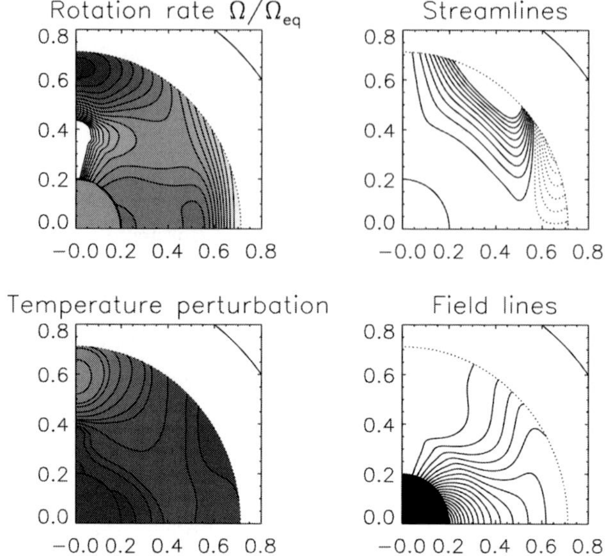

Figure 7.3. Numerical results of Equations (7.13)–(7.18) in a steady-state calculation for $f_\nu = f_\kappa = 5 \times 10^8$ and $f_\eta = 5 \times 10^6$. Each quadrant shows the solution in the radiative zone only, and the dotted line represents the edge of the convection zone. The rotation rate contours (from darker to lighter shading) range from $0.6\Omega_{eq}$ to Ω_{eq}. The streamlines are shown with dotted lines for clockwise flows and solid lines for anti-clockwise flows. The temperature perturbations range from 0 K to +50 K.

$\nabla^2 T = 0$ at the outer boundary. The main consequence of this new set of boundary conditions is to eliminate spurious Ekman flows and let the force balance within the tachocline dictate the flow amplitude and geometry.

The background state used was derived from a realistic solar model (Christensen-Dalsgaard *et al.* 1991) where, for numerical purposes, the thermal conductivity, viscosity and magnetic diffusivity are artificially increased by the factors f_k, f_ν and f_η respectively; this is necessary, since viscous and magnetic diffusion layers on the artificial outer boundary are otherwise too thin to be resolved. Typical values of f achieved in preliminary simulations are of the order of 10^7, with corresponding Ekman numbers of the order of 10^{-6}; when $f_\nu = f_\eta = f_\kappa$ the solar values of the magnetic and thermal Prandtl numbers are respected.

In the absence of strong magnetic fields the amplitude and geometry of the meridional flow satisfy the expectations from the Gough & McIntyre model: the steady-state solutions appear to depend on the thermal conductivity only, confirming that the weak flows that may be driven by the artificial stresses on the outer boundary are negligible compared to the baroclinic flows. These numerical results do therefore provide a good insight into the slow tachocline dynamics.

A thorough quantitative study of the numerical solutions is currently being performed, but preliminary qualitative results are found to be very sensitive to the thermal and magnetic diffusion parameters f_k and f_η. According to the scalings obtained in Section 7.2.3, the magnetic Reynolds number corresponding to the tachocline ventilation flow is

$$Rm \sim \frac{\kappa}{\eta}\frac{\Omega_0^2}{N^2}\frac{r_{\rm cz}^2}{\Delta^2}\frac{\tilde{\Omega}}{\Omega_0} \sim 0.01\frac{r_{\rm cz}^2}{\Delta^2}, \tag{7.36}$$

for solar values of the diffusion and rotation parameters. Hence provided there exists a confining mechanism for the tachocline and $\Delta \ll r$ then $Rm \gg 1$, confirming the nonlinear interaction between the field and the flow; on the other hand, $Rm \ll 1$ if the tachocline is not confined. Again, this dual structure suggests either a very strong sensitivity of the equilibrium solution to the input parameters, or even the existence of multiple equilibria.

For most parameter values (in the low-diffusivity limit) numerical simulations show that the internal field retains a mainly dipolar structure with field lines connecting to the convective region. The interior rotation profile is close to a state of isorotation, and no tachocline is observed in this limit.

For carefully chosen parameters, however, it is possible to obtain solutions that are encouragingly close to what may be expected from a slow tachocline (see Figure 7.3). Meridional flows burrow into the radiative zone and confine the field to the interior except within the upwelling region. The width of the upwelling region is always of the order of the depth of the tachocline, and the flow direction within the upwelling region is roughly parallel to the rotation axis. Contrary to the incompressible simulations, field confinement also occurs in the polar regions. Interestingly, a thermal boundary layer appears to be present in the polar regions, but not in the equatorial regions.

7.2.4.2 Time-dependent calculations

The first numerical time-dependent simulations of a slow solar tachocline following the idea of Gough & McIntyre were presented by Sacha Brun and Jean-Paul Zahn at the workshop. The numerical algorithm used is the ASH code (Glatzmaier 1984; Clune *et al.* 1999; Miesch *et al.* 2000; Brun *et al.* 2004), which performs a spectral decomposition of the governing MHD anelastic equations into spherical harmonics and Chebyshev polynomials in the horizontal and vertical directions respectively. The massively parallel numerical algorithm achieves significant resolution in all three directions. It is ideally suited for studying the radiative–convective interface.

Brun & Zahn (2006) study numerically the dynamical evolution of the radiative zone when subject to shearing from the overlying convective region, and in the presence of a large-scale embedded primordial field. Their computational domain includes the radiative zone only, and they model the radiative–convective interface as an impermeable, electrically and thermally conducting, sheared boundary. Various initial magnetic field configurations are studied, ranging from deeply embedded fields to open field configurations. Furthermore, the assumption of axial symmetry is dropped, which enables them to study the emergence of all the possible non-axisymmetric MHD and baroclinic instabilities that have recently been discussed (see Chapter 10, by Gilman & Cally, and Chapter 11, by Hughes, in this book), as well as the angular momentum transport from the associated Reynolds and turbulent Maxwell stresses (see Section 7.2.5.1).

The numerical values of the viscous, thermal and magnetic transport coefficients (ν, η and κ) used in the ASH code are far greater than the microscopic solar values; however, by respecting their hierarchy (i.e. by respecting the hierarchy of all expected boundary layer widths and all dynamical timescales), Brun & Zahn attempt to capture the essential dynamical balance in the tachocline, if not quantitatively at least qualitatively.

Their main result could reshape our view of the slow tachocline: none of the simulations appears to reach the steady-state balance suggested by the Gough & McIntyre model. Instead, the system is observed to evolve in time following the diffusion of the magnetic field out of the radiative zone. In consequence, the dynamical evolution of the interior depends crucially on the initial magnetic configuration.

For initially open field lines, isorotation is rapidly achieved, as suspected from the results of MacGregor & Charbonneau (1999). The meridional flows are strongly suppressed by the Lorentz force exerted by the mostly radial field lines, and fail to confine the field (the magnetic Reynolds number associated with their flows is of order of unity). After a rapid transient period (roughly, one Alfvén time), the system continues evolving as a result of the slow global field dissipation, whilst remaining in a Ferraro state. There is no evidence for the presence of a tachocline in this case.

When the field is initially in a configuration close to what one may expect from the Gough & McIntyre steady state (corresponding to the marginally confined field configuration of Rüdiger & Kitchatinov 1997), one could expect that the meridional flows, not being hindered by the field, would act in such a way as to confine it (see the incompressible analogue discussed in the Section 7.2.4.1). However, Brun & Zahn find that in this case also, the field lines quickly diffuse across the initially existing tachocline, connect to the convection zone and from there ensues Ferraro isorotation within a short Alfvénic timescale. It appears that although meridional flows of the kind predicted by Gough & McIntyre are indeed observed in the simulation, they do not have enough time to achieve dynamical balance in the magnetic diffusion layer before the field diffuses and connects with the convective zone.

Only for a deeply confined initial field does the outward diffusion occur slowly enough to allow for the formation of the tachopause. In that case, magnetic field lines are indeed seen to be confined to the radiative interior by the meridional flows, except in the polar regions which retain a modest amount of latitudinal shear. This simulation appears to reproduce the Gough & McIntyre view of the slow tachocline, save for a very important difference: the Ohmic diffusion of the internal field is only partially reduced by the tachocline dynamics, so that the field amplitude steadily decreases on a magnetic diffusion timescale. As this happens, the position and width of the tachocline and tachopause slowly change (the tachopause rises, and the tachocline becomes correspondingly thinner).

The absence of a stable steady state implies a direct relationship between the observed tachocline structure and the initial field configuration. This result, should it be confirmed, has important implications for dynamo action during the pre-Main Sequence phase of solar evolution. A primordial centrally condensed magnetic field configuration can presumably only be achieved by a timely switch from a steady-state or largely irregular dynamo to a cyclic dynamo, which must happen before the convection zone has entirely retreated to its present radius. This idea is plausible given that the timescale for the evolution of the convective–radiative interface ($\sim 10^7$ yr) is much shorter than the magnetic diffusion timescale ($\sim 10^{10}$ yr). In addition, the Mount Wilson Ca II program has found strong observational evidence for a transition from irregular dynamo action in very young stars to periodic dynamos for older stars (Saar *et al.* 1994). This trend has been associated with the transition between young, very rapid rotators and older, slower rotators, and interestingly, the timescale for magnetic braking of very young stars is also of the order of 10^7 yr. Schüssler (1975), Parker (1981) and Mestel & Weiss (1987) studied the typical magnetic fields that are likely to remain from dynamo action during the pre-Main Sequence stage; perhaps it is time to revisit their results in the light of Brun & Zahn's simulations using modern dynamo models, numerical algorithms and recent observations.

However, the numerical results obtained by Brun & Zahn pose another important problem. In all simulations, even for the most centrally condensed initial field configurations, the tachopause eventually reaches the outer boundary and, as field lines connect with the convective region, the system switches to the usual Ferraro state of isorotation. Using a rough scaling argument to compensate for the large diffusivities used in the simulations, Brun & Zahn estimate that this state is likely to be achieved *before* the present solar age regardless of the initial conditions. This striking result is difficult to reconcile with helioseismic observations; if confirmed, it could shed serious doubts on the relevance of the current slow tachocline model to the solar radiative zone. However, I will discuss in Section 7.2.5.3 how a better understanding of the outer boundary conditions to be applied to the slow tachocline model may rescue the situation.

7.2.5 Discussion and prospects for slow tachocline models

In recent years, slow tachocline models have come under increased scrutiny and criticism. By design, they ignore phenomena occurring on rapid timescales, concentrating instead on the secular dynamical interaction between slow meridional flows and the internal field. As such, they neglect three important effects that are likely to have a significant impact on the fragile balance described above: the potential axisymmetric and non-axisymmetric instabilities of the calculated equilibria, the combined effects of all possible rapid-timescale angular-momentum transporters known to exist in the tachocline, and the effect of an overlying dynamo field. In addition, the typical boundary conditions used to model the interface with the convective zone are highly idealized and may distort our view of the tachocline. I shall now discuss briefly the consequences of these effects on our understanding of the tachocline dynamics.

7.2.5.1 Stability of the slow tachocline models

Slow tachocline models may be subject to a wide variety of instabilities, including purely hydrodynamical shear and baroclinic instabilities, MHD instabilities of the large-scale primordial field, magnetic buoyancy instabilities, magneto-rotational instabilities and magneto-shear instabilities. Detailed investigations in the context of the slow tachocline model are only just beginning.

Linear and weakly nonlinear stability analyses of an idealized purely hydrodynamical tachocline shear flow in the non-diffusive limit have been performed by Watson (1981), Charbonneau *et al.* (1999b), Dikpati & Gilman (2001) and myself (Garaud 2001). The tachocline latitudinal shear is found to be close to marginal stability. The observed radial shear is stabilized by the very strong stratification (the typical Richardson number is of the order of a thousand). However, as Schatzman *et al.* (2000) point out, the standard Richardson criterion for stratified shear instability must be corrected to account for thermal diffusion in the tachocline; in that case, the radial shear is again close to marginal stability. In addition, Petrovay (2003) suggests that independent shellular fluid motions create much stronger small-scale radial shear layers, which could lead to secondary shear instabilities in the tachocline. This interesting proposal has not been confirmed numerically yet, but would correspond to a scenario close to that proposed by Spiegel & Zahn (1992), and have important consequences for all slow and fast tachocline models alike.

In any case, the addition of magnetic fields changes the nature and stability of non-axisymmetric perturbations; reviews of the stability of tachocline flows in the presence of strong fields and of the effects of magnetic buoyancy are given in Chapters 10 and 11. The magneto-rotational instability (see Chapter 12 by Ogilvie) could operate in regions of the Sun where angular velocity decreases outward from the rotation axis (as it does in the polar regions). Balbus & Hawley (1994) showed that the strong local stratification of the tachocline limits displacements to horizontal surfaces, as expected; this could provide a source of latitudinal momentum mixing in the polar regions.

Even more problematic for slow tachocline models are the well-known non-axisymmetric field instabilities of a mostly dipolar field in stellar interiors. Early works by Wright (1973), Markey & Tayler (1973, 1974) and Pitts & Tayler (1985) already suggested that a purely dipolar structure deep in the interior (as assumed in the above slow tachocline models) was subject to adiabatic perturbations near its neutral points (any confined field structure necessarily has such points). These are known to be stabilized by the presence of toroidal fields, but the current slow tachocline field structures are indeed found to be unstable (Brun & Zahn 2006). A new method for finding possible stable field structures in stellar interiors was developed by Braithwaite & Spruit (2004). It would be interesting to see how the slow tachocline models may be modified by the additional constraint that the underlying primordial field should be in a stable configuration.

Self-consistent studies of the model and of its stability have tentatively been performed. The Newton–Raphson relaxation algorithm I have used to calculate steady-state solutions of the slow tachocline equations cannot find unstable equilibria. Therefore the solutions found for the range of diffusion parameters studied are known to be stable to all axisymmetric perturbations. However, it provides no information on the evolution of non-axisymmetric perturbations. The numerical algorithm used by Brun & Zahn (2006), on the other hand, is ideally suited for the study of three-dimensional instabilities of all kinds. They observe the growth of non-axisymmetric instabilities associated with the primordial dipolar field, but do not detect any other intrinsic instabilities in the tachocline region. This result is interesting in the light of the local and global analyses mentioned above, but could be consistent with instabilities that only develop at high Reynolds and magnetic Reynolds numbers.

In conclusion, there are clear signs that the slow tachocline model might be unstable to a variety of non-axisymmetric instabilities. These could play an important role in redistributing chemical species, angular momentum and thermal energy within the tachocline, and must therefore be analysed. Various clues to the relative lack of mixing below the tachocline also suggest that any derived model should be constructed in such a way as to maximize stability below the tachocline; this constrains the geometry of the assumed primordial field.

7.2.5.2 Gravity waves as angular momentum transporters

The tachocline is known to host a wide spectrum of gravity waves, excited by overshooting convective plumes pounding on the stably stratified interior. These waves transport and deposit angular momentum further down in the radiative zone; the differential damping between prograde and retrograde waves is known to accentuate shearing flows and can be likened to some kind of anti-diffusion mechanism (McIntyre 2003, and Chapter 8 of this book; Kumar *et al.* 1999; Kim & MacGregor 2001, 2003; Talon & Charbonnel 2005). Quantitative estimates for the amplitude of the gravity waves thus generated, as well as their damping rate as a result of nonlinear interactions (mode–mode interaction or critical layer interaction) are difficult to obtain, although numerical simulations provide a new promising route for resolving this problem (Rogers & Glatzmaier 2006a). To complicate matters, dynamical interactions between the gravity waves and magnetic fields in the tachocline transfer energy into a wider spectrum of Alfvén waves, with correspondingly different propagation and damping mechanisms (Kim & MacGregor 2003). The global action of gravity and Alfvén waves on the background fluid generates large-scale dynamical structures that can have a radial extent much larger than the overshoot layer. Moreover, although the total flux of angular momentum transported is small, it is nonetheless important on the secular timescales considered for the slow tachocline models. Thus in this case again, significant modifications to the existing slow tachocline models could be required.

7.2.5.3 Boundary conditions

One of the most difficult problems faced by all tachocline models (including the fast tachocline, see Section 7.3) is the choice of boundary conditions used to describe the convective-radiative interface. The problem is exacerbated in the case of the slow tachocline, where meridional flows play an important role in redistributing angular momentum, preserving the thermal-wind balance and confining the internal field. Indeed, artificial flows generated on the boundary of the computational domain are an inevitable consequence of any attempt to impose stresses locally. Two situations may arise.

If the boundary is assumed to be impermeable, Ekman and Ekman–Hartmann layers form (the layer structure is modified for stress-free boundaries, but does not disappear); numerical models must monitor the amplitude of these boundary layer flows and ensure that they are only a small perturbation to the baroclinic flows of interest. This constraint places upper limits on the values of the Prandtl (ν/κ) and inverse Roberts (η/κ) numbers. However, even in a limit where Ekman flows can be neglected, the presence of an impermeable outer boundary constrains the geometrical structure of the meridional flow cells by limiting their upper radial extent, and by mass conservation, their latitudinal geometry. This numerical artefact is inevitable in the case of impermeable boundaries, and will affect the latitudinal force balance within the tachocline.

Another option is to relax the condition of impermeability. In that case, continuity of radial stresses replaces the condition of impermeability, but the problem is then merely transposed into a Reynolds stress modelling problem for turbulent convection. In addition, associated with the thought that it is possible to approximate the radiative–convective interface with simple 'boundary conditions' is the underlying assumption that the structure and dynamics of the convective region are independent of the tachocline dynamics. However, the recent works of Miesch (2003) and Rempel (2005) refute this hypothesis. The differential rotation near the convective–radiative interface is related to the differential rotation in the convective region, which results from the angular momentum balance between Reynolds stresses and large-scale meridional flows; these flows burrow into the tachocline and advect entropy to create a latitudinal entropy gradient which strongly constrains differential rotation through the thermal-wind balance. Thus the radiative–convective system is inseparably coupled. It is to be hoped that in the next few years, models will pay particular attention to modelling the convective zone *and* the tachocline simultaneously.

The role of the convection zone as a boundary condition on the magnetic field is even more ambiguous. Even while leaving aside the possible presence of a dynamo field in the outer layers of the tachocline (see Section 7.3 for a review of the effect of the dynamo field on the tachocline dynamics), currently used boundary conditions could be warping our conclusions on the slow tachocline dynamics. All models thus far assume the convective zone to be nearly perfectly insulating ($\eta \to \infty$) and match a potential field to the internal field. By assumption, field lines are smoothly anchored to the convective zone (i.e. to the outer boundary). However, we know that this is very far from the true situation: overshooting plumes interact with the magnetic field lines, stirring and shaking them, advecting them into large horizontal excursions, promoting reconnection as well as regeneration (the dynamo effect). In fact, it is more likely that the combined effect of convection is to confine the interior field (at least, its long-term averaged component) somewhat below the overshoot region. Indeed, flux expulsion and magnetic pumping by the convective plumes (Tobias *et al.* 2001; Dorch & Nordlund 2001) is sometimes thought of as being the principal reason for the lack of overlap between the internal primordial field and the dynamo field (as discussed by Boruta 1996). By contrast, the underlying assumption that field lines *can* be smoothly anchored into the convective zone leads to the ubiquitous emergence of a Ferraro state of isorotation in most numerical simulations of the slow tachocline. It will be interesting to know whether this conclusion holds should a more realistic model of the effect of the overshooting plumes on the primordial field be used.

In any case, the presence of a dynamo field may entirely change our view of the solar tachocline; the next section reviews recent models that explicitly involve the solar dynamo in the tachocline dynamics.

7.3 Dynamo field confinement: the fast tachocline

The solar dynamo field is observed through the regular emergence of strong flux concentrations at the solar surface, which appear in the form of active regions composed of dark sunspots and bright faculae. In Chapter 13, Tobias & Weiss review current observational knowledge of the solar dynamo and the potential role of the tachocline in its generation. Some important models favour the radiative–convective interface as the optimal location for the solar dynamo (Parker 1993): field stretching by the strong shear in the azimuthal flow can generate large-scale toroidal fields, accumulating in the tachocline until buoyancy instabilities trigger their rise into the convective region. From there, part of the flux emerges coherently through the surface, while the rest is promptly distorted into small-scale fields in all directions. Non-zero mean flow helicity results in non-zero mean poloidal flux generation, which is then pumped back down into the tachocline by convective overshooting. Many alternative models of the solar dynamo exist (see the review by Ossendrijver 2003), in some of which dynamo action is independent of the tachocline shear and relies only on turbulent and large-scale motions within the convective zone (Glatzmaier 1984; Brun *et al.* 2004; Brandenburg 2005). In reality, however, magnetic flux is necessarily pumped into the tachocline by overshooting convective plumes (Tobias *et al.* 2001).

The inevitable presence of strong dynamo-generated magnetic fields in the tachocline naturally raises many questions. What are the consequences for the tachocline dynamics? How far down into the tachocline does the dynamo field penetrate? Could the dynamo field be entirely, or partly, responsible for the observed rotation profile below the convective zone?

Contrary to the primordial field confinement models described above, the dynamics arising from the interaction of the tachocline shear with the dynamo field occurs on much shorter timescales. The intrinsic field variability is of the order of 11 yr, with a much larger amplitude than the assumed primordial field (and a correspondingly much shorter Alfvén time). Shear and magneto-shear instabilities operate on timescales typical of the rotation rate and Alfvén timescales (see Chapter 10). Finally, where overshoot is implied, the flow turnover timescale is of the order of a month. For obvious reasons, this new view of the tachocline was loosely called the fast tachocline (Gilman 2000).

7.3.1 Fast tachocline diffusion models

How deep is the fast tachocline? A quick answer associates the thickness of the fast tachocline with the dynamo field penetration depth. The dynamo field is pumped into the overshoot layer by downward penetrating plumes (Tobias *et al.* 2001) and diffuses downward into the radiative zone. However, the regular field polarity reversal plays an important role in limiting the field diffusion, since each cycle nearly cancels out the previous one (Mestel & Weiss 1987); as a result the field is strongly suppressed within a skin depth $\delta_{\rm SD} \sim (\tau_{\rm D}/\tau_\eta)^{1/2} r_{\rm cz}$ (assuming the dynamo is exactly periodic with a period $\tau_{\rm D}$ and where $\tau_\eta = r_{\rm cz}^2/\eta$ is the Ohmic diffusion timescale). For a laminar tachocline with microscopic diffusivity $\eta \sim 400\,{\rm cm^2\,s^{-1}}$ the skin depth is less than a few kilometres. This figure can be increased to a few megametres should one consider eddy diffusion in a turbulent tachocline with $\eta_{\rm t} \sim 10^{10}\,{\rm cm^2\,s^{-1}}$ (Forgács-Dajka & Petrovay 2001). Whether turbulence in the tachocline does indeed act as an 'eddy diffusivity' should be kept in mind throughout the following section, and is discussed in more detail in Section 7.3.2 and by Diamond *et al.* in Chapter 9.

A promising way of confining the tachocline was first suggested and later developed by Forgács-Dajka & Petrovay (2001, 2002; Forgács-Dajka 2004). They consider the structure of a turbulent tachocline pervaded by an oscillatory dynamo field. The field diffuses downward into the radiative zone and interacts with the tachocline shear. By construction, within the dynamo skin depth the magnetic diffusion timescale is of the order of the dynamo period. The Alfvén crossing time, on the other hand, depends on the imposed field amplitude and can be assumed to be much smaller than the dynamo period for fields of the order of several kilogauss or larger. Ferraro isorotation along the poloidal field lines is therefore rapidly achieved.

In their first paper on the dynamics of the fast tachocline, Forgács-Dajka & Petrovay assume a given poloidal field structure within the dynamo skin-depth and impose a sheared angular velocity profile at the interface with the convective zone (see Equation (7.4)). These are equivalent to the assumption that all meridional motions are negligible within the tachocline; indeed, in that case the equations governing the poloidal and toroidal components of the field decouple. The poloidal component satisfies a simple diffusion equation with periodic forcing, which has a spatially damped oscillatory solution. Here for simplicity the poloidal field B_p is assumed to have the functional form

$$B_p(r,\theta,t) = \overline{B}_p(r,\theta)\cos(\omega_{\rm D} t), \tag{7.37}$$

where $2\pi/\omega_{\rm D} = \tau_{\rm D} = 22\,{\rm yr}$. Under those conditions, the azimuthal component of the momentum and induction equations can be integrated to obtain the profiles of angular velocity and toroidal field as functions of time.

An approximate analytical solution to the governing equations can be derived in the limit of large poloidal field strength (i.e. in the limit where there is a clear separation between the Alfvén time and the dynamo period), and thin tachocline. Let $v_{\rm A}$ be the typical Alfvén velocity of the imposed poloidal field; then by assumption $\epsilon = r\omega_{\rm D}/v_{\rm A} \ll 1$. Following Forgács-Dajka & Petrovay (2001), the equations are for simplicity written in a local Cartesian system (with $\theta \leftrightarrow x$ and $r \leftrightarrow z$). In units of the Alfvén timescale and the radius of the convective zone the non-dimensional governing equations are

$$\partial_t u_\phi = \cos(2\pi\epsilon t)\partial_x B_\phi + \frac{\tau_{\rm A}}{\tau_\nu}\nabla^2 u_\phi, \tag{7.38}$$

$$\partial_t B_\phi = \cos(2\pi\epsilon t)\partial_x u_\phi + \frac{\tau_{\rm A}}{\tau_\eta}\nabla^2 B_\phi. \tag{7.39}$$

In the limit $\epsilon \ll 1$ it is possible to perform a two-timescale analysis and seek solutions on the slow timescale $\tau = \epsilon t$ (which evolves on the timescale of the cyclic dynamo field). The slow solutions satisfy the reduced equation

$$\cos(2\pi\tau)\partial_x B_\phi = -\frac{\tau_{\rm A}}{\tau_\nu}\nabla^2 u_\phi, \tag{7.40}$$

$$\cos(2\pi\tau)\partial_x u_\phi = -\frac{\tau_{\rm A}}{\tau_\nu}\nabla^2 B_\phi, \tag{7.41}$$

and, should one assume that $\partial_z \gg \partial_x$, can be found analytically; they display an oscillatory temporal structure with the timescale of the imposed field $\tau_{\rm D}$, and an oscillatory spatially damped structure below the convective–radiative interface on a typical lengthscale δ_D, where

$$\frac{\delta_{\rm D}}{r} = \left(\frac{4\tau_{\rm A}^2}{\tau_\nu\tau_\eta\cos^2(2\pi\tau)L^2}\right)^{1/4} \tag{7.42}$$

and L is the latitudinal wavenumber of the imposed poloidal field. Not surprisingly, this estimate is equivalent to the depth of a Hartmann layer for an imposed field with field lines parallel to the boundary and amplitude $B_0\cos(2\pi t/\tau_{\rm D})$. This fast tachocline model therefore predicts the same tachocline thickness scalings as a function of the imposed field as had been obtained by Rüdiger & Kitchatinov (1997).[3] By extension, there is a natural generalization of the result should the imposed dynamo field have a strong radial component.

For the model assumptions to be consistent, it is important to verify that $\delta_{\rm SD} \gg \delta_{\rm D}$. This places lower limits on the imposed field strength for a given turbulent

[3] Forgács-Dajka & Petrovay (2001) derive other scaling laws between the confining field strength and the tachocline thickness in the limit where the dynamo frequency is higher (which could be applicable for stars other than the Sun).

diffusivity. In addition, if the field is much weaker than about a kilogauss, the simple two-timescale analysis fails and interactions between the dynamo forcing and the Alfvén waves could lead to the excitation of modes with new frequencies. This has not been investigated yet.

Numerical solutions have been computed by Forgács-Dajka & Petrovay (2001) for a dipolar poloidal field of varying amplitude. They show a clear confinement of the imposed latitudinal shear for large enough field strength (typically, $|B_{\rm p}| \sim 0.2$T for $\eta_{\rm t} \sim 10^6\,{\rm m}^2\,{\rm s}^{-1}$). The latitudinal variation of the field amplitude leads to a significant latitudinal variation of the tachocline depth, which is consistent with the above estimates. Observations, however, reveal only a weak latitudinal variation of the tachocline position and width (Charbonneau *et al.* 1999a) which could in principle set strong constraints on the poloidal field geometry diffusing from the overlying dynamo. As expected also from the analysis, there is a significant temporal variability of the depth and aspect of the tachocline on a period of 11 years (the differential rotation is independent of the field polarity). Both results confirm and quantify common expectations that there must exist some variability in the tachocline angular velocity profile on the dynamo timescale. However, precise helioseismic observations by MDI/SOI on board SoHO have only been available for slightly less than one solar cycle, and little to no tachocline variability on the dynamo timescale has yet been detected (Corbard *et al.* 2001). Definite answers on this topic are impatiently awaited: so far, only 1.3 yr torsional oscillations have been found (Howe *et al.* 2000).

In following works, Forgács-Dajka & Petrovay (2002) and Forgács-Dajka (2004) study various improvements to the model, including the effects of a large-scale (imposed) meridional flow, of a radially varying magnetic diffusivity and varying magnetic Prandtl number. The background state is derived from the solar model of Guenther *et al.* (1992). They also calculate the poloidal component of the dynamo field self-consistently from the poloidal component of the advection–diffusion equation: in these new simulations the poloidal field is advected by the imposed meridional flows in addition to diffusion. Finally, they impose a realistic description of latitudinal and temporal variation of the migrating dynamo field as a boundary condition, which is derived from the observations of Stenflo (1994). The modelled meridional flows are poleward near the solar surface with a velocity of about 10–20 $\rm m\,s^{-1}$, in accordance with direct observations of the motion of small magnetic features (e.g. Komm *et al.* 1993) or inferences from local helioseismology (Giles *et al.* 1997). Two geometries are studied: a single-cell structure with an equatorward return flow in the tachocline, and a double-cell structure with a poleward return flow in the tachocline and a null node at about $r = 0.85R_\odot$. Note that numerical simulations of turbulent convection do not appear to favour the view of stable long-lived circulation cells deep in the convective zone; meridional flows

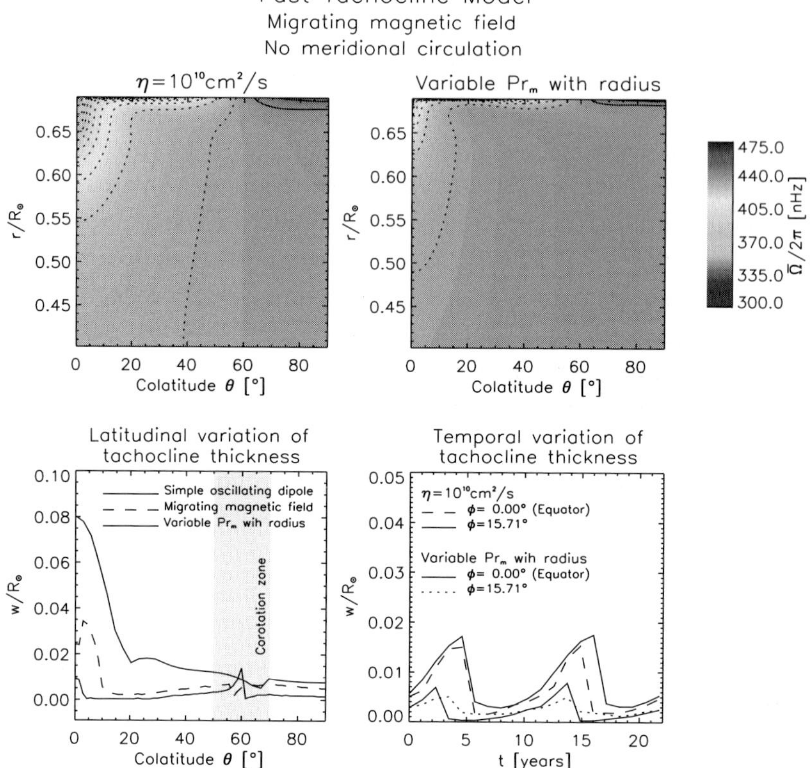

Figure 7.4. Numerical solutions for the fast tachocline model of Forgács-Dajka & Petrovay. This simulation includes a realistic representation of the poloidal field extracted from the butterfly diagram, but neglects meridional motions. *Upper panels:* resulting differential rotation spreading into the radiative interior in two cases. In the left panel $\eta = \nu = 10^6\ \mathrm{m^2\,s^{-1}}$ throughout the domain (in which case $Pm = \nu/\eta = 1$). In the right panel Pm is varied with depth between 0.024 and 0.1. In this case the variations of η and ν are: $\log_{10}\eta = 3.5 - 6$ and $\log_{10}\nu = 2 - 5$. *Bottom panels:* corresponding latitudinal variation (left) and temporal variation (right) of the tachocline thickness.

are instead very intermittent, with strongly variable geometries (Brun & Toomre 2002).

The results, illustrated in Figure 7.4, can be summarized as follows. The role of the meridional flows as transporters of angular momentum naturally aids the tachocline confinement process in the case of the modelled two-cell circulation pattern (by transporting angular momentum poleward) and hinders it in the case of the single-cell circulation pattern. The numerical simulations confirm these expectations, and suggest that flows as slow as a few centimetres per second in the tachocline have a significant impact on the observed differential rotation profile.

A strong decrease in turbulent magnetic diffusivity with depth beneath the tachocline is expected from the steep increase in the background stratification. Note that the decrease in turbulent mixing below the tachocline is clearly constrained by independent observations of the light element depletion fraction (see Chapter 3). In that case again, the imposed convection zone shear is still easily quenched by the fast tachocline fields. However, across the turbulent/laminar transition the dynamo field penetration is abruptly suppressed and is therefore not able to reduce any deep-seated residual shear related to solar spin-down; this is an intrinsic problem of all fast tachocline models. One possible solution stems from the fact that the solar dynamo cycle is not exactly periodic. Mestel & Weiss (1987) suggested that the (apparently) random component of the dynamo field could diffuse much deeper into the radiative zone than its periodic counterpart. I investigated this possibility in detail (Garaud 1999), and found that an internal field with an rms value of about $10^{-4}B_0$ (where B_0 is the amplitude of the poloidal component of the dynamo field) could build up deeper in the interior.

7.3.2 Discussions and prospect for the dynamo confinement model

In comparison with slow tachocline models, the idea proposed by Forgács-Dajka & Petrovay has the advantage of being based on a robust balance of forces, which holds even in the presence of instabilities (it does in fact rely on the presence of instabilities), and can be tuned to compensate any additional angular momentum transport from convective plumes, gravity waves or meridional flows. The spatial variation of the tachocline depth observed in the numerical simulations can be reconciled with observations for specific poloidal field structures, and the strong temporal variation observed could still be consistent with observations should the tachocline be in fact a little bit shallower than current estimates (this statement is mostly based on the resolution of helioseismic inversions).

One must nonetheless bear in mind the three assumptions inherent in the model: the tachocline is turbulent, the turbulence leads to an eddy diffusivity greater than 10^9 cm^2 s^{-1} in the tachocline and, finally, the dynamo generation mechanism does not rely on the detailed tachocline structure.

If we assume that the tachocline has a width $\sim 0.02R_\odot$, then turbulent motions at the level required by the fast tachocline model cannot result from overshooting plumes only. The stability of the tachocline to hydrodynamic and magnetohydrodynamic instabilities was discussed in Section 7.2.5.1 and is reviewed in Chapters 10 and 11; magneto-shear instabilities offer a promising route for the maintenance of turbulent motions. In fact, these instabilities are so ubiquitous that the maintenance of large-scale fields in the tachocline appears to be the more relevant problem. Nonetheless, the first of the three governing assumptions is not much under dispute.

However, the role of turbulent motions in 'diffusing' large-scale fields is a far more difficult issue. Although very commonly used in astrophysical MHD models, the physical basis for turbulent diffusivity, as well as its parametrization, is still ambiguous. The concept of turbulent diffusion is typically derived from heuristic arguments on the vectorial form of the averaged electromotive force due to small-scale fields and flows (see Chapter 13):

$$\overline{(\mathbf{u} \times \mathbf{b})}_i = \alpha_{ij} B_j - \beta_{ijk} \partial_j B_k \ldots \tag{7.43}$$

This expression naturally emphasizes the tensorial nature of the turbulent diffusivity β; assuming that $\beta \sim \eta_{\rm t}$ is a scalar is a largely unjustified (but commonly used) simplification.

The turbulent diffusivity is known to be quenched when the magnetic fields start having a strong effect on the turbulent flow (near energy equipartition); at the largest scales in the tachocline, this effect is relevant for fields upward of a few thousand gauss, which already has implications for fast tachocline models. But the situation may in fact be much worse. In the tachocline, the magnetic Prandtl number is of the order of $\nu/\eta \sim 10^{-2}$; if a small-scale dynamo indeed operates at these values of the magnetic Prandtl number (Boldyrev & Cattaneo 2004) magnetic energy accumulates somewhere on the turbulent inertial range and reaches equipartition well before the larger scale field does. This process could quench the turbulent magnetic diffusivity for much lower field strengths (Cattaneo & Vainshtein 1991). Catastrophic η-quenching is shown to occur in two-dimensional flows through numerical simulations (Cattaneo 1994) and quasi-linear closure (Gruzinov & Diamond 1995). The situation is still unclear in the case of three-dimensional flows. The η-quenching process could pose serious threats to the fast tachocline models: using the scalings proposed by Cattaneo & Vainshtein (1991), large-scale fields as low as a few gauss would suffice to quench the turbulent diffusivity of the fast tachocline by several orders of magnitude. This creates an intrinsic contradiction within the model.

In any case, the current fast tachocline model neglects all effects of the turbulent motions except for their role in enhancing the magnetic diffusivity. However, other macroscopic effects are known to occur and are likely to play an important role in the tachocline dynamics. Turbulent flux expulsion has been observed in a wide variety of systems where turbulent and laminar regions coexist (Tao *et al.* 1998; Tobias *et al.* 2001). Field generation by small-scale turbulent motions, the α-effect, has also been predicted by turbulence closure models (Krause & Rädler 1980) and observed in numerical simulations (e.g. Brandenburg *et al.* 1990). Finally, non-isotropic Reynolds stresses and turbulent Maxwell stresses may be as important as the large-scale Lorentz forces in reducing the imposed shear. In other words, a consistent model for the fast tachocline will require a consistent description of the effects of turbulent motions on the large-scale flows and fields.

Building on this idea, another natural step in the study of fast tachocline confinement models is to calculate self-consistently the temporal evolution of the field and the flow, using for instance a mean-field dynamo model. Indeed, current mean-field models calculate the field evolution assuming a given angular velocity profile in the tachocline, whereas current tachocline confinement models study the effect of an assumed dynamo field on the shear. In an integrated model, can dynamo action be sustained if the radial shear is quenched by the dynamo itself? This could indeed happen should dynamo action rely more on the latitudinal shear than the radial shear, or if the solar dynamo is more of an α^2-dynamo than an $\alpha\Omega$-dynamo. Reproducing simultaneously the tachocline profile and the solar cycle is an interesting challenge which could provide much insight into the correct parametrization of the α- and β-effects.

7.4 Discussion and prospects

We have now reached a stage in the process of studying the tachocline dynamics where there exists a large enough variety of studies, models and observations to support critical discussions. What are the next steps in the study of the tachocline magnetohydrodynamics? The few points that I believe will have a significant impact on our understanding of the tachocline in the next few years are the following.

Coexisting 'fast' and 'slow' tachoclines? In the light of the discussions outlined in Sections7.2 and 7.3, is it still possible to consider the idea of coexisting 'fast' and 'slow' tachoclines? The only way to do this would be to construct a complex layered structure starting from the bottom of the convective zone with a turbulent, magnetic overshoot region, which gradually quietens downward to give way to a more laminar region where the large-scale (dynamo) fields are pumped, stored and stretched. Slightly further down, the low magnetic diffusivity forbids the oscillating field from penetrating very far down and thus appear successively the well-ventilated, magnetic free region of the Gough & McIntyre tachocline, the magnetic diffusion layer and finally the magnetically constrained interior. And most of the above must be packed, according to observations, within a total width spanning no more than 2%–4% of the solar radius. This scenario can only work if fluid motions in the tachocline are to a very large degree two-dimensional. However, there are doubts that this may be the case at all times despite the strong stratification. Numerical simulations suggest there are occasional very strong overshooting events with large radial extent. Can the slow tachocline balance survive these mixing events? More precisely if, as suspected, the Gough & McIntyre model indeed harbours multiple equilibria, mixing events extending between the interior and the overshoot region could dredge out interior field lines and drag them into the convective region, triggering the transition from a confined interior field to the open field configuration. Should this happen, there is no simple mechanism capable of returning the system to the confined field configuration of the Gough & McIntyre model within the typical timescale of occurrence of the mixing events. Furthermore, magnetic buoyancy and other instabilities (see the discussions in Chapters 10, 11 and 12) are intrinsically three-dimensional. In particular, the 'tipping' instabilities discussed by Spruit (1999, 2002; Braithwaite & Spruit 2004) may change the picture drastically, as explained by McIntyre in Chapter 8.

The role of the interaction between overshooting plumes and an internal primordial field. As discussed elsewhere, this interaction is likely to play a dominant role in the tachocline dynamics. Tamara Rogers and I have begun studying this phenomenon to determine whether this may indeed be a sufficient, self-consistent way of confining an internal field while bypassing the need for baroclinic meridional flows. We hope to show for instance that the ubiquitous emergence of Ferraro rotation in laminar models is in fact an artefact of the simplified interface conditions; in fact, we believe that the interaction between overshoot and an internal field may form the basis for a minimalist model of the tachocline and the radiative interior.

The role of gravity waves. Talon & Charbonnel (2005) have recently claimed that the continuous adjustment of the angular velocity of the radiative core to that of the convection zone could in fact be entirely attributed to gravity wave mixing. This would suppress the need for an internal primordial field. An important task for the near future is to test the Talon & Charbonnel model for angular momentum transport against direct numerical simulations of gravity waves in the solar interior (Rogers & Glatzmaier 2006b), and to investigate ways in which observations (combining asteroseismology, surface light-element abundances and magnetic activity measurements) may help distinguish between the magnetic and non-magnetic scenarios.

The early evolution of the Sun and its relation to the internal primordial field. Given its likely dominant role in the interior dynamics, it is perhaps disappointing that we know so little about the interior field. How much of the collapsing cloud's magnetic flux survives the fully convective phase of stellar evolution? What happens to this flux as the convective zone finally retreats? The Mount Wilson observations of the magnetic activity of very young solar type stars now permit a more comprehensive study of the correlation between dynamo action, rotation and internal structure: can we construct a model of the early solar magnetism that would include these new data and enable us to predict the current internal field strength and geometry?

Self-consistent mean-field hydrodynamics and dynamo models. Current mean-field dynamo models assume a given differential rotation profile, while current fast tachocline models assume a given magnetic field profile. Rempel (2005) showed that it is now possible to use mean-field hydrodynamics to model simultaneously the tachocline and the convection zone; the extension of this work to include magnetic stresses as well as mean-field dynamo processes might provide a model of rotation and dynamo action in the Sun. This would be a significant advance in the field, since the self-consistent determination of rotation (which can be measured by helioseismology) and meridional flows (which appear to constrain the equatorward sunspot drift throughout the cycle in many types of dynamos) may help distinguish between various competing dynamo models. Comparison with the rotation profile and magnetic activities of other stars would also help refine our understanding of this exceedingly complex system. In fact, such an approach may be the only route towards a better understanding of interior dynamics: it is becoming increasingly clear that we have very little hope of reaching the asymptotic values of the Reynolds and Rayleigh numbers in three-dimensional simulations of the whole Sun that would permit a trustworthy study of the convection zone and the tachocline. However, numerical simulations in a local box are on the other hand much closer to solar values, and may help constrain the parametrizations to be used in mean-field models.

Acknowledgments

I thank all of the Isaac Newton workshop participants for enlightening and stimulating discussions about this fabulous subject. The completion of this manuscript would not have been possible without the help and support of Nic Brummell and Douglas Gough. I also thank Fausto Cattaneo, Gary Glatzmaier, Chris Jones, Michael McIntyre, Bob Rosner, Steve Tobias and Nigel Weiss for clarifying many of the complex scientific issues discussed here.

References

Acheson, D. J. & Hide, R. (1973). *Rep. Prog. Phys.*, **36**, 159.
Balbus, S. A. & Hawley, J. F. (1994). *Astrophys. J.*, **266**, 769.
Boldyrev, S. & Cattaneo, F. (2004). *Phys. Rev. Lett.*, **92**, 144 501.
Boruta, N. (1996). *Astrophys. J.*, **458**, 832.
Boyer, D. W. & Levy, E. H. (1984). *Astrophys. J.*, **277**, 848.
Braithwaite, J. & Spruit, H. C. (2004). *Nature*, **431**, 819.

Brandenburg, A. (2005). *Astrophys. J.*, **625**, 539.

Brandenburg, A., Tuominen, I., Nordlund, A., Pulkkinen, P. & Stein, R. F. (1990). *Astron. Astrophys.*, **232**, 277.

Brummell, N. H., Clune, T. L. & Toomre, J. (2002). *Astrophys. J.*, **570**, 825.

Brun, A. S. & Toomre, J. (2002). *Astrophys. J.*, **570**, 865.

Brun, A. S. & Zahn, J.-P. (2006). *Astron. Astrophys.*, submitted.

Brun, A. S., Miesch, M. S. & Toomre, J. (2004). *Astrophys. J.*, **614**, 1073.

Cattaneo, F. (1994). *Astrophys. J.*, **434**, 200.

Cattaneo, F. & Vainshtein S. I. (1991). *Astrophys. J.*, **376**, L21.

Charbonneau, P., Christensen-Dalsgaard, J., Henning, R. *et al.* (1999a). *Astrophys. J.*, **527**, 445.

Charbonneau, P., Dikpati, M. & Gilman, P. A. (1999b). *Astrophys. J.*, **528**, 523.

Christensen-Dalsgaard, J., Gough D. O. & Thompson, M. J. (1991). *Astrophys. J.*, **378**, 413.

Clune T. L., Elliott, J. R., Glatzmaier, G. A., Miesch, M. S. & Toomre, J. (1999). *Parallel Comp.*, **25**, 361.

Corbard, T., Jiménez-Reyes, S. J., Tomczyk, S., Dikpati, M. & Gilman, P. A. (2001). In *Helio- and Astero-Seismology at the Dawn of the Millennium*, ed. A. Wilson. ESA Publications, p. 265.

Cowling, T. G. (1945). *Mon. Not. Roy. Astron. Soc.*, **105**, 166.

Cowling, T. G. (1957). *Magnetohydrodynamics*. New York: Interscience.

Dikpati, M. & Gilman, P. A. (2001). *Astrophys. J.*, **551**, 536.

Dorch, S. B. F. & Nordlund, Å. (2001). *Astron. Astrophys.*, **365**, 562.

Dormy, E., Cardin, P. & Jault, D. (1998). *Earth Planet. Sci. Let.*, **160**, 15.

Dormy, E., Jault, D. & Soward, A. M. (2002). *J. Fluid Mech.*, **462**, 263.

Elliott, J. R. & Gough, D. O. (1999). *Astrophys. J.*, **516**, 475.

Ferraro, V. C. A. (1937). *Mon. Not. Roy. Astron. Soc.*, **97**, 458.

Forgács-Dajka, E. (2004). *Astron. Astrophys.*, **413**, 1143.

Forgács-Dajka, E. & Petrovay, K. (2001). *Sol. Phys.*, **203**, 195.

Forgács-Dajka, E. & Petrovay, K. (2002). *Astron. Astrophys.*, **389**, 629.

Garaud, P. (1999). *Mon. Not. Roy. Astron. Soc.*, **304**, 583.

Garaud, P. (2001). *Mon. Not. Roy. Astron. Soc.*, **324**, 68.

Garaud, P. (2002). *Mon. Not. Roy. Astron. Soc.*, **335**, 707.

Giles, P. M., Duvall, T. L., Scherrer, P. H. & Bogart, R. S. (1997). *Nature*, **390**, 52.

Gilman, P. A. (2000). *Sol. Phys.*, **192**, 27.

Glatzmaier, G. A. (1984). *J. Comp. Phys.*, **55**, 461.

Gough, D. O. & McIntyre, M. E. (1998). *Nature*, **394**, 755.

Gruzinov, A. V. & Diamond, P. H. (1995). *Phys. Plasmas*, **2**, 1941.

Guenther, D. B., Demarque, P., Kim, Y.-C. & Pinsonneault, M.H. (1992). *Astrophys. J.*, **387**, 372.

Howe, R., Christensen-Dalsgaard, J., Hill, F. *et al.* (2000). *Science*, **287**, 2456.

Kim, E. J. & MacGregor, K. B. (2001). *Astrophys. J.*, **556**, L117.

Kim, E. J. & MacGregor, K. B. (2003). *Astrophys. J.*, **588**, 645.

Komm, R. W., Howard, R. F. & Harvey, J. W. (1993). *Sol. Phys.*, **147**, 207.

Krause, F. & Rädler, K.-H. (1980). *Mean-field Magnetohydrodynamics and Dynamo Theory*. Berlin: Akademie-Verlag.

Kumar, P., Talon, S. & Zahn, J.-P. (1999). *Astrophys. J.*, **520**, 859.

MacGregor, K. B. & Charbonneau, P. (1999). *Astrophys. J.*, **519**, 911.

Markey, P. & Tayler, R. J. (1973). *Mon. Not. Roy. Astron. Soc.*, **163**, 177.

Markey, P. & Tayler, R. J. (1974). *Mon. Not. Roy. Astron. Soc.*, **168**, 505.

McIntyre, M. E. (2003). In *Stellar Astrophysical Fluid Dynamics*, ed. M. J. Thompson & J. Christensen-Dalsgaard. Cambridge: Cambridge University Press, p. 111.
Mestel, L. (1953). *Mon. Not. Roy. Astron. Soc.*, **113**, 716.
Mestel, L. & Weiss, N. O. (1987). *Mon. Not. Roy. Astron. Soc.*, **226**, 123.
Miesch, M. S. (2003). *Astrophys. J.*, **586**, 663; *Astrophys. J.*, **611**, 268.
Miesch, M. S., Elliott, J. R., Toomre, J. *et al.* (2000). *Astrophys. J.*, **532**, 593.
Ossendrivjer, M. A. J. H. (2003). *Astron. Astrophys. Rev.*, **11**, 287.
Parker, E. N. (1981). *Geophys. Astrophys. Fluid Dyn.*, **18**, 175.
Parker, E. N. (1993). *Astrophys. J.*, **407**, 342.
Petrovay, K. (2003). *Sol. Phys.*, **215**, 17.
Pitts, E. & Tayler, R. J. (1985). *Mon. Not. Roy. Astron. Soc.*, **216**, 139.
Rempel, M. (2005). *Astrophys. J.*, **622**, 1320.
Rogers, T. M. & Glatzmaier, G. A. (2005). *Astrophys. J.*, **620**, 432.
Rogers, T. M. & Glatzmaier, G. A. (2006a). *Astrophys. J.*, submitted.
Rogers, T. M. & Glatzmaier, G. A. (2006b). *Astrophys. J.*, in preparation.
Rüdiger, G. & Kitchatinov, L. L. (1997). *Astron. Nachr.*, **318**, 273.
Rüdiger, G. & Pipin, V. V. (2001). *Astron. Astrophys.*, **375**, 149.
Saar, S. H., Brandenburg, A., Donahue, R. A. & Baliunas, S. L. (1994). In *Cool Stars, Stellar Systems and the Sun.* ed. J.-P. Caillault, Astron. Soc. Pac. Conf. Ser. **64**, p. 468.
Schatzman, E., Zahn, J.-P. & Morel, P. (2000). *Astron. Astrophys.*, **364**, 876.
Schüssler, M. (1975). *Astron. Astrophys.*, **38**, 263.
Schüssler, M., Caligari, P., Ferriz-Mas, A. & Moreno-Insertis, F. (1994). *Astrophys. J.*, **422**, 652.
Spiegel, E. A. & Zahn, J.-P. (1992). *Astron. Astrophys.*, **265**, 106.
Spruit, H. C. (1999). *Astron. Astrophys.*, **349**, 189.
Spruit, H. C. (2002). *Astron. Astrophys.*, **381**, 923.
Stenflo, J. O. (1994). In *Solar Surface Magnetism*, ed. R. J. Rutten & C. J. Schrijver. Dordrecht: Kluwer, p. 365.
Talon, S. & Charbonnel, C. (2005). *Astron. Astrophys.*, **440**, 981.
Tao, L., Proctor, M. R. E. & Weiss, N. O. (1998). *Mon. Not. Roy. Astron. Soc.*, **300**, 907.
Tobias, S. M. (2005). In *Fluid Dynamics and Dynamos in Astrophysics and Geophysics*, ed. A. M. Soward, C. A. Jones, D. W. Hughes & N. O. Weiss. London: CRC Press, p. 193.
Tobias, S. M., Brummell, N. H., Clune, T. L. & Toomre, J. (2001). *Astrophys. J.*, **549**, 1183.
Watson, M. (1981). *Geophys. Astrophys. Fluid Dyn.*, **161**, 285.
Wright, G. A. E. (1973). *Mon. Not. Roy. Astron. Soc.*, **162**, 339.

8

Magnetic confinement and the sharp tachopause

Michael E. McIntyre

The discovery by Spruit of a new small-scale turbulent dynamo has significantly changed the tachocline model proposed by Gough & McIntyre (1998). The small-scale dynamo is shear driven, is characteristic of stably stratified flows, and is mediated by the kink or 'tipping' instability elucidated for such flows by R. J. Tayler. The dynamo works best in high latitudes and supports turbulent Maxwell stresses large enough to dominate the angular momentum transport, taking over from the pure mean meridional circulation (MMC) proposed by Gough & McIntyre (1998). What survives from the Gough & McIntyre proposal is the laminar thermomagnetic boundary layer at the tachopause, essential for the confinement of the interior field $\mathbf{B}_\mathrm{i}$ by high-latitude downwelling. That downwelling is, however, itself confined within a double boundary layer at the tachopause. The thermomagnetic boundary layer sits just underneath a modified Ekman layer, in which the turbulent Maxwell stress of the small-scale dynamo diverges.

The effects of compositional stratification in the helium settling layer under the tachopause are considered. It is concluded that Gough & McIntyre's (1998) 'polar pits' to burn lithium are dynamically impossible and that the tachopause is not only sharp but globally horizontal. That is, the tachopause, as marked by the top of the helium settling layer, follows a single heliopotential to within a very tiny fraction of a megametre from equator to pole. Therefore the stably stratified tachocline, defined in high latitudes as the layer of dynamically significant shear beneath the convection zone, must be thick enough to burn lithium. This is consistent with the helioseismic evidence because the high-latitude shear, even though crucial to the maintenance of the dynamo action, is held down in magnitude, by the dynamo's turbulent Maxwell stresses, to values too small to be visible.

The Solar Tachocline, D. W. Hughes, R. Rosner and N. O. Weiss.
Published by Cambridge University Press.

8.1 Introduction

Following Spiegel & Zahn (1992) and others, I start from the assumption that the fluid dynamics of the tachocline is a multi-timescale problem. Specifically, in order to understand the structure of the present tachocline I assume, and will argue in what follows, that one has to consider fluid-dynamical processes over the full range of timescales from the gigayear or secular timescale of solar evolution to the months and years of convection-zone overshoot and upper-tachocline MHD instabilities and turbulence, all touched on in other chapters in this book. That many timescales are important should hardly need saying, but does, perhaps, need saying here if only to counter the false dichotomy 'slow versus fast' that seems to have taken hold in the literature. Indeed it seems possible, now, that even so basic a quantity as the tachocline thickness Δ may depend on the gigayear-timescale history, as well as on a variety of turbulent processes over a large range of timescales.

In what follows I assume it unnecessary to repeat my old arguments (1994, 2003a) against the Spiegel–Zahn horizontal-eddy-viscosity hypothesis – which arguments, in turn, point toward the inevitable existence of a global-scale magnetic field $\mathbf{B}_{\rm i}$ in the radiative interior, whether of fossil or dynamo origin (Gough & McIntyre 1998), as the only way to account not only for the interior's solid rotation but also for the smallness of Δ, at most several tens of megametres according to helioseismology (see Chapter 3 by Christensen-Dalsgaard & Thompson). The argument for inevitability still seems significant in itself, given the far greater uncertainties about the origin, and the viability, of magnetic fields in the radiative interior. There, the gigayear-timescale escapology of magnetic fields has Houdini-like possibilities (see Chapter 11 by Hughes) involving the nonlinear effects of instabilities and Parker flux-tube buoyancy in combination.

The argument for inevitability of a magnetic interior can be summarized in two parts. First, a non-magnetic interior cannot be held in solid rotation by real stratified, layerwise-two-dimensional turbulence. Such turbulence, if it were to be excited, would tend to be 'anti-frictional' – to drive the system away from solid rotation and not toward it (McIntyre 1994, 2003a,b and references therein). The effect would be qualitatively unlike that of the hypothesized horizontal eddy viscosity. Second, a non-magnetic interior would be incapable of withstanding another fluid-dynamical process that would also drive it away from solid rotation and that would, furthermore, as originally pointed out by Spiegel & Zahn (1992), make Δ values significantly larger than permitted by the helioseismic evidence. That process – the downward 'radiative spreading' or 'Haynes–Spiegel–Zahn burrowing', into a non-magnetic interior, of mean meridional circulations (MMCs) and differential rotation – will be revisited here together with the concomitant notion of 'gyroscopic pumping'. As well as making Δ values too large, the downward-burrowing MMCs would prevent a helium settling layer from forming at the top of the interior, as well as probably burning too much beryllium.

The arguments against non-magnetic horizontal eddy viscosity will prove robust, I believe (a) because of their clearcut basis in the fundamental principles of non-magnetic, stratification-constrained eddy motion, especially potential-vorticity conservation and invertibility (see McIntyre 2003a,b and references therein), and (b) because of the comprehensive testing and vindication of those fundamental principles by high-resolution observations and modelling of, especially, the Earth's stratosphere.[1] So the main focus of this chapter will not be on those arguments, but rather on how, if the existence of the global-scale interior $\mathbf{B}_\mathrm{i}$ is accepted as practically certain, the scenario of Gough & McIntyre (1998) (already discussed by Garaud in Chapter 7) now needs to be modified in the light of advances in our knowledge of MHD turbulence. The focus is not now on asking *whether* it is a global-scale $\mathbf{B}_\mathrm{i}$ that limits Δ, but on understanding more clearly *how* it does so. The MHD-turbulent aspects will force a re-examination of how azimuthal stresses are supported between the interior and the overlying turbulent layers, and how they fit in with the contributions of MMCs to angular momentum exchange.

Despite radical changes, one important feature of the Gough & McIntyre (1998) scenario seems to have survived so far, with a little help from Occam's razor. This is the prediction of a ventilated (helium-poor) tachocline terminated by a sharp tachopause, across which there is a strong jump in compositional or heavy-element abundance gradients, from zero in the tachocline to a finite value in the helium settling layer just beneath, corresponding to a contribution N_μ^2 to the buoyancy frequency squared that is a significant fraction of the typical thermal value $N^2 \sim 10^{-6}\,\mathrm{s}^{-2}$. This points toward the validity of helioseismic calibrations of the kind attempted in Elliott & Gough (1999).

A feature that does not, on the other hand, survive from Gough & McIntyre (1998) in any form at all is the large-scale, laminar, field-free ($\mathbf{B} \equiv 0$) downwelling throughout high latitudes, occupying a substantial fraction of the thickness of the tachocline. The original Gough & McIntyre (1998) scenario relied entirely on Reynolds and Maxwell stresses in the convection zone to produce (by gyroscopic pumping) the downwelling MMC needed (a) to confine the interior field $\mathbf{B}_\mathrm{i}$ in high latitudes as well as (b) to transfer angular momentum as necessary and (c) to ventilate the tachocline. But the hypothesis of large-scale field-free downwelling in high latitudes, pumped entirely by convection-zone stresses, is now untenable – whether or not we include overshoot-layer stresses – because it has been convincingly shown by Spruit (1999, 2002), building on the classic work of R. J. Tayler

[1] A striking observational example, illustrating the detail in which the stratosphere is now observed, can be quickly found by googling "`gyroscopic pump in action`". There is also a vast literature of published papers in leading journals; see, for example, Manney *et al.* (1994) and Riese *et al.* (2002).

in the 1970s, that most if not all of the high-latitude downwelling region, even if initially field-free, could not remain so. Because of its vertical shear, the region would be MHD-unstable in such a way as to evolve into a small-scale dynamo.

The dynamo action is mediated by what Spruit (1999, 2002) conveniently calls a 'Tayler instability' – a stratification-modified pinch or kink-type ('tipping') instability – of the toroidal field wound up by the shear, on large horizontal scales but on radial scales small enough for thermal diffusion to counteract the stable stratification. The ability of the Tayler instability to close the dynamo loop has been verified by numerical experiments (Braithwaite & Spruit 2006). This small-scale dynamo seems likely to be most effective in latitudes within the poleward half of the range, and possibly also in some lower-latitude band or bands not too close to the equator.

The implication (see Section 8.4) turns out to be that the MMC, or at least the high-latitude downwelling branch most critically needed to confine $\mathbf{B}_\mathrm{i}$, is gyroscopically pumped by turbulent Maxwell stresses that diverge not in the convection zone but, rather, near the base of an MHD-turbulent tachocline. This region will be referred to as the *lowermost tachocline* in high latitudes. The orders of magnitude dictate that the stress divergence and consequent MMC are confined to within a fraction of a megametre of the tachopause, where a double boundary-layer structure must exist. The turbulence and gyroscopic pumping could be continuous or intermittent, depending on $|\mathbf{B}_\mathrm{i}|$ values.

Before developing these ideas it is necessary to deal with one fundamental question that was raised at the workshop. Gough & McIntyre's (1998) inevitability argument and its further developments just sketched rely, of course, on the physical reality of the gyroscopic-pumping and burrowing mechanisms for MMCs penetrating a non-magnetic interior. Those mechanisms are well understood and have been carefully studied. They show up most plainly in thought-experiments in which the Reynolds and Maxwell stress divergences in the overlying turbulent layers are replaced by an artificially prescribed, azimuthally symmetric, azimuthally directed force field $\bar{F}$ (Haynes *et al.* 1991). If that force field pushes fluid retrogradely, for instance, then the Coriolis effect tries to turn the fluid poleward. As detailed analysis confirms, this amounts to a systematic mechanical pumping action that drives MMCs. Ekman pumping is the special case in which the force happens to be frictional. But any azimuthal force will do, hence the generic term 'gyroscopic pumping'. Persistent gyroscopic pumping in some layer of any stratified, rotating, thermally relaxing and *non-magnetic* system with a finite pressure scale height generates MMCs that continually burrow downward. This was first clearly shown in the detailed, and complementary, independent investigations by Haynes *et al.* (1991) and Spiegel & Zahn (1992). The burrowing mechanism is so fundamental – to any attempt to understand the tachocline and to assess magnetic versus non-magnetic scenarios – that I find it convenient to give the mechanism a distinctive name, 'Haynes–Spiegel–Zahn burrowing' or 'HSZ burrowing' for brevity, whenever verbal precision is necessary.[2]

The question raised at the workshop was whether HSZ burrowing is a real physical phenomenon. It was claimed, in effect, that the two studies just cited are qualitatively in error and that there is no such thing as HSZ burrowing, even in the absence of the interior magnetic field $\mathbf{B}_i$. The claim was based on a recent study of MMCs using non-magnetic equations (Gilman & Miesch 2004) whose results appear to imply that MMCs driven from above cannot penetrate downward more than a negligible distance, probably less than the vertical resolution of helioseismic inversions. If that were correct then most of the arguments in this chapter, and in its predecessors including Gough & McIntyre (1998), would fail utterly. Therefore Section 8.2 revisits the problem studied in Gilman & Miesch (2004), using the same formulation and notation. It turns out that through a quirk of formulation the solutions obtained by Gilman & Miesch make up an incomplete set. They are a special subset of solutions, *for each of which the gyroscopic pumping exactly vanishes at each latitude*. No gyroscopic pumping implies no burrowing! There is, after all, no conflict. Indeed the analysis in Section 8.2, based on an idealized slab model, provides the simplest possible illustration of the pumping and burrowing mechanisms, supplementing the original analytical and numerical work of Haynes *et al.* (1991) and Spiegel & Zahn (1992).

Section 8.3 goes on to argue that turbulence in the interior, below the tachopause, must be exceedingly sporadic. Thus, within the gigayear perspective a random snapshot of the Sun is almost certain to show an interior that is entirely laminar or very nearly so. Broadly speaking this is consistent with the standard solar modelling assumption of a microscopically diffusive helium settling layer, though it remains possible that the layer is somewhat thickened, and indeed its heavy-element contrast somewhat increased, by the sporadic interior mixing.

Section 8.4 examines what Spruit's (2002) arguments then imply about tachocline and tachopause structure and high-latitude downwelling. As already mentioned, such downwelling is critical to the confinement of $\mathbf{B}_i$, a point underlined by recent numerical studies (Garaud 2002; Braithwaite & Spruit 2004; Brun & Zahn 2006) showing the tendency for the dipolar poloidal part of an internal field to diffuse its lines upward and outward through a substantial high-latitude region. That tendency is, however, easily held in check by the downwelling within the double-boundary-layer structure of the lowermost tachocline. Thus the double boundary layer appears well able to confine $\mathbf{B}_i$ in high latitudes.

[2] As already mentioned, it has also been called 'spreading' but, with the Sun's gravitational field pulling hard on my imagination, I prefer 'burrowing' because it unambiguously connotes downwardness.

Intriguingly, if frustratingly, the mean shears within the double boundary layer turn out to be far too small to be helioseismically visible. Moreover, the same appears true of shears throughout the bulk of the high-latitude, stably stratified tachocline. Therefore the visible shear must, in high latitudes, reside wholly in the lower convection zone and overshoot layer. As will be seen shortly this is consistent with the helioseismic evidence. A similar situation may be expected in any low-latitude band that goes turbulent via the shear driven, Tayler-mediated small-scale dynamo action, though, even if such a band exists, the properties of the Tayler instability – favoured by a poleward decrease in the toroidal field wound up by the shear – suggest that the band would have limited latitudinal extent. It might also exhibit unsteady behaviour, such as a life cycle involving poleward migration on timescales perhaps $\sim 10^6$ yr or more.

Section 8.5 extends the idealized analysis of Section 8.2 to allow for compositional gradients in the underlying helium settling layer, in order to reassess Gough & McIntyre's (1998) 'lithium-burning polar pit' hypothesis. It appears that N_μ values, acting in concert with the surrounding $\mathbf{B}_\mathrm{i}$, are more than enough to inhibit the formation of such pits and, indeed, to constrain the tachopause – defined as the bottom of the ventilated layer, equivalently the top of the helium settling layer – to be very close to the horizontal.

Moreover, this constraint on tachopause slope holds tightly even on a global scale. It appears that the tachocline, assuming it is sufficiently ventilated, must have not only an approximately constant chemical composition but also constant depth over all latitudes. More precisely, the tachopause has to follow an effective gravitational–centrifugal potential, globally, to within a very tiny fraction of a megametre.

If this picture is anywhere near correct then the only way to burn lithium is simply for Δ, defined in terms of tachopause depth, to be large enough. A careful lithium-burning modelling study by Christensen-Dalsgaard *et al.* (1992) suggests a need for Δ values close to 65 Mm measured downwards from the helioseismic bottom of the convection zone at $0.713R_\odot$. This puts the tachopause at $0.62R_\odot$. If one superposes the $0.62R_\odot$ circle on to Figure 3.7 of Chapter 3, then especially in high latitudes one sees what looks like a substantial shear-free region beneath the overshoot layer, consistent with the earlier statement that the visible shear must, in high latitudes, reside wholly in the lower convection zone and overshoot layer. This of course is very different from the scenario of Gough & McIntyre (1998). Section 8.6 offers some concluding remarks, mainly on some uncertainties regarding tachocline ventilation.

It might be thought that the terminology should be changed if, as I am now suggesting, the ventilated tachocline is distinctly deeper than the tachocline defined by shears visible in a helioseismic inversion. But observational invisibility does not imply dynamical insignificance. And indeed, in the scenario to be developed, the invisible shear has a crucial role in the high-latitude dynamics and ventilation of the tachocline, all the way down to the tachopause – defined, as here, to mean the ventilated layer and its lower boundary, or equivalently the top of the helium settling layer.

8.2 Gyroscopic pumping and HSZ burrowing

Consider the thought experiment of Haynes *et al.* (1991), performed on Gilman & Miesch's (2004) non-magnetic, linearized Cartesian slab model. We take coordinates (x, y, z) respectively eastward, northward and upward as in Gilman & Miesch, with corresponding velocity components (u, v, w) and Coriolis vector idealized as $(0, 0, 2\Omega)$. For definiteness the top of the model, $z = z_{\text{top}}$ say, is taken to be isothermal, stress-free, and impermeable to mass. The prescribed azimuthal force field $\{\bar{F}(y,z), 0, 0\}$ is applied to an upper layer $z_{\text{f}} < z < z_{\text{top}}$. The force $\bar{F}$ is assumed weak enough for linearization to remain valid. We ask to what extent the response to $\bar{F}$ penetrates downward into the unforced region $z < z_{\text{f}}$, where we take the buoyancy frequency N of the stratification to be constant as in Gilman & Miesch (2004). As in that work we ignore compositional gradients, as would be an appropriate idealization if the interior were non-magnetic and HSZ burrowing active over the gigayear timescale. For then no helium settling layer would have a chance to form (further discussion in Section 8.5), and the thermal stratification would dominate. Steady-state solutions of the type found by Gilman & Miesch (2004) should, of course, be valid in the unforced region.

The profile of N within the forcing layer $z_{\text{f}} < z < z_{\text{top}}$ will be left unspecified. In the original scenario of Gough & McIntyre (1998), in which the forcing layer was identified as the convection zone, we would have $N \equiv 0$ for $z_{\text{f}} < z < z_{\text{top}}$. But there is no difficulty in including the overshoot layer, and indeed an entire turbulent tachocline, as part of the forcing layer. The thought experiment is meant to imitate the effect of any overlying layer, stratified in any way, in which the turbulent Reynolds and Maxwell stresses in an x-averaged description diverge to give the force field $\bar{F}(y,z)$. Any such force field, arising from internal stresses, must have a domain integral that vanishes,

$$\iint \bar{F}(y,z)\,\mathrm{d}y\,\mathrm{d}z = 0, \tag{8.1}$$

even though its vertical integral $\bar{\mathcal{F}}(y) = \int_{z_{\text{f}}}^{z_{\text{top}}} \bar{F}(y,z)\,\mathrm{d}z$ need not vanish.

Again following Gilman & Miesch (2004) we use the Boussinesq equations and describe thermal relaxation toward radiative equilibrium by a constant thermal diffusivity $\kappa \approx 10^7\,\text{cm}^2\,\text{s}^{-1}$. Some aspects of the problem depend on non-Boussinesq effects, which in a doubly infinite domain select downward penetration at the expense of upward, as illustrated by Haynes *et al.*'s analysis. Here we have replaced those effects by the artifice of cutting off the fluid domain at $z = z_{\text{top}}$.

Defining the buoyancy–acceleration anomaly ϑ in the standard way as gravity times the fractional temperature anomaly on a pressure surface, $\vartheta = gT/\bar{T}$ in Gilman & Miesch's notation, we have, for axisymmetric dynamics $\partial/\partial x = 0$,

$$\frac{\partial v}{\partial y} + \frac{\partial w}{\partial z} = 0, \tag{8.2a}$$

$$\frac{\partial u}{\partial t} - 2\Omega v - \nu \frac{\partial^2 u}{\partial z^2} = \bar{F}(y, z), \tag{8.2b}$$

$$\frac{\partial^2 v}{\partial z \partial t} + 2\Omega \frac{\partial u}{\partial z} + \frac{\partial \vartheta}{\partial y} - \nu \frac{\partial^3 v}{\partial z^3} = 0, \tag{8.2c}$$

$$\frac{\partial \vartheta}{\partial t} + N^2 w - \kappa \frac{\partial^2 \vartheta}{\partial z^2} = 0. \tag{8.2d}$$

As in Gilman & Miesch (2004), we have included viscous terms, with constant momentum diffusivity ν. The third equation, (8.2c), may be called the generalized thermal-wind equation. It is formed by eliminating the pressure between the hydrostatic equation and the vertical derivative of the meridional momentum equation. As is realistic for the solar tachocline we assume that Ω is effectively large (rapidly rotating system, small Rossby number), so that in (8.2c) there is a powerful tendency toward thermal-wind balance, $2\Omega\, \partial u/\partial z \approx -\partial\vartheta/\partial y$.

If a system like this is started from an undisturbed initial state with u, v, w, and ϑ all zero then, as Haynes *et al.* showed in an essentially similar problem, the typical behaviour within the forcing layer is robustly as follows. First, u accelerates in response to $\bar{F}$, followed by Coriolis turning of (u, v). The system then approaches a locally steady or nearly steady state in which thermal-wind balance prevails, and in which $-2\Omega v$ has come into approximate balance with $\bar{F}$ in Equation (8.2b). The effect of the ν term in (8.2b) is equivalent to a slight redistribution of $\bar{F}$, leaving the qualitative picture unaffected. The balance

$$-2\Omega v \approx \bar{F} \tag{8.3}$$

describes the persistent gyroscopic pumping of meridional flow v by the steady azimuthal force field $\bar{F}$. Note that $\bar{F} < 0$ implies $v > 0$, confirming that a retrograde force pumps fluid poleward.

Now Gilman & Miesch's results should apply to the unforced region $z < z_{\text{f}}$. They assume a steady state with $\bar{F} \equiv 0$, leading to a single equation that applies in the unforced region,

$$\frac{\partial^6 v}{\partial z^6} + \frac{4\Omega^2}{\nu^2}\frac{\partial^2 v}{\partial z^2} + \frac{N^2}{\nu\kappa}\frac{\partial^2 v}{\partial y^2} = 0. \tag{8.4}$$

(This comes from assuming N constant, taking $\partial^3/\partial z^3$ of (8.2c), then successively eliminating u, ϑ and w.) Gilman & Miesch (2004) consider solutions of the form $v \propto \mathrm{e}^{kz}\sin(y/\ell)$, where ℓ is a suitable latitudinal lengthscale and k is a complex constant satisfying the characteristic equation

$$k^6 + \frac{4\Omega^2}{\nu^2}k^2 - \frac{N^2}{\nu\kappa}\ell^{-2} = 0, \qquad (8.5)$$

of whose six roots three correspond to downward evanescence. Consideration of the scale $(\mathrm{Re}\,k)^{-1}$ for evanescence when the latitudinal lengthscale ℓ takes reasonable values $\sim 10^2$ Mm gives vertical scales of the order of a few tens of megametres at most, even when ν and κ are both taken to have large eddy values $\sim 10^{12}\,\mathrm{cm}^2\,\mathrm{s}^{-1}$. Microscopic values give a small fraction of a megametre. If these were the only possible solutions then they would certainly imply what was claimed at the workshop, namely that there is no such thing as HSZ burrowing into a non-magnetic interior. Gough & McIntyre's (1998) inevitability argument would then fail.

Let us ask, however, what a boundary-layer solution of this kind in the unforced region $z < z_\mathrm{f}$ would imply about the forcing function $\bar{F}(y,z)$ in the layer above. All variables are downward evanescent. Therefore, by integrating (8.2a) over all z and invoking the assumption that the upper boundary $z = z_\mathrm{top}$ is impermeable to mass, we may deduce that the y derivative of $\int_{-\infty}^{z_\mathrm{top}} v\,\mathrm{d}z$ vanishes, so that

$$\int_{-\infty}^{z_\mathrm{top}} v\,\mathrm{d}z = C, \qquad (8.6)$$

where C is a constant, provided also that the upper boundary is stress-free, i.e. that $\nu\,\partial(u,v)/\partial z = 0$. So integrating (8.2b) for the steady state gives

$$\bar{\mathcal{F}}(y) = \int_{z_\mathrm{f}}^{z_\mathrm{top}} \bar{F}(y,z)\,\mathrm{d}z = -2\Omega\int_{-\infty}^{z_\mathrm{top}} v\,\mathrm{d}z = -2\Omega C. \qquad (8.7)$$

Now this is compatible with (8.1) only if $C = 0$; therefore

$$\bar{\mathcal{F}}(y) = -2\Omega\int_{-\infty}^{z_\mathrm{top}} v\,\mathrm{d}z = 0 \qquad (8.8)$$

at each y. In other words, there is no net gyroscopic pumping – no vertically integrated azimuthal force, and no vertically integrated meridional mass flux and volume flux in the boundary layer – at any y. Gilman & Miesch's solutions are all solutions for which the net gyroscopic pumping exactly vanishes at each y.

That fact is not obvious from Gilman & Miesch's perspective, in which only the unforced layer $z \leqslant z_\mathrm{f}$ is considered. We may note, however, that all of their solutions satisfy the special relation $\nu\,\partial^2 u/\partial y\partial z = 2\Omega w$. This can be straightforwardly verified either from the solutions, or from the variant of (8.8) obtained by integrating (8.2b) from $-\infty$ to any $z \leqslant z_\mathrm{f}$, noting that $\bar{F} = 0$. Even though it might appear that w is being arbitrarily prescribed at the top – and should therefore represent any gyroscopic pumping from above – the u field, invisible in Gilman & Miesch's formulation, has cunningly organized itself in such a way that the boundary $z = z_{\rm f}$ exerts an azimuthal viscous stress on the fluid beneath that just cancels[3] the pumping effect of the prescribed w.

To double-check this we look at an explicit solution that includes the upper forcing layer $z_{\rm f} < z < z_{\rm top}$. For simplicity we set $N \equiv 0$ and $\bar{F} \propto \sin(y/\ell)$, independent of z within the forcing layer. Then within that layer we see that Equations (8.2) admit a simple solution of the form $v = -\bar{F}/2\Omega$, with $u \propto \sin(y/\ell)$, both independent of z, the remaining variables being given by $\vartheta \equiv 0$ and $w = -(z_{\rm top} - z)(d\bar{F}/dy)/2\Omega \propto (z_{\rm top} - z)\cos(y/\ell)$. But a boundary-layer solution in which u is a continuous function of z satisfies (8.2b) at $z = z_{\rm f}$ only if (consistently with $\nu\partial^2 u/\partial y\partial z = 2\Omega w$ for $z \leqslant z_{\rm f}$) we add a delta function to $\bar{F}$ whose strength is precisely $-(z_{\rm top} - z_{\rm f})\bar{F}$. That is, to get a solution of Gilman & Miesch's boundary-layer form we must choose this extra contribution to $\bar{F}$ such that the total force integrates to zero, $\bar{\mathcal{F}}(y) = 0$, as already seen from (8.8). For this particular solution we also need a delta-function heat source and sink $\propto \cos(y/\ell)$ at $z = z_{\rm f}$, that is, where N^2 is discontinuous, but such an artifice does not affect the issue of gyroscopic pumping.

For the generic case in which $\bar{\mathcal{F}}(y)$ does not, by contrast, vanish, we must expect to find additional solutions that do not have the boundary-layer character implied by (8.5). Even within the steady-state framework, we do not have to look far to find them. In place of (8.4) consider the corresponding equation for u. Substituting $v \propto \partial^2 u/\partial z^2$ from (8.2b), we have

$$\frac{\partial^8 u}{\partial z^8} + \frac{4\Omega^2}{\nu^2}\frac{\partial^4 u}{\partial z^4} + \frac{N^2}{\nu\kappa}\frac{\partial^4 u}{\partial y^2 \partial z^2} = 0, \tag{8.9}$$

with characteristic equation

$$k^8 + \frac{4\Omega^2}{\nu^2}k^4 - \frac{N^2}{\nu\kappa}\ell^{-2}k^2 = 0. \tag{8.10}$$

This has two more roots, both zero, signalling the existence of two extra solutions, $u =$ constant and $u \propto z$. So if we leave the y-origin arbitrary the general solution in the unforced region is

$$u = \left(\sum_{1}^{6} C_j e^{k_j z} + C_7 + C_8 z\right)\sin(y/\ell), \tag{8.11}$$

[3] The Spiegel–Zahn eddy viscosity similarly cancels the pumping from above, through horizontal rather than vertical transmission of azimuthal stress. In effect one has two gyroscopic pumps, an upper pump producing a certain mass flux, and a lower one negating it by producing an equal and opposite mass flux.

where the k_j are the six roots of (8.5) and the C_j are arbitrary constants. Such solutions are applicable when, for instance, we take

$$\bar{\mathcal{F}}(y) = \mathcal{F}_0 \sin(y/\ell) \tag{8.12}$$

with constant $\mathcal{F}_0$.

To see what (8.11) means physically, it is simplest to consider first a problem with an artificial lower boundary, say $z = 0$, far beneath the forcing layer, where far means many evanescence height scales $(\mathrm{Re}\,k)^{-1}$. On $z = 0$ we impose $u = v = w = 0$ (impermeable and no-slip) and $\kappa\, \partial\vartheta/\partial z = 0$ (heat flux held to its background value). Then with $\bar{\mathcal{F}}(y) = \mathcal{F}_0 \sin(y/\ell)$ it is a straightforward exercise to prove that $C_7 = 0$ and $C_8 = \nu^{-1}\mathcal{F}_0$, with exponentially small error, and that the solution in the unforced region is

$$u = \left(\sum_{j=1}^{3} C_j \mathrm{e}^{k_j z} + \nu^{-1} \mathcal{F}_0\, z \right) \sin(y/\ell), \tag{8.13}$$

where k_1, k_2, k_3 are the downward-evanescent roots of (8.5). They are needed to describe details within a thin layer near the top. Beneath that layer, we have $v = w = 0$ and $\vartheta = 2\Omega\, \ell\, \nu^{-1} \mathcal{F}_0 \cos(y/\ell)$. The upward-evanescent roots k_4, k_5, k_6 are absent because the solution just described satisfies the four lower boundary conditions as it stands, with exponentially small error. The coefficient of z, $C_8 = \nu^{-1}\mathcal{F}_0$, is determined regardless of details near the top, because in the steady state (8.8) is replaced by

$$\bar{\mathcal{F}}(y) = \nu \left.\frac{\partial u}{\partial z}\right|_{z=0} . \tag{8.14}$$

This comes from integrating (8.2a) and (8.2b) from $z = 0$ to z_{top}, then using (8.1) and the bottom boundary conditions. In order to have a steady state in this linear model, the net applied force $\bar{\mathcal{F}}$ must be balanced at each y by the stress on the bottom. This pins down the coefficient of z. The existence of the steady-state solution (8.13) is an easy way to see that, in the original time-dependent thought experiment, the influence of $\bar{\mathcal{F}}$ must have burrowed all the way to the bottom – regardless of how far down the bottom may be. If we take the bottom down toward $z = -\infty$ then the time to reach the steady state increases without bound.

Notice that this solution describes another situation in which the gyroscopic pumping has been cancelled by a viscous stress. Before that, as the burrowing proceeds, the pumping drives an MMC whose Coriolis force accelerates u values up to such extremes, $\propto \nu^{-1}$ when ν is considered small, that the viscous stress spanning the entire depth $0 < z < z_{\mathrm{f}}$ comes into balance with the force applied to the overlying layer $z_{\mathrm{f}} < z < z_{\mathrm{top}}$. Of course such extremes could violate the original linearization. But the real significance of the foregoing is that the response to $\bar{\mathcal{F}}(y)$

in the *absence* of an artificial lower boundary must be inherently time-dependent, as originally shown by HSZ.

On the long timescale of the burrowing process, and when $\kappa \gg \nu$, the time derivative in (8.2d) may be neglected as well as that in (8.2c), where, moreover, thermal-wind balance is an excellent approximation. So by taking $\partial^2/\partial y^2$ of (8.2b) and then successively eliminating v through (8.2a), w through (8.2d) with $\partial/\partial t$ neglected, then finally ϑ through thermal-wind balance $\partial\vartheta/\partial y = -2\Omega\,\partial u/\partial z$ in place of (8.2c), we get

$$\left(\frac{\partial}{\partial t} - \nu\frac{\partial^2}{\partial z^2}\right)\frac{\partial^2 u}{\partial y^2} - \frac{4\Omega^2\kappa}{N^2}\frac{\partial^4 u}{\partial z^4} = \frac{\partial^2 \bar{F}}{\partial y^2}, \tag{8.15}$$

recovering Spiegel & Zahn's result that when thermal relaxation is diffusive and ν sufficiently small then the burrowing behaviour is hyperdiffusive, with hyperdiffusivity $4\Omega^2\ell^2\kappa/N^2$ for latitudinal lengthscale ℓ. This may be compared with Haynes *et al.*'s result in the Boussinesq limit, $H \to \infty$ in their notation: when the thermal relaxation is Newtonian with timescale $\kappa_{\text{Newt}}^{-1}$ then the burrowing behaviour is diffusive with diffusivity $4\Omega^2\ell^2\kappa_{\text{Newt}}/N^2$. Notice that the timescale for burrowing is sensitive to the latitudinal scale ℓ, behaving as ℓ^{-2}.

Before leaving this topic we note for completeness the steady-state solution of (8.2a)–(8.2d) that idealizes the laminar-downwelling scenario of Gough & McIntyre (1998), within a non-magnetic tachocline of nominal thickness $\Delta = z_{\text{f}}$ at the bottom of which, $z = 0$, there is a thermomagnetic boundary layer able to accept a certain volume flux $-w_0\cos(y/\ell)$, say, per unit area. That flux is governed by the magnitude of the global-scale interior field $\mathbf{B}_{\text{i}}$, and so the overlying layers must adjust themselves so as to pump exactly that much flux, which flux Gough & McIntyre estimated to scale as $|\mathbf{B}_{\text{i}}|^{1/3}$.

With microscopic values of ν and κ we may take $\Delta \gg (\text{Re}k)^{-1}$. Then, apart from details near $z = z_{\text{f}} = \Delta$, the solution in $0 < z < \Delta$ is as follows. It confirms the Gough & McIntyre (1998) result that, for given $u|_{z=\Delta}$, $w_0 \propto \Delta^{-3}$ implying $\Delta \propto |\mathbf{B}_{\text{i}}|^{-1/9}$:

$$u = \frac{N^2 w_0}{24\,\Omega\ell\kappa}\, z^2(3\Delta - 2z)\sin(y/\ell), \tag{8.16a}$$

$$v = 0, \tag{8.16b}$$

$$w = -w_0\,\cos(y/\ell), \tag{8.16c}$$

$$\vartheta = \frac{N^2 w_0}{2\kappa}\, z(\Delta - z)\cos(y/\ell). \tag{8.16d}$$

The relation $w_0 \propto \Delta^{-3}$ follows from (8.16a) with $z = \Delta$. Also $\mathcal{F}_0 = 2\Omega\ell w_0$ in (8.12), from integrating (8.2a) and (8.2b) as before. Following Gough & McIntyre (1998), we have assumed isothermal conditions $\vartheta = 0$ at $z = 0$ as well as at $z = z_{\text{top}}$, and $N \equiv 0$ in the forcing layer $z_f < z < z_{top}$ to make it into an idealized convection zone. The model tachocline described by (8.16) is frictionless, with angular momentum exchange across it mediated solely by the MMC and handed over to the Maxwell stress in the thermomagnetic boundary layer. Gough & McIntyre estimated that $w \sim -10^{-5}\,\mathrm{cm\,s^{-1}}$, more than enough to ventilate the tachocline and to confine $\mathbf{B}_i$ in high latitudes.

8.3 The nearly laminar magnetic interior

Following Gough & McIntyre's (1998) inevitability argument we now take for granted the existence of the global-scale interior field $\mathbf{B}_i$, and expand our timeframe to the gigayear perspective of solar spin-down. Let us accept, in particular, that the present-day interior is close to solid rotation essentially because spin-down was, and presumably still is, Ferraro-constrained – in other words constrained by the Alfvénic elasticity of a sufficiently strong poloidal component of $\mathbf{B}_i$.

This is almost the same thing as saying that $\mathbf{B}_i$ was, and is, strong enough to stop HSZ burrowing, allowing a helium settling layer to form. The burrowing depends on the sustained gyroscopic pumping of an MMC, whose Coriolis force accelerates a deepening layer of differential rotation in thermal-wind balance. It is the resulting baroclinicity, together with thermal diffusion, that allows the MMC to persist and to continue burrowing. If the Ferraro constraint is strong enough to stop the differential rotation (with the help of MHD shear instabilities as necessary, see below), then it also stops the baroclinicity[4] and therefore the burrowing. In other words it impedes the response to the pumping, almost as if the interior were solid. Thus the response is limited to being an MMC such that its entire mass flux can be accepted by the thermomagnetic boundary layer at the top of the interior, as in Gough & McIntyre (1998).

The estimates of Mestel & Weiss (1987) and the detailed numerical experiments of Charbonneau & MacGregor (1993) suggest that the order of magnitude required to impose the Ferraro constraint is $|\mathbf{B}_i| \gtrsim 10^{-2}\,\mathrm{G}$. The inevitability argument of Gough & McIntyre (1998) then implies that $\mathbf{B}_i$ must be at least this strong, in reality, and furthermore, as already mentioned, that in high latitudes $\mathbf{B}_i$ must be largely confined to the interior by the gyroscopic pumping from above, as required in spin-down scenarios like those of Charbonneau & MacGregor (1993). If the poloidal field were not so confined then its lines would diffuse upward and outward through a substantial high-latitude region, such that the Sun's differential rotation

[4] I use the term 'baroclinicity' in its most fundamental sense, meaning the non-vanishing of the $\nabla p \times \nabla \rho$ term in the three-dimensional vorticity equation, where ρ is density and p is total pressure including the hydrostatic background. In the case of thermal-wind balance this in turn implies the non-vanishing of the axial derivative of angular velocity Ω and hence, usually, violation of the Ferraro constraint. In a perfect gas the non-vanishing of $\nabla p \times \nabla \rho$ is equivalent to the non-vanishing of $\nabla p \times \nabla \vartheta$ and of $\nabla p \times \nabla T$, where T is temperature, and is therefore equivalent to having non-vanishing isobaric gradients of T.

would differ from that observed. Such scenarios are illustrated in various ways by the numerical experiments of Garaud (2002), Braithwaite & Spruit (2004), and Brun & Zahn (2006) mentioned earlier.

In spin-down scenarios like those of Charbonneau & MacGregor (1993) there are poloidal-field tori within the interior, surrounding the neutral ring, that do not thread the convection zone or tachocline. In order to spin those tori down, avoiding a 'dead zone' of super-rotation surrounding the neutral ring and hence a contradiction with the helioseismic evidence, Charbonneau & MacGregor (1993) had to use an artificial viscosity ν far greater than the actual microscopic viscosity. In the real Sun, therefore, some kind of turbulent eddy viscosity must be involved.

Now Spruit (1992, 2002) cogently argues that, when shear develops in the interior, the first turbulent process to kick in will be a small-scale dynamo mediated by Tayler instabilities – stratification-modified pinch or kink-type ('tipping') instabilities – of the toroidal field wound up by the shear. See also the numerical verification of dynamo action by Braithwaite & Spruit (2006). The dynamo is shear-driven and, arguably, has the robustness of an interchange instability. One may therefore reasonably assume that it will act to reduce shear through the Maxwell stresses produced by windup. In this respect it is somewhat like the better-known magneto-rotational instability in hot accretion disks (see Chapter 12 by Ogilvie), which, however, has a much higher shear threshold (Spruit 1999). Therefore the Tayler-mediated small-scale dynamo appears likely to be the main mechanism inhibiting the formation of super-rotating 'dead zones' in the real Sun.

The existence of the Ferraro constraint, aided by the rapid damping of global torsional oscillations by phase mixing, implies that instability need only occur at one location on each torus. This point is significant since, unlike the magneto-rotational instability, the Tayler instability tends to be ineffective near the equator and so needs the help of the Ferraro constraint, if it is to bring about uniform spin-down. Without the Ferraro constraint, there would be nothing to stop global-scale sub-threshold shears from building up. Near the equator, above the neutral ring, magnetorotational instability may have a role as well. Recall that by adopting sufficiently small scales the instabilities can make use of thermal diffusivity, κ, to release the constraint due to thermal stratification (e.g. Townsend 1958; Fricke 1969; Zahn 1974; Acheson 1978).

Now because spin-down is so slow, we may expect the instabilities to kick in very sporadically in space and time, and certainly not uniformly throughout the interior. The Ferraro constraint is needed for that reason as well. Such sporadic or intermittent behaviour is generic for any high-Reynolds-number fluid system whose coarse-grain shear is well below all instability thresholds. In this respect the Sun's interior must be somewhat like sheared, stably stratified terrestrial fluid systems at high Richardson number. In all such systems it is well known that turbulence occurs sporadically, the more so the higher the Richardson number. The terrestrial lower stratosphere is a case in point. The sporadic occurrence of turbulence there is familiar to everyone in these days of universal air travel. Most of the time the seat belt sign is off, and the ride almost perfectly smooth. Since coarse-grain Richardson numbers Ri are large, shear-instability thresholds $Ri \lesssim \frac{1}{4}$ cannot be exceeded over large volumes.

We are forced to conclude – because of the extreme slowness of spin-down – that the Sun's interior, even more than the terrestrial stratosphere, must be laminar at most times and locations. And, as already remarked, the whole picture is consistent with the presence of a distinct helium settling layer in perhaps the top 100 Mm or so of the radiative envelope, for which there is some helioseismic support (see Section 3.5 in this book).

Corresponding estimates for the stably stratified tachocline (Spruit 2002) point toward the opposite conclusion. A coarse-grain view of the tachocline puts it well above threshold, in high latitudes at least (see Equations (8.26)–(8.28) below). The high-latitude tachocline seems therefore likely to be in some sense much more turbulent than the interior. For a turbulent tachocline we need to consider how convection-zone stresses are handed over to the interior. This involves understanding how an MHD-turbulent flow goes over into a laminar, Ferraro-constrained flow. It is this problem that is considered next.

8.4 The high-latitude tachocline and its invisible shear

How then is the stress handed over? More precisely, what is the pattern of angular momentum transport, from some combination of MMCs and turbulent stresses, that transmits to each latitude of the mostly laminar, Ferraro-constrained interior any torque that arises from the convection zone's propensity to rotate differentially? And could that pattern include an MMC capable of confining $\mathbf{B}_\mathrm{i}$ in a band of high latitudes – let us say something like latitudes 50°–80° or colatitudes 10°–40° – holding the field lines of $\mathbf{B}_\mathrm{i}$ nearly horizontal there against magnetic diffusion, as required to bring about Ferraro-constrained spin-down in most of the interior?

Now it happens that the Tayler instability is likely to be effective in something like the same latitude band, as well as in the neighbourhood of the pole. To assess this more closely one would need to consider the latitudinal gradients of the actual toroidal field produced by the small-scale dynamo – which is why a low-latitude band might also be unstable, from time to time at least – and one would need to consider the possible shear-induced modifications of the Tayler instability itself (see Chapter 10 by Gilman & Cally).

For the moment, however, I simply assume that there is an 'active band' of high latitudes, probably something like the nominal 50°–80°, where the vertical shear is enough to drive the Tayler-mediated small-scale dynamo and for Spruit's (2002) order-of-magnitude estimates to apply. I ignore horizontal shear, in effect supposing that the tachocline is in shellular solid rotation in the active band of latitudes. The angular-velocity contours in Figure 3.7 of Chapter 3 hint that this may not be too bad an approximation.[5] Thus the focus is on the vertical structure. I further simplify by assuming a single latitudinal scale $\ell \sim 10^2$ Mm, in a formal sense staying with slab-model thinking for the moment. It will be convenient to stay with the slab-model notation as well; thus, z will still be the upward, i.e. radial, coordinate, and $\partial u/\partial z$ will be the vertical shear of the mean azimuthal velocity. (Strictly it is $\partial/\partial z$ of the mean *angular* velocity that is relevant, but the difference is unimportant for present purposes.) It will be convenient to define a non-dimensional shear

$$q = \Omega^{-1}\partial u/\partial z \tag{8.17}$$

(Spruit's (2002) notation); the threshold value of $|q|$ will be denoted by q_{crit}. Its order of magnitude is given by (8.26) below.

The key points are listed next, followed by the order-of-magnitude relations that underpin them. It will emerge that the processes involved cover practically the entire range of timescales from gigayears down to the months and years of convective overshoot and the solar-cycle dynamo. The latter, being self-evidently a large-scale, low-latitude dynamo as well as a relatively fast one, is a different beast altogether from the small-scale, stably stratified dynamo presently under discussion. The small-scale dynamo will turn out to be vastly slower, yet still fast in comparison with gigayears. To avoid confusion it will need to be remembered that 'small-scale' refers not to horizontal scales but only to the vertical scale of the eddy motion.

(i) The small-scale dynamo has plenty of headroom, given any of the current estimates of tachocline thickness Δ. This would be so even if the real high-latitude tachocline were as thin as the $\Delta \approx 13\,\text{Mm} \approx 0.019R_\odot$ estimated by Elliott & Gough (1999), let alone the $\Delta \approx 65\,\text{Mm} \approx 0.09R_\odot$ now anticipated in connection with lithium burning. The vertical scale δ_κ of the eddy motion, governed here by the thermal diffusivity κ acting to release the stratification constraint, is of the order of 10^{-1} Mm, see Equation (8.25) below.

(ii) The dominant azimuthal stress across horizontal area elements is the turbulent Maxwell stress. Its mean value is proportional to the local vertical shear $\partial u/\partial z$ with

[5] Shellular solid rotation in the active band is plausible, in any case, because of the shear-reducing propensities of the Tayler-mediated small-scale dynamo pointed out by Spruit (2002). Indeed, unlike non-magnetic turbulence, the small-scale dynamo could have taken on the role of the Spiegel–Zahn horizontal eddy viscosity had it not been for the dependence on the latitudinal gradients of toroidal field. That dependence precludes the small-scale dynamo from being effective across all latitude bands, implying that the inevitability argument of Gough & McIntyre (1998) still holds good. In order to enforce solid rotation in the manner of the Spiegel–Zahn theory, a horizontal eddy viscosity would need to support a stress that transmits azimuthal torques horizontally across all latitudes. It would need to produce, respectively, prograde and retrograde torques in high and low latitudes, in just such a way as to cancel the gyroscopic pumping from above.

a proportionality coefficient ν_e, see Equation (8.24) below, that is approximately constant like an ordinary viscosity. In particular, ν_e is independent of shear for any supercritical shear $|\partial u/\partial z| > \Omega q_{crit}$. This shear-independence of ν_e is remarkable for a fully developed turbulent flow. Spruit (2002) aptly calls it a 'coincidence'. For the stably stratified tachocline, $\nu_e \sim 1.6 \times 10^8\,\mathrm{cm^2\,s^{-1}}$.

(iii) So powerful is the Maxwell stress that it dominates the angular momentum transport in the bulk of the high-latitude, stably stratified tachocline. This statement holds over a vast range of possible $|\mathbf{B}_i|$ values. It dominates even when $|\partial u/\partial z|$ is much smaller than typical coarse-grain shear values estimated from helioseismology. That is part of why the shear $|\partial u/\partial z|$ in the lower, stably stratified portion of a 65 Mm deep ventilated tachocline may be expected to be helioseismically invisible in high latitudes. We shall see that the magnitude of ν_e is large enough to bring $|\partial u/\partial z|$ down to values close to threshold, Ωq_{crit}, in high latitudes. Such values are about an order of magnitude less than the visible shear.

(iv) The simplest version of the implied scenario is for $|\partial u/\partial z|$ to stay just above threshold, $|\partial u/\partial z| \sim \Omega q_{crit}$. We shall see that this is possible if tachopause $|\mathbf{B}_i|$ values are large enough, $\gtrsim 10^2\,\mathrm{G}$. There are other possible versions, for lower $|\mathbf{B}_i|$ values, in which time-averaged $|\partial u/\partial z|$ values are sub-threshold and the dynamo action intermittent. In such cases ν_e takes $|\partial u/\partial z|$ below threshold and switches off, $|\partial u/\partial z|$ then builds up through gyroscopic pumping (temporarily like an unsteady version of the scenario in Gough & McIntyre (1998)), then ν_e switches on again, and so on cyclically. Possible cycle times could be anywhere in the range from $\sim 10^6$ yr upward, depending on $|\mathbf{B}_i|$.

(v) In the bulk of the stably stratified tachocline, thermal-wind balance holds robustly. There, the weak vertical shear constrains baroclinicity *qua* latitudinal buoyancy gradients $|\partial\vartheta/\partial y|$ to be weak as well. Furthermore, the dynamo turbulence leaves unaffected both the N value of the subadiabatic thermal stratification itself and the value of κ felt by mean motions (H. C. Spruit, private communication). This is because of the way the turbulent motion depends on κ to release the stratification constraint. So MMCs are still tied to $|\partial\vartheta/\partial y|$ via the microscopic κ value, $\sim 10^7\,\mathrm{cm^2\,s^{-1}}$, just as if the turbulence were absent, i.e. in just the same way as in Gough & McIntyre (1998). The upshot is that in the bulk of the stably stratified tachocline there is no MMC, to a first approximation, and that even with the weakened $|\partial u/\partial z|$ the angular momentum transport, there, is mediated predominantly by ν_e. To a higher approximation, one might expect an MMC like that of Gough & McIntyre except that there is now no impediment to weak equatorial downwelling.

(vi) One peculiar consequence is that, in stark contrast with Gough & McIntyre (1998), the present scenario, as developed so far, appears to leave Δ values almost completely unconstrained. This opens the possibility already mentioned that Δ is large enough, ≈ 65 Mm, to explain lithium burning, even with no 'polar pits'. It seems that Δ is determined in a rather subtle and delicate way, not amenable to simple order-of-magnitude analysis. Indeed it may well be that Δ is not determined by quasi-steady dynamics but, rather, depends on the history of convection-zone retreat and helium settling layer formation, as well as on $|\mathbf{B}_i|$ values. (Thus the scatter in lithium abundance found in samples of solar-type stars might be related to a scatter in $|\mathbf{B}_\text{i}|$ values as well as to rotation histories.)

(vii) The dynamo begins to lose headroom in a lowermost turbulent layer of thickness $\sim \delta_\kappa \sim 10^{-1}$ Mm. Notice from Equation (8.25) below that the scale δ_κ is, like ν_e, independent of shear, as long as the dynamo is switched on. As we enter the lowermost turbulent layer, vertical eddy scales and ν_e values must decrease downward. Shear values $|\partial u/\partial z|$ increase, but not enough to stop the turbulent Maxwell stress from diverging and giving rise to an azimuthal force $\bar{F}$, hence gyroscopic pumping.

(viii) A slight extension of Spruit's (2002) arguments suggests that $\nu_\text{e} \propto z^2$ within the lowermost turbulent layer, joining continuously to the constant value $\nu_\text{e} \sim 1.6 \times 10^8\,\text{cm}^2\,\text{s}^{-1}$in the bulk of the tachocline, where z is measured from some virtual origin near the bottom of the lowermost layer. Further analysis suggests that the azimuthal and meridional turbulent Maxwell stress components

$$\nu_\text{e}\left(\frac{\partial u}{\partial z}, \frac{\partial v}{\partial z}\right) = (\sigma, \tau), \tag{8.18}$$

say, take on a modified Ekman-layer structure, breaking the thermal-wind constraint as well as gyroscopically pumping an MMC in the form of a poleward Ekman mass flux. Note that this pumping is entirely due to the fluctuating Maxwell stresses described by the eddy viscosity ν_e, and nothing whatever to do with the sort of quasi-steady Maxwell stresses that would characterize a laminar Hartmann or Ekman–Hartmann layer, or the thermomagnetic boundary layer of Gough & McIntyre (1998).

(ix) To the extent that we have shellular solid rotation $\Omega(z)$ in the active band of latitudes, and the dynamo is switched on, the poleward Ekman mass flux must converge so as to produce an approximately uniform downwelling, $w_{\text{Ek}} < 0$. To see this one has to depart from slab-model geometry and substitute spherical or polar cylindrical geometry. The vertically integrated mass-flux convergence is approximately uniform for the same reasons as in ordinary laminar spin-down in a laboratory cylinder. It is only the vertical structure, not the vertically integrated mass flux, that is changed by the vertically variable eddy viscosity within the modified Ekman layer. Indeed we have the simple formula

$$w_{\text{Ek}} = \nu_\text{e}\left.\frac{\mathrm{d}(\ln\Omega)}{\mathrm{d}z}\right|_{\text{bulk}}, \tag{8.19}$$

implying $w_{\text{Ek}} < 0$ since $\mathrm{d}(\ln\Omega)/\mathrm{d}z\,|_{\text{bulk}} < 0$ in the high-latitude tachocline. The value of ν_e in (8.19) is just the constant bulk value $\nu_\text{e} \sim 1.6 \times 10^8\,\text{cm}^2\,\text{s}^{-1}$ outside the layer. The formula is readily derived by assuming incompressible flow together with the gyroscopic-pumping relation (8.3), setting $\bar{F} = \partial\sigma/\partial z$ in (8.3), then integrating across the modified Ekman layer and computing the horizontal volume-flux convergence in polar geometry. So (8.19) depends only on the fact that $\sigma(z)$ drops from $\nu_\text{e}\,\partial u/\partial z|_{\text{bulk}}$ down to zero across the modified Ekman layer, as the small-scale dynamo finally runs out of headroom. It does not depend at all on the detailed vertical structure within the modified Ekman layer.

(x) The downwelling described by (8.19) is prevented by $\mathbf{B}_\mathrm{i}$ from burrowing into the interior, as noted in Section 8.3. Having nowhere else to go, the mass flux must recirculate through a laminar thermomagnetic boundary layer of thickness $\delta_{\kappa\eta}$, say, like that proposed by Gough & McIntyre (1998), lying immediately beneath the modified Ekman layer and forming with it a tight double-boundary-layer structure. Values of $\delta_{\kappa\eta}$, from Equation (8.30) below, go like $|\mathbf{B}_\mathrm{i}|^{-1/3}$ but are typically a fraction of a megametre. Thus we have convergent poleward flow in the lowermost turbulent layer, and divergent equatorward flow in the laminar thermomagnetic boundary layer just beneath. It is in this way that the stress transmitted by ν_e, i.e. by the averaged *fluctuating* Maxwell stress in the bulk of the stably stratified tachocline, is handed over via the MMC in the lowermost tachocline to the *quasi-steady* Maxwell stress in the outermost fringe of the laminar interior – which fringe is just the thermomagnetic boundary layer. That boundary layer therefore has a dual role: it serves both as the laminar sublayer of the turbulent lowermost tachocline, and also as the outermost fringe of the laminar, Ferraro-constrained interior. It is here that the Ferraro constraint begins to make itself felt directly, through the downwelling and advective-diffusive balance in the boundary layer as discussed by Gough & McIntyre (1998). And it is this same downwelling and advective–diffusive balance that brings about the high-latitude confinement of $\mathbf{B}_\mathrm{i}$, in the same way as in Gough & McIntyre's work.

The order-of-magnitude relations on which the foregoing statements are based are now summarized. The relations are equivalent to those in Spruit (2002) except that I revert to formal slab-model thinking and use $\ell \sim 10^2$ Mm as the latitudinal scale instead of the tachocline radius r used by Spruit; the scale ℓ roughly corresponds to what Gough & McIntyre (1998) called r/L. Like Spruit (2002) I also ignore factors like $2\cos\theta$ in front of Ω, where θ is colatitude, and factors like π.

The formal assumption of a single latitudinal scale ℓ may not be as bad as it sounds, despite the importance of the real polar geometry for the pattern of mass transport in the MMC, as noted in point (ix) above. The Tayler instability, as such, has a large horizontal reach because of its kink or tipping-type kinematics dominated by azimuthal wavenumber $m = 1$. It is certainly able to reach across the pole – one might say more aptly 'slide across the pole', as suggested in Figure 1 of Spruit (1999) – and will probably do so even though the mean shear defined by azimuthal averaging must, technically speaking, vanish at the pole. The instability is a physical process with no respect for coordinate singularities. Indeed, it tends to use as much horizontal space as is available to it, and Spruit's (2002) estimates assume that it does so. As in Gough & McIntyre (1998), the scale ℓ is meant to be no more than a rough way of characterizing the magnitudes of horizontal derivatives constrained by the available horizontal space.

Let η be the microscopic magnetic (Ohmic) diffusivity and Ω_A the typical toroidal field strength produced by the small-scale dynamo within the tachocline, measured as angular Alfvén speed, i.e. as the number of radians of longitude per unit time travelled by the phase of an Alfvén wave. We assume that the microscopic diffusivities satisfy

$$\kappa \gg \eta \gg \nu, \tag{8.20}$$

consistent with typical numerical orders of magnitude $\kappa \sim 1.4 \times 10^7\ \text{cm}^2\,\text{s}^{-1}$, $\eta \sim 4 \times 10^2\ \text{cm}^2\,\text{s}^{-1}$, $\nu \sim 3 \times 10^1\ \text{cm}^2\,\text{s}^{-1}$ near the top of the tachocline, at $0.7R_\odot$, and $\kappa \sim 1 \times 10^7\ \text{cm}^2\,\text{s}^{-1}$ and $\eta \sim 3 \times 10^2\ \text{cm}^2\,\text{s}^{-1}$ at $0.62R_\odot$ (see the estimates of Gough in Chapter 1, particularly Table 1.1). Following Spruit (1999, 2002) we assume

$$N \gg \Omega \gg \Omega_{\text{A}}, \tag{8.21}$$

the first of which is well satisfied with thermal buoyancy frequency $N \sim 10^{-3}\,\text{s}^{-1}$, and $\Omega \sim 3 \times 10^{-6}\,\text{s}^{-1}$. The second is also well satisfied because, defining the dimensionless thermal diffusivity and Prandtl–Rossby ratio by

$$K = \kappa/N\ell^2 \sim 10^{-10}, \qquad P = \Omega/N \sim 3 \times 10^{-3}, \tag{8.22}$$

with $\kappa = 1\times10^7\ \text{cm}^2\,\text{s}^{-1}$ and $\ell = 10^2$ Mm $= 10^{10}$ cm, we have[6] from Equation (19) of Spruit (2002) that

$$\Omega_{\text{A}}/\Omega = q^{1/2}(KP)^{1/8} \ll 1, \tag{8.23}$$

since the dimensionless shear $q \lesssim 1$ even with extreme assumptions, as will emerge shortly. Now a slight rearrangement of Equations (10) and (32) of Spruit (2002) produces[7]

$$\nu_{\text{e}} = \ell^2\Omega(KP)^{1/2} = \Omega\,\delta_\kappa^2 \sim 1.6 \times 10^8\ \text{cm}^2\,\text{s}^{-1}, \tag{8.24}$$

where

$$\delta_\kappa = \ell(KP)^{1/4} \sim 0.7 \times 10^{-1}\ \text{Mm}. \tag{8.25}$$

The second formula for ν_{e} in (8.24) shows at once why the lowermost turbulent layer of thickness $\sim \delta_\kappa$ will have the characteristics of an Ekman layer, point (vii) ff.

[6] As long as the small-scale dynamo's toroidal magnetic field is expressed as the Alfvén angular velocity Ω_{A}, the spherical and cylindrical radii, as such, do not enter any of the formulae being quoted from Spruit (2002). The significance of the symbol r in Spruit (2002) is always that of the available latitudinal lengthscale. That is why the formulae are written here using ℓ in place of r.

[7] All these expressions depend on Equation (49) of Spruit (1999), after correcting a typographic error: the last occurrence of N should be Ω; see Equation (A29) of Spruit (1999) and the footnote on p. 927 of Spruit (2002). We may note also that the statement on p. 194*b* of Spruit (1999) that 'rotation does not by itself remove the instability' is made in the wrong context, that of zero diffusivities. The statement is correct for the real-world *diffusive* problems of interest here, but incorrect for a diffusionless problem. This latter point is illustrated by Equation (10.15) in Chapter 10, which is for the kink or tipping mode, azimuthal wavenumber $m = 1$, of a diffusionless Tayler instability in the case of solid background kinetic rotation Ω and Alfvénic rotation Ω_{A}. In that diffusionless case, $\Omega \geqslant \Omega_{\text{A}}$ implies stability.

above, since not only is δ_κ independent of q, the 'coincidence' mentioned in point (ii) above, but also, by a further coincidence, δ_κ is the same as the Ekman thickness scale $(\nu_e/\Omega)^{1/2}$.

The dimensionless shear threshold or critical shear for the small-scale dynamo to operate is, from Equation (27) of Spruit (2002),

$$q_{\text{crit}} = K^{1/4} P^{-7/4} (\eta/\kappa) \sim 2.5 \times 10^{-3} \tag{8.26}$$

at $0.62R_\odot$. Reading Ω values from the horizontal contours in Figure 3.7 of Chapter 3, we see that Ω goes from about 390 to 430 nHz, corresponding to a fractional change

$$\alpha = (430 - 390)/410 = 1 \times 10^{-1}, \tag{8.27}$$

from which we may derive a nominal q value, with the conservative choice $\Delta =$ 65 Mm,

$$q = \alpha\ell/\Delta \sim 1.5 \times 10^{-1}. \tag{8.28}$$

Even with such a large Δ this nominal shear is nearly two orders of magnitude greater than q_{crit}. However, as already noted, the stress and therefore the actual shear, in the bulk of the stably stratified tachocline, is tightly linked by (8.19) to the downwelling velocity w_{Ek}, which must equal the downwelling velocity $w_{\kappa\eta}$ that can be accepted by the thermomagnetic boundary layer. That is why the actual shear, in the stably stratified tachocline, is likely to be far smaller than the nominal shear just computed – though still dynamically significant, sharply distinguishing the tachocline from the interior – and why a tachocline 65 Mm deep could be consistent with the high-latitude Ω contours in Figure 3.7 of Chapter 3, despite appearances.

We assume that the estimate of $w_{\kappa\eta}$ by Gough & McIntyre (1998) is correct in order of magnitude:

$$|w_{\kappa\eta}| \sim \eta/\delta_{\kappa\eta} \propto |\mathbf{B}_\text{i}|^{1/3} \propto V_{\text{Ai}}^{1/3}, \tag{8.29}$$

where V_{Ai} is the interior Alfvén speed corresponding to $|\mathbf{B}_\text{i}|$, about $0.4\,\text{cm s}^{-1}$ per gauss near $0.62R_\odot$, with density $\rho \sim 0.42\,\text{g cm}^{-3}$, and where the boundary-layer thickness scale is

$$\delta_{\kappa\eta} = K^{1/3} \left(\frac{\eta}{\kappa}\right)^{1/6} \left(\frac{\Omega\,\ell}{V_{\text{Ai}}}\right)^{1/3} \ell, \quad \propto (\kappa\eta)^{1/6}. \tag{8.30}$$

Equating w_{Ek} to $w_{\kappa\eta}$ and using (8.17) and (8.19), with $\partial u/\partial z \sim \ell\,\partial\Omega/\partial z$, we have

$$q \sim \ell \left.\frac{\mathrm{d}(\ln\Omega)}{\mathrm{d}z}\right|_{\text{bulk}} = \ell\frac{w_{\text{Ek}}}{\nu_e} \sim K^{-1/3} \left(\frac{\eta}{\nu_e}\right) \left(\frac{\kappa}{\eta}\right)^{1/6} \left(\frac{V_{\text{Ai}}}{\Omega\,\ell}\right)^{1/3}, \tag{8.31}$$

equivalently

$$V_{\text{Ai}} \sim \Omega\,\ell\,q^3 K \left(\frac{\nu_e}{\eta}\right)^3 \left(\frac{\eta}{\kappa}\right)^{1/2}. \tag{8.32}$$

For an extreme value $|q| \sim 1$ this would imply an impossibly large $|\mathbf{B}_\mathrm{i}|$ of the order of thousands of megagauss, again suggesting that $|q| \ll 1$ and further supporting our earlier assumption (8.23). It should be cautioned, however, that Gough & McIntyre's (1998) scaling relation (8.29) has yet to be verified by a full analysis of the boundary-layer structure, and indeed $\delta_{\kappa\eta}$ and therefore Equations (8.31)–(8.32) might well change at high $|\mathbf{B}_\mathrm{i}|$ values, because Maxwell stresses then modify the meridional momentum balance assumed by Gough & McIntyre (see also Chapter 7). For $|q| = q_\mathrm{crit}$ we have a more reasonable value $V_\mathrm{Ai} = V_\mathrm{Ai(crit)}$, say, corresponding to $|\mathbf{B}_\mathrm{i}| \sim 10^2\,\mathrm{G}$. This follows from (8.24), (8.26) and (8.32):

$$\begin{aligned} V_\mathrm{Ai(crit)} &= \Omega^4 \ell^7 K^{13/4} P^{-15/4} \eta^{1/2} \kappa^{-7/2} \\ &= \Omega\, \ell K^{1/4} P^{-3/4} (\eta/\kappa)^{1/2} \sim 0.4 \times 10^2\,\mathrm{cm}^2\,\mathrm{s}^{-1}, \end{aligned} \tag{8.33}$$

implying in turn that $\delta_{\kappa\eta} \sim 0.7 \times 10^{-1}\,\mathrm{Mm}$ and $w_{\kappa\eta} = w_{\mathrm{Ek}} \sim -4 \times 10^{-5}\,\mathrm{cm\,s}^{-1}$. This magnitude $V_\mathrm{Ai(crit)} \sim 0.4 \times 10^2\,\mathrm{cm}^2\,\mathrm{s}^{-1}$ or $|\mathbf{B}_\mathrm{i}| \sim 10^2\,\mathrm{G}$ represents the critical order of magnitude of $\mathbf{B}_\mathrm{i}$ above which the stably stratified, high-latitude tachocline can continuously sustain small-scale dynamo action and below which the dynamo action would have to be intermittent, point (iv) above.

A curious aspect of the scaling (8.33) is the implication that $V_\mathrm{Ai(crit)} = \ell\,\Omega_\mathrm{A}$ at threshold. One may see this by substituting (8.26) into (8.23). Therefore the critical magnitude of $\mathbf{B}_\mathrm{i}$ – whose most important component for this purpose is the *poloidal* component, as explained by Gough & McIntyre (1998) – coincides with the order of magnitude of the Tayler-unstable eddy *toroidal* field of the small-scale dynamo. Furthermore, we see from (8.30) and (8.33) that $V_\mathrm{Ai} = V_\mathrm{Ai(crit)}$ implies $\delta_{\kappa\eta} \sim \delta_\kappa$. It seems that, just above threshold, the scaling for the small-scale dynamo eddies is the same as Gough & McIntyre's scaling for the thermomagnetic boundary layer. This is perhaps not unreasonable since both structures have shallow aspect ratios δ_κ/ℓ, $\delta_{\kappa\eta}/\ell$ and both, at threshold, feel not only a strong Coriolis effect but also the magnetic as well as the thermal diffusivity. We may also note from Equation (3) of Spruit (2002) that, under these threshold conditions, the eddy timescale for the small-scale dynamo, i.e. the growth time for the Tayler instability, is $\Omega/\Omega_\mathrm{A}{}^2 \sim 10^4\,\mathrm{yr}$ – fast from some viewpoints and slow from others.

A full analysis of the double-boundary-layer structure is beyond our scope here, and awaits further investigation. However, in the lower portion of the modified Ekman layer, where we are provisionally supposing that the eddy viscosity falls off like z^2 as the small-scale dynamo runs out of headroom, points (vii) and (viii) above, the Ekman-layer equations have complex power-law solutions that give some idea of the structure. Figure 8.1 shows some possible profiles of σ, u, τ and v. These satisfy Equation (8.18) with $\nu_\mathrm{e} \propto z^2$ together with the standard Ekman-layer

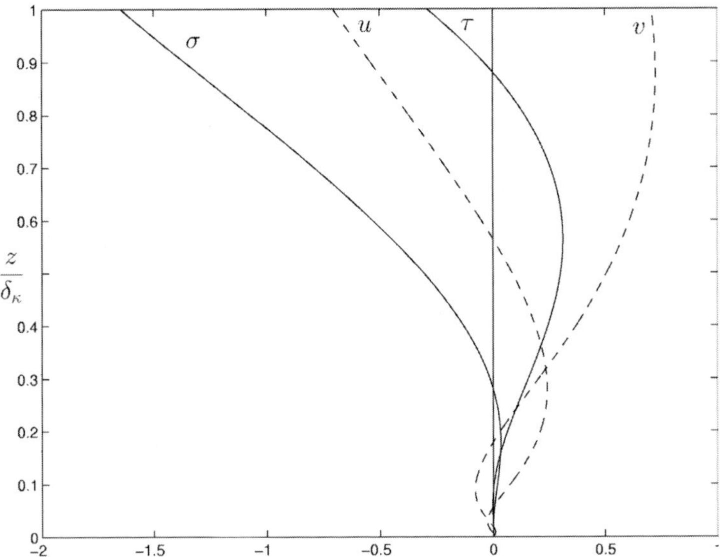

Figure 8.1. Solutions of Equations (8.18) and (8.34) regarded as an idealized model of the lower portion of the modified Ekman layer where the eddy viscosity $\propto z^2$. Somewhat arbitrarily, this lower portion is assumed to occupy a layer of thickness δ_κ, in dimensionless coordinates $0 \leqslant z \leqslant 1$. Again somewhat arbitrarily, the eddy viscosity in (8.18) is taken to have reached the value $\frac{1}{2}\nu_{e\infty}$ at $z = 1$, where $\nu_{e\infty}$ denotes the asymptotic value $\sim 1.6 \times 10^8\,\mathrm{cm^2\,s^{-1}}$ in the bulk of the stably stratified tachocline above. The solutions finite at $z = 0$ are then $(u, v) = (\mathrm{Re}, \mathrm{Im})\, e^{3i\pi/4} z^a$ and $(\sigma, \tau) = z^2 \partial(u, v)/\partial z = (\mathrm{Re}, \mathrm{Im})\, a e^{3i\pi/4} z^{a+1}$, in dimensionless units, where $a = \frac{1}{2}\{-1 + \surd(1 + 16\mathrm{i})\} = 0.9591 + 1.3707\mathrm{i}$. In fact there is a one-parameter family of solutions with $a = \frac{1}{2}\{-1 + \surd(1 + 4\mathrm{i}C)\}$, where $C = 4z_{1/2}^2$ with $z_{1/2}$ the dimensionless altitude at which the eddy viscosity reaches the value $\frac{1}{2}\nu_{e\infty}$. Such solutions cannot describe the upper portion of the layer where the viscosity profile approaches its asymptotic value $\nu_{e\infty}$, nor can they correctly describe the fine details near the bottom of the real modified Ekman layer where it interfaces with the thermomagnetic boundary layer. This is because the dynamo runs out of headroom somewhere *above* $z = 0$, depending on $|\mathbf{B}_\mathrm{i}|$ values. The saving grace, however, is that the mass-flux relation (8.19) depends only on σ going to zero somehow, and not on the detailed vertical structure.

equations

$$-2\Omega v = \partial\sigma/\partial z, \qquad 2\Omega u = \partial\tau/\partial z; \tag{8.34}$$

see caption for further details. The profiles give what seems to be a qualitatively reasonable description of the lower portion of the modified Ekman layer, showing how the shears can stay finite and the stresses go to zero as $z \to 0$. There is another set of complex power-law solutions, rejected as unphysical, for which the shears and stresses go to infinity as $z \to 0$.

In the upper portion of the layer, not shown, where the power-law solutions cease to apply, as the eddy–viscosity profile departs from its z^2 dependence and begins to approach its asymptotic value $\nu_{e\infty} \sim 1.6 \times 10^8\,\mathrm{cm^2\,s^{-1}}$ in the bulk of the stably stratified tachocline above, we can imagine the profiles being smoothly continued upward with τ and v making an oscillatory approach to zero in the usual manner of Ekman profiles. The azimuthal stress σ must continue toward its asymptotic negative value $\nu_{e\infty}\,\partial u/\partial z|_{\mathrm{bulk}}$, and the azimuthal shear $\partial u/\partial z$ toward a corresponding negative value, smaller in magnitude than in the portion of the u profile visible in Figure 8.1, point (vii) above. Again one expects an oscillatory approach toward these asymptotes. A consistent description of the upward continuation requires the second of (8.34) to be replaced by an equation corresponding to a steady, variable-viscosity version of (8.2c) with a small but significant thermal-wind term, as already hinted by the scaling relation $\delta_{\kappa\eta} \sim \delta_\kappa$. That is a further sense in which the Ekman layer is 'modified'. The v profile and its upward continuation describe, of course, the gyroscopically-pumped poleward flow.

The issue of tachocline ventilation turns out to involve subtleties that depend on the effects of compositional stratification N_μ. So we discuss the latter first.

8.5 The effects of compositional stratification N_μ

As already emphasized, the helium settling layer just beneath the tachocline owes its existence to the suppression of global-scale HSZ burrowing by the interior field $\mathbf{B}_\mathrm{i}$. Once the settling layer has formed, the vertical gradient of mean molecular weight μ adds a contribution

$$N_\mu^2 = -g\,\partial \ln \mu/\partial z \tag{8.35}$$

to the buoyancy frequency squared that is a significant fraction of the typical thermal value $N^2 \sim 10^{-6}\,\mathrm{s^{-2}}$. For instance, a standard solar model (Figure 3.4 of Chapter 3) gives a fractional contrast $\mathrm{d}\ln\mu = 0.014$ across the settling layer and a corresponding reduced gravity $g' = 0.014g \sim 1 \times 10^3\,\mathrm{cm\,s^{-2}}$. Measuring the slope shown in the inset to Figure 3.4a, one gets $\partial/\partial z \sim (0.05R_\odot)^{-1} \sim (35\,\mathrm{Mm})^{-1}$; so $N_\mu^2 \sim g'/0.05R_\odot \sim 0.3 \times 10^{-6}\,\mathrm{s^{-2}}$, or $N_\mu \sim 0.5 \times 10^{-3}\,\mathrm{s^{-1}}$. However, neither $\mathrm{d}\ln\mu$ nor N_μ^2 can really be said to be known precisely, because the helioseismic evidence is undergoing revision, as discussed in Chapter 3, though still generally supporting the existence of the settling layer. It is possible that the real settling layer may be somewhat deepened, with $\partial/\partial z$ perhaps more like $(100\,\mathrm{Mm})^{-1}$, by the weak and highly sporadic interior turbulent mixing discussed in Section 8.3. Furthermore, the overall μ contrast across the layer could be somewhat bigger than indicated by the number $\mathrm{d}\ln\mu = 0.014$, if the same weak mixing were even slightly effective in bringing up helium-rich gas from the core, on the gigayear timescale.

Fortunately, however, the following arguments depend only on very rough orders of magnitude for $\mathrm{d}\ln\mu$ and N_μ^2.

The main issue is whether HSZ burrowing can penetrate the interior near the polar weak spots in $\mathbf{B}_\mathrm{i}$, as speculated by Gough & McIntyre (1998). These are the zero points or 'hairy-sphere defects' of the vector field formed by the horizontal projection of $\mathbf{B}_\mathrm{i}$. If such burrowing were possible, then it could create 'polar pits' or 'cauldrons', in which lithium could be burned even if Δ were less than 65 Mm. The most favourable conditions for such burrowing would be that $|\mathbf{B}_\mathrm{i}|$ is altogether negligible near the poles. We ask whether, in that most favourable case, the burrowing could locally penetrate the helium settling layer in those neighbourhoods. It will appear that the answer is a clear 'no'.

The non-magnetic slab model of Section 8.2 is sufficient to reveal the essential effects, which turn out to be insensitive to the choice of horizontal scale ℓ. The only changes needed are to replace the thermal buoyancy acceleration ϑ by the total buoyancy acceleration $\vartheta + \vartheta_\mu$ in (8.2c), and to append an equation for the compositional buoyancy acceleration ϑ_μ. In the latter equation we may safely neglect all diffusive effects, which are tiny.[8] By analogy with the thermal buoyancy acceleration we define ϑ_μ as g times the fractional departure of μ from its background stratification, so that the equation for ϑ_μ is

$$\frac{\partial \vartheta_\mu}{\partial t} + N_\mu^2 w = 0, \tag{8.36}$$

in which we idealize by taking N_μ = constant. Simplifying (8.2c) as before, we have the appropriate form of the thermal-wind equation,

$$2\Omega\,\frac{\partial u}{\partial z} + \frac{\partial(\vartheta + \vartheta_\mu)}{\partial y} = 0, \tag{8.37}$$

and readily find that (8.15) is replaced by

$$\left\{\left(\frac{\partial}{\partial t} - \kappa_\mu \frac{\partial^2}{\partial z^2}\right)\frac{\partial^2}{\partial y^2} - \frac{4\Omega^2\kappa}{N^2}\frac{\partial^4}{\partial z^4}\right\}\frac{\partial u}{\partial t} = \left(\frac{\partial}{\partial t} - \kappa_\mu \frac{\partial^2}{\partial z^2}\right)\frac{\partial^2 \bar{F}}{\partial y^2}, \tag{8.38}$$

where $\kappa_\mu = \kappa N_\mu^2/N^2$. The microscopic viscosity ν has been neglected, since it is nearly as small as the helium self-diffusivity, $\chi \sim 10^1\ \mathrm{cm^2\,s^{-1}}$, which has already been neglected in (8.36).

Now the key point is that (8.38) is an equation for $\partial u/\partial t$ and not for u. The $\partial/\partial t$ is a crucial and essential feature, coming from the need to eliminate ϑ_μ between (8.36) and (8.37). By contrast, ϑ is eliminated as in the derivation of (8.15), via (8.2d) with its $\partial/\partial t$ neglected. That is appropriate because of the enormous magnitude of $\kappa \sim 10^7\ \mathrm{cm^2\,s^{-1}}$ relative to χ.

[8] For instance $\chi \sim 10^1\ \mathrm{cm^2\,s^{-1}}$, where χ is the self-diffusivity of helium anomalies in the appropriate hydrogen–helium mixture (see Chapter 1, Table 1.1).

With a relatively small horizontal scale ℓ – as before, we consider slab-model solutions sinusoidal in y/ℓ – one might think at first that the new quasi-diffusive term in κ_μ signals the possibility of burrowing straight down into the helium settling layer. But appearances are deceptive here.

Consider a thought-experiment in which the forcing is switched on at time zero. The time-dependent solutions of (8.38) below the forcing layer, right-hand side zero, describe burrowing that commences in just the same way as with the Spiegel–Zahn equation (8.15). The hyperdiffusive term dominates the new quasi-diffusive term in the earliest stages, in which the vertical scale increases from zero. As the disturbance penetrates more deeply, however, the quasi-diffusive term comes into balance with the hyperdiffusive term. Thus $\partial u/\partial t$ reaches a steady state with vertical structure $\exp(z/h)$, where the vertical scale h is given by $h = (2\Omega\ell/N)(\kappa/\kappa_\mu)^{1/2} = 2\Omega\ell/N_\mu$. This is just the (non-diffusive) Rossby height belonging to the horizontal scale ℓ, and the response, from then onward, is nothing but the well-known Eliassen response to gyroscopic pumping in a non-diffusive stratified fluid. Its most important feature, for our purposes, is that $\partial u/\partial t$ and $\partial\vartheta_\mu/\partial t$ are steady, not u and ϑ_μ. The other variables ϑ, v and w are all steady. The response consists of perpetual spin-down, with u and ϑ_μ asymptotically proportional to t.

This means, of course, that the response is self-limiting, in one of two possible ways. The first way is for the spin-down to continue – with the u and ϑ_μ terms in (8.37) asymptotically proportional to t – until the compositional stratification surfaces are overturned and the stratification is wiped out. That is what would have taken place on a global scale, preventing the helium settling layer from forming at all, had there been no $\mathbf{B}_\text{i}$ and no Ferraro constraint. Such a response is a nonlinear response, outside the scope of our linearized equations.

The second way, which is the one relevant here, is well within the scope of the equations. If, as here, the gyroscopic pumping is ultimately the result of the convection zone's propensity to rotate differentially, then there is a saturation value beyond which the spin-down cannot proceed, having taken up all the available differential rotation and thus killed off the gyroscopic pumping. We may say that the underlying layers are fully spun-down. Just what the final saturation value might be is difficult to say, but one may reasonably suppose that spin-down cannot proceed beyond limits governed by the value of α in Equation (8.27), $\alpha \sim 10^{-1}$, the fractional angular-velocity increment across the whole tachocline. It is easy to verify (see the Margules slope estimate below) that such limits are essentially zero for present purposes. They tell us that the self-limiting of the Eliassen response would take place with hardly any tilting of the compositional isopleths.

In other words, for realistic α the helium settling layer spanning the poles presents an almost perfect barrier against HSZ burrowing. That is why the polar pits cannot be dug.

Two further points need comment. The first is that Equation (8.38) also admits perpetual-spin-down solutions with a linear dependence on z, such as $u \propto t\ \sin(y/\ell)$ and $u \propto zt\ \sin(y/\ell)$. These, however, fail to satisfy physically reasonable boundary conditions. For instance the first of them requires both $|\vartheta|$ and $|\kappa\,\partial\vartheta/\partial z|$ to increase like t at the top boundary, if $w = 0$ at some bottom boundary. This is because ϑ has to be asymptotically proportional to $zt\ \cos(y/\ell)$ in order to avoid violating (8.37), in which $\partial u/\partial z = 0$ despite the perpetual tilting of μ-surfaces, implying $\vartheta = -\vartheta_\mu$. The second solution has a similar pathology and, furthermore, does not even permit $w = 0$ at the bottom, since it can be shown to imply a z-independent contribution to w, $\propto t\ \cos(y/\ell)$. Also, both types of solution would disappear if any horizontal heat diffusion were allowed. So we need not consider them further.

The second point is that the tilting of compositional isopleths or stratification surfaces is so small, in fact, that it tightly constrains vertical displacements of the tachopause even on a global scale. We have just found that HSZ burrowing is ineffective even at the polar weak spots of $\mathbf{B}_\mathrm{i}$. Still less is it effective in the rest of the interior where the Ferraro constraint has control. There is no MMC to tilt the thermal, or overturn the compositional, stratification surfaces. The implication is that those surfaces must be accurately horizontal, in the sense that they accurately follow the gravitational–centrifugal heliopotentials.

As a check on that assertion, and to get some idea of its error bar, let us calculate the tilting of compositional stratification surfaces that would occur if they alone were tilted and if the Ferraro constraint were artificially relaxed, to permit a thermal-wind shear across the helium settling layer of the same order as the shear across the whole of the high-latitude tachocline. The slope can be obtained from the thermal-wind equation (8.37) evaluated with ϑ_μ alone, or equivalently and more directly from the Margules slope formula $2\Omega U/g'$, where $U \sim \Omega\alpha\ell$, the velocity increment across the layer, and α is the fractional angular-velocity increment as before. Even with the extreme value $\alpha \sim 10^{-1}$ we have $U \sim 3 \times 10^3\,\mathrm{cm\,s^{-1}}$ and a Margules slope $2\Omega^2\alpha\ell/g' \sim 2 \times 10^{-5}$. The nominal elevation change over a distance $r \sim 500\,\mathrm{Mm}$ is only $10^{-2}\,\mathrm{Mm}$. The real elevation change from pole to equator, with the Ferraro constraint brought back into play, is therefore far, far smaller still – a very tiny fraction of a megametre indeed.

8.6 Concluding remarks

The main issue not yet addressed is that of the tachocline's ventilation timescale. This turns out to be by far the most delicate issue, and crude order-of-magnitude arguments are unable to decide it directly. Taken at face value, the threshold numbers used in Section 8.4 imply gigayear ventilation times. This is because the main ventilation mechanism is now turbulent mixing by the small-scale dynamo. From

Equations (15) and (19) or the first of (43) in Spruit (2002), we have an eddy diffusivity D for vertical material transport of the order of

$$D \sim q\,\Omega\,\ell^2\,P^{3/4}K^{3/4}, \tag{8.39}$$

in the notation of Section 8.4. If q is close to its threshold, $q_{\rm crit}$, then one may verify by substitution from (8.26) that D is of the same order as the microscopic magnetic diffusivity $\eta \sim$ 300–400 $\rm cm^2\,s^{-1}$. With $\Delta \sim 65$ Mm the nominal ventilation time Δ^2/η is then about the same as the Sun's Main Sequence lifetime, ~4 Gyr. Perhaps this is not an accident: could it be that the thickness of the tachocline is such that it can only just stay ventilated?

One can imagine playing games with factors like π^2 or taking one or two tens of megametres off the Δ value by assuming a deep overshoot layer, or one could suppose that the small-scale dynamo in the stably stratified tachocline is well above threshold, with the implication from (8.32) that $|\mathbf{B}_{\rm i}| \gg 10^2\,\rm G$. And when a full analysis of the double-boundary-layer structure becomes available, including a quantitative numerical model, then the net effect of the numerical factors might go one way or the other. As regards large $|\mathbf{B}_{\rm i}|$, there seems no reason why the Sun should not have an interior field as strong as that of ordinary (non-neutron) magnetic stars, which should allow us to consider $|\mathbf{B}_{\rm i}|$ values perhaps into the hundreds of kilogauss, magnetic escapology permitting. If, despite the cautionary remark below (8.32), the 1/3 power in (8.31) were to apply over the whole range of $|\mathbf{B}_{\rm i}|$ and $|V_{\rm Ai}|$ values, then we would be able to use $D \sim 10\eta$.

However, we may also invoke Occam's razor, appealing to the effects of compositional stratification discussed in Section 8.5. The key point is again the dynamical impossibility of significantly tilting the compositional isopleths in the helium settling layer. This presents a powerful barrier not only against the burrowing of MMCs but also against the turbulent erosion of heavy elements *into* the tachocline. Erosion rates must be severely limited by that circumstance alone. They will be further limited by the diffusive leakage of $\mathbf{B}_{\rm i}$ across the tachopause, and into the tachocline, in those latitude bands equatorward of the active high-latitude band where there is either no confining downwelling, or very weak downwelling such as might occur over the equator (point (v) in Section 8.4). In such latitude bands the Ferraro constraint will reach across the tachopause, now defined as the top of the helium settling layer, and will tend to suppress shear across it and protect it from any kind of erosion. So the tachocline could be helium-poor, therefore, not so much because of fast ventilation from above, as in the scenario of Gough & McIntyre (1998), but because of minuscule erosion rates of heavy elements across such a heavily protected compositional tachopause.

There remains, however, the lithium problem, which of itself still argues for substantial ventilation. Further discussion must await detailed solutions of the nonlinear equations for the double-boundary-layer structure, as well as a more quantitative description of the small-scale dynamo.

One final twist in the tail of this tale. The *visible* shear at the top of the high-latitude tachocline – visible, for instance, in Figure 3.7 of Chapter 3 and already indicated by helioseismology to occupy mainly the lower convection zone and its overshoot layer (see Chapters 1 and 3) – would be dynamically impossible in the presence of the small-scale dynamo. That is very clear from Spruit's order-of-magnitude relations as used in Section 8.4, to the extent that Equation (8.29) correctly indicates the thermomagnetic boundary layer's mass-carrying capability. At first sight it might seem paradoxical: 'surely the lower convection zone and overshoot layer are much more turbulent?' But that would be to underestimate the power of the Maxwell stresses in the small *vertical* scale dynamo, arising from the large *horizontal* reach of its eddy structures via the Tayler instability's kink or tipping-type kinematics, dominated by azimuthal wavenumber $m = 1$, and reflected in the large vertical eddy viscosity ν_e. So the suggestion must be that the convective plumes break up that horizontal structure, disconnecting and reconnecting the wound-up field lines in such a way as to drastically reduce the eddy viscosity and permit much larger shears.

Acknowledgments

I thank the organizers, in alphabetical order Pascale Garaud, Douglas Gough, David Hughes, Bob Rosner, Nigel Weiss, and Jean-Paul Zahn, for inviting me to participate in this interesting and stimulating workshop. Jørgen Christensen-Dalsgaard kindly tutored me on everything about solar models, Pascale Garaud stimulated me to re-examine the thermomagnetic boundary-layer scaling in Gough & McIntyre (1998), leading to the cautionary remarks about large $|\mathbf{B}_i|$, Nigel Weiss patiently helped me to improve my general knowledge of MHD and other astrophysical processes, and at the eleventh hour Douglas Gough kindly shared with me his careful calculations of diffusivity values as well as offering wise and helpful comments on many aspects of the problem. Above all, however, on this occasion, I am grateful to Henk Spruit for an extensive correspondence that helped me to improve my understanding of the Tayler-mediated small-scale dynamo and of magnetic stability, instability, and escapology.

References

Acheson, D. J. (1978). *Phil. Trans. R. Soc. Lond.*, A**289**, 459.
Braithwaite, J. & Spruit, H. C. (2004). *Nature*, **431**, 819.
Braithwaite, J. & Spruit, H. C. (2006). *Astron. Astrophys.*, in press.

Brun, A. S. & Zahn, J.-P. (2006). *Astron. Astrophys.*, submitted.
Charbonneau, P. & MacGregor, K. B. (1993). *Astrophys. J.*, **417**, 762.
Christensen-Dalsgaard, J., Gough, D. O. & Thompson, M. J. (1992). *Astron. Astrophys.*, **264**, 518.
Elliott, J. R. & Gough, D. O. (1999). *Astrophys. J.*, **516**, 475.
Fricke, K. (1969). *Astron. Astrophys.*, **1**, 388.
Garaud, P. (2002). *Mon. Not. Roy. Astron. Soc.*, **329**, 1.
Gilman, P. A. & Miesch, M. S. (2004). *Astrophys. J.*, **611**, 568.
Gough, D. O. & McIntyre, M. E. (1998). *Nature*, **394**, 755.
Haynes, P. H., Marks, C. J., McIntyre, M. E., Shepherd, T. G. & Shine, K. P. (1991). *J. Atmos. Sci.*, **48**, 651. See also *J. Atmos. Sci.*, **53**, 2105.
Manney, G. L., Zurek, R. W., O'Neill, A. & Swinbank, R. (1994). *J. Atmos. Sci.*, **51**, 2973.
McIntyre, M. E. (1994). In *The Solar Engine and its Influence on the Terrestrial Atmosphere and Climate* (Vol. **25** of NATO ASI Subseries I, Global Environmental Change), ed. E. Nesme-Ribes. Heidelberg: Springer, p. 293.
McIntyre, M. E. (2003a). In *Stellar Astrophysical Fluid Dynamics*, ed. M. J. Thompson & J. Christensen-Dalsgaard. Cambridge: Cambridge University Press, p. 111.
McIntyre, M. E. (2003b). In *Perspectives in Fluid Dynamics*, ed. G. K. Batchelor, H. K. Moffatt & M. G. Worster. Cambridge: Cambridge University Press, p. 557.
Mestel, L. & Weiss, N. O. (1987). *Mon. Not. Roy. Astron. Soc.*, **226**, 123.
Riese, M., Manney, G. L., Oberheide, J. *et al.* (2002). *J. Geophys. Res.*, **107** (D23), Art. No. 8179.
Spiegel, E. A. & Zahn, J.-P. (1992). *Astron. Astrophys.*, **265**, 106.
Spruit, H. C. (1999). *Astron. Astrophys.*, **349**, 189.
Spruit, H. C. (2002). *Astron. Astrophys.*, **381**, 923.
Townsend, A. A. (1958). *J. Fluid Mech.*, **4**, 361.
Zahn, J.-P. (1974). In *Stellar Instability and Evolution*, ed. P. Ledoux, A. Noels & A. W. Rodgers. Dordrecht: Reidel, p. 185.

9

β-Plane MHD turbulence and dissipation in the solar tachocline

Patrick H. Diamond, Sanae-I. Itoh, Kimitaka Itoh & Lara J. Silvers

The physical processes causing the turbulent dissipation and mixing of momentum and magnetic fields in the solar tachocline are discussed in the context of a simple model of two-dimensional MHD turbulence on a β-plane. The mean turbulent resistivity and viscosity for this model are calculated. Special attention is given to the enhanced dynamical memory induced by small scale magnetic fields and to the effects of magnetic fluctuations on nonlinear energy transfer. The analogue of the Rhines scale for β-plane MHD is identified. The implications of the results for models of the solar tachocline structure are discussed.

9.1 Introduction

The tachocline is a thin, stably stratified layer of the solar interior situated in the radiative zone, immediately below the convection zone (Miesch 2005; Tobias 2005). This layer connects the latitudinal differential rotation of the solar convection zone to the expected solid body rotation of the solar interior (Schou *et al.* 1998; see also Chapter 3 in this book by Christensen-Dalsgaard & Thompson). Thus, flows in the tachocline are sheared (both poloidally and radially), with the predominant structure being that of a radially sheared toroidal flow. The stratification of the tachocline is strongly stable (with Richardson number $Ri \gg 1$), and the magnetic field strength is significant, though magnetic pressure is still much smaller than thermal pressure, consistent with hydrostatic equilibrium, i.e. $B^2/8\pi \ll p$.

In addition to its intrinsic interest, the tachocline has received considerable attention recently on account of its pivotal role in the proposed interface dynamo of the Sun (Parker 1993; see also Chapter 13 by Tobias & Weiss). Interest in the interface dynamo has been sparked by the many fundamental questions recently raised concerning the mean field $\alpha\omega$ theory of the solar dynamo and its traditional constituents,

The Solar Tachocline, D. W. Hughes, R. Rosner and N. O. Weiss.
Published by Cambridge University Press.

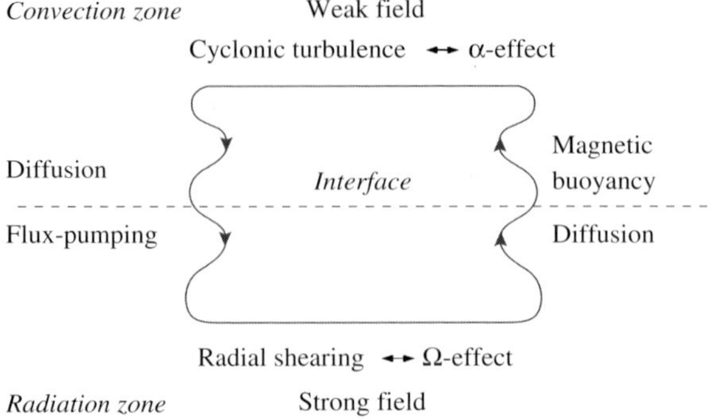

Figure 9.1. A schematic of the interface dynamo cycle, which operates near the boundary of the convection and radiation zones. In this cartoon the tachocline sits just below the convection zone. Magnetic field is amplified by an α-effect in the convection zone, 'pumped' into the tachocline by convective overshoot, amplified by shearing in the tachocline, and returned to the convection zone by magnetic buoyancy.

the alpha (α) and turbulent diffusivity (η_T) effects.[1] In this regard, possible quenching of η_T and α in inverse proportion to the product of magnetic Reynolds number (Rm) and mean magnetic field intensity (B_0^2), (i.e. $\sim(1+RmB_0^2)^{-1}$), is a particularly strong motivation to consider alternative dynamo scenarios (Cattaneo & Vainshtein 1991; Gruzinov & Diamond 1994; Diamond *et al.* 2005a). The interface dynamo concept proposes an escape from quenching by separating the location of cyclonic turbulence (which drives the α-effect) from the site of shearing (ω-effect) and the consequent strong magnetic field build-up. In the interface dynamo, weak poloidal fields are amplified by α in the convection zone, then transported to the tachocline by either turbulent diffusion or entrainment by convectively overshooting plumes (a process referred to as 'magnetic flux pumping'; Tobias *et al.* 2001). Once in the tachocline, the magnetic field is amplified by flow shearing to form strong toroidal loops. A cartoon of the interface dynamo scenario is shown in Figure 9.1. The toroidal field remains stored below the convection zone, until it erupts upward into the convection zone, and eventually into the solar atmosphere (Parker 1966; Hughes 1991). Also, the nature of field amplification in the interface dynamo is such that the overall sensitivity to the value of α is reduced. In the interface scenario α is required only to convert toroidal field to poloidal, while toroidal field is actually amplified by tachocline shear. This state brings to mind the familiar contrast between α^2 and $\alpha\omega$ dynamos. In α^2 dynamos, the field growth rate $\gamma \sim \alpha$ while for $\alpha\omega$ dynamos $\gamma \sim \sqrt{\alpha\Omega}$ with the reduced sensitivity to α evident. Given the many attractive features of the interface dynamo, it is no stretch to say that understanding

[1] Note that the latter is often denoted by β. To avoid confusion in this chapter we use η_T.

the tachocline is an important prerequisite for a theory of the solar dynamo, solar cycle, etc.

Perhaps the most basic and important questions concerning the tachocline are:

(i) why does it exist?
(ii) where is it located and why is it so sharply localized?

Regarding existence, the tachocline is formed by the penetration of a shear into the stably stratified radiation zone. This process of sheared flow penetration is ultimately driven by solar spin-down, i.e. the loss of angular momentum from the Sun on account of its coupling to the outgoing, rotating, solar wind (Bretherton & Spiegel 1968; Spiegel & Zahn 1992), or, alternatively, by the stresses maintaining latitudinal differential rotation in the convection zone (see Chapter 1 in this volume by Gough). Spin-down or stresses in turn generate meridional circulation cells, which drive the inward penetration of shear. Alternatively put, approximate *thermal wind balance* tightly links radiative heat transport-driven baroclinic torques to the vertical variation of the centrifugal force (Mestel 1999). Any small deviation from exact thermal wind balance necessarily implies meridional circulation, which must exist to balance the torque budget. Thus, it may be said that meridional cells are spawned by the competition between stratification (N^2) and rotation (Ω^2), in the sense that these two effects directly compete against one another in the thermal wind balance. This process of meridional flow-driven 'burrowing' is sometimes referred to as *gyroscopic pumping* (McIntyre 2000, 2003), and is similar to the well-known mechanism of the Eddington–Sweet circulation in convectively stable stellar interiors (Eddington 1926; Sweet 1950). Indeed, the basic timescale of tachocline penetration is the Eddington–Sweet time scale $\tau_{\mathrm{ES}} = (N/2\Omega)^2 r_0^2/\kappa$, where r_0 is the radius of the tachocline boundary, κ is the thermal diffusivity, N is the Brunt–Väisälä frequency and Ω is the rotation frequency.

The second question above, i.e. what limits or localizes the tachocline, is the more challenging one, by far. Radial mixing is ineffective, on account of the severe limitation imposed on it by the strong stable stratification in the solar radiation zone. It would just slightly enhance the Eddington–Sweet penetration rate, so making little or no difference in the outcome of that theory. Thus, one must turn to turbulent viscous mixing, as in the Spiegel–Zahn scenario (Spiegel & Zahn 1992), in which tachocline burrowing is balanced by 'horizontal' turbulent momentum transport, or to magnetic field effects, as in the Gough–McIntyre scenario (Gough & McIntyre 1998). In the latter scenario, tachocline penetration is opposed by a hypothetical fossil dipolar magnetic field in the solar radiation zone. The fossil dipole field is separated from the tachocline and convection zone by a magnetic separatrix or 'tachopause', at which nonlinear dissipative MHD processes (i.e. magnetic reconnection) are both crucially important and excruciatingly difficult to calculate. One key element in determining the tachocline thickness according to the Gough–McIntyre model is the balance of shearing of poloidal fields with resistive dissipation of toroidal magnetic fields. This magnetic dissipation can be either resistive (i.e. related to radial diffusion, as actually assumed by Gough & McIntyre) or turbulent, and so related to poloidal transport and mixing of magnetic fields. Thus, it is interesting to note *that in both tachocline formation scenarios, turbulent transport and dissipation play central roles*. Finally, we note here that other tachocline models exist but are beyond the scope of this discussion (see Chapter 7 by Garaud). For an alternative magnetic model see Rüdiger & Kitchatinov (1977).

Tachocline turbulence is quasi-geostrophic turbulence in a spherical shell, and is excited primarily by forcing due to convective overshoot. The forcing may also be thought of as a surrogate for the input of energy to smaller scales as a result of large scale instability. As we argue below, tachocline turbulence almost certainly has a strong magnetic component and thus should be thought of as quasi-geostrophic MHD turbulence, the *simplest* incarnation of which is β-plane MHD turbulence. Here, we consider the β-plane model, rather than a spherical surface or spherical shell model, for reasons of simplicity. Predictive understanding of turbulent transport and dissipation of *both* momentum (closely related to vorticity in two dimensions) *and* magnetic fields in β-plane MHD turbulence is necessary in order to construct tachocline formation models. *Such turbulent transport provides the key element of dissipation, which limits or offsets the meridional cell-driven 'burrowing'* (McIntyre 2003). This chapter discusses the physics of transport and dissipation in β-plane MHD. Both momentum and magnetic field transport are considered.

There are at least three specific reasons why understanding β-plane MHD turbulence is an interesting and challenging task. These are as follows.

(i) The simultaneous presence and coexistence of eddies, Rossby waves and Alfvén waves at different scales.

(ii) The freezing of magnetic potential and field into the fluid at high magnetic Reynolds number Rm. 'Freezing in' has been shown to severely limit turbulent diffusion of magnetic fields in two dimensions.

(iii) The tendency of even two-dimensional turbulence to stretch magnetic fields and thus to 'Alfvénize' the turbulence, producing a high intensity spectrum of small scale magnetic fields. Also, in two-dimensional MHD, energy forward cascades, rather than inverse cascades as in a two-dimensional fluid, once the magnetic field intensity exceeds a weak minimal level (Kraichnan 1965; Pouquet 1978).

These three observations in turn suggest the following.

(i) As in the case of ordinary geostrophic turbulence, a scale emerges in β-plane MHD turbulence that demarks the boundary between an 'eddy turbulence' range of scales

Table 9.1. *Elements of Tachocline Formation Scenarios*

Element	Spiegel–Zahn model	Gough–McIntyre model
Drive of tachocline formation	Spin-down, meridional circulation cells	Spin-down, meridional circulation cells
Tachocline limiter	Horizontal viscosity and momentum mixing	Fossil poloidal magnetic field in radiative core
Turbulent dissipation mechanism	Turbulent *viscosity* in quasi-geostrophic fluid or MHD	Turbulent *resistivity* in quasi-geostrophic two-dimensional MHD
Critical balance	$\nu_{\mathrm{h}}\nabla_{\mathrm{h}}^2\mathbf{v}$ vs. 'burrowing' of tachocline	$(\eta\partial_r^2 + \eta_{\mathrm{h}}\partial_{\mathrm{h}}^2)B_\phi$ vs. shearing of B_p into B_ϕ
Critical issues in turbulence physics	(1) Cascade direction (2) Rhines scale in β-plane MHD (3) Momentum transport	(1) Turbulent resistivity, quenching of η_{T} in β-plane MHD (2) $\langle A^2\rangle$ spectral transport direction

and a 'wave turbulence' range. *We shall show that in β-plane MHD, this analogue of the well-known Rhines scale can be Rm dependent* (Rhines 1975).

(ii) For scales $\ell < \ell_{\mathrm{RM}}$, where ℓ_{RM} is the β-plane MHD 'Rhines scale', the dynamics is essentially that of two-dimensional MHD turbulence. So, turbulence tends to 'Alfvénize', thus enhancing memory and quenching turbulent diffusion and dissipation of both momentum and field. Moreover, energy *forward* cascades, even in two-dimensional MHD.

(iii) For scales $\ell > \ell_{\mathrm{RM}}$, the dynamics is that of a gas of Rossby waves. In particular, turbulent transport is controlled by *wave interaction*, so the simplified conventional wisdom about two-dimensional MHD turbulence is not applicable.

The upshot of all this is that the actual *dynamics* of turbulent mixing of momentum and magnetic field in the tachocline is quite unclear, and that horizontal turbulent viscosity (ν_{h}) and resistivity (η_{h}) are either quenched or significantly reduced! This discussion indicates the need to seriously consider the micro-physics of turbulent transport in the tachocline environment when constructing models of tachocline formation (see Table 9.1).

The remainder of this chapter is organized as follows. Section 9.2 introduces the model and discusses some basic aspects of β-plane MHD, including the important Zeldovich theorem. The effective Rhines scale for β-plane MHD is discussed in Section 9.3. In particular, we show that in β-plane MHD at high Rm, the effective Rhines scale ℓ_{RM} increases with Rm. Spectral transfer and turbulent dissipation are discussed in Section 9.4. The eddy and Alfvén wave forward cascade range at scales $\ell < \ell_{\mathrm{RM}}$ and the Rossby wave dominated range at $\ell > \ell_{\mathrm{RM}}$ are dealt with separately. We show that for $\ell < \ell_{\mathrm{RM}}$, both ν_{h} and η_{h} are significantly reduced by the effects of small scale magnetic fields. We also discuss the effects of Rossby wave interactions on turbulent transport at larger scales. Section 9.5 discusses the implications of these results for tachocline structure formation scenarios.

9.2 Some basic aspects of two-dimensional MHD turbulence on a β-plane

Though virtually all previous studies of turbulence and turbulent transport in the tachocline have been in the context of neutral fluid models, tachocline turbulence is very likely *MHD turbulence*, or turbulence with a substantial magnetic component. This assumption is natural, given the strong toroidal magnetic field of the tachocline *and* the presence of magnetic field sources both above and below the tachocline. Specifically, the tachocline magnetic field is fuelled from above by overshooting plumes, which originate in the convection zone and which entrain convection zone magnetic fields while they fall into the stably stratified tachocline below the convection zone. Likewise, small elements or loops of the fossil field thought to reside in the solar radiation zone (as in the Gough–McIntyre scenario) may enter the tachocline following reconnection events, which occur at the tachopause, and which thus fuel the tachocline magnetic field from *below*. Thus, the readily available sources of magnetic field as well as the stable stratification and apparent minute thickness of the layer suggest that turbulence in the tachocline is *'shellular' MHD turbulence*, which is two-dimensional in character. Here we make the simplest of approximations and neglect the thickness of the shell, thus taking the dynamics to be two-dimensional. Since shellular turbulence in a layer of finite thickness can exhibit complex vertical couplings (as in multi-layer models; Pedlosky 1987), further simplification is desirable. Thus, we focus on the *absolutely minimal model* of such shellular MHD turbulence, *namely two-dimensional MHD turbulence on a β-plane* (Bracco *et al.* 1998). This system includes constituents of both geostrophic turbulence (i.e. Rossby waves, vortices) and two-dimensional MHD turbulence (i.e. Alfvén waves etc). Despite the simplicity of this minimalist model, developing a theory of β-plane MHD turbulence is still useful for tachocline modelling, since such a theory can constrain and elucidate the physics of the turbulent transport and dissipation coefficients (i.e. turbulent viscosity and resistivity) which (partially) determine the structure and the thickness of the tachocline in either the Spiegel–Zahn or the Gough–McIntyre scenario. In particular, both horizontal turbulent viscosity ν_{h} and horizontal turbulent resistivity η_{h} are set by β-plane MHD turbulence. The former is central to the Spiegel–Zahn scenario, as discussed earlier. Horizontal resistivity is important to the Gough–McIntyre scenario since, in this model, shearing of poloidal, radiation-zone magnetic fields is presumed to be balanced by resistive dissipation of the toroidal, tachocline field.

Up until now, only Ohmic, vertical diffusion of magnetic fields has been considered by Gough & McIntyre. Turbulent horizontal diffusion is also possible and is, very likely, a stronger effect (i.e. $-B_0 \sin\theta\,\partial\bar{\Omega}/\partial\theta \sim (\eta\partial_r^2 + \eta_{\mathrm{T}}\partial_{\mathrm{h}}^2)B_\phi$, with η_{T} dominant).

The model of β-plane MHD turbulence that we employ is simply two-dimensional MHD with the β effect added to the vorticity equation. Using $\mathbf{B} = \nabla A \times \widehat{\mathbf{z}}$ and $\mathbf{V} = \nabla\psi \times \widehat{\mathbf{z}}$, the governing equations can be written as:

$$\partial_t \nabla^2\psi + \nabla\psi \times \widehat{\mathbf{z}} \cdot \nabla\nabla^2\psi + \beta\partial_x\psi = \nabla A \times \widehat{\mathbf{z}} \cdot \nabla\nabla^2 A + \nu\nabla^2\nabla^2\psi + \widetilde{f}, \quad (9.1)$$

$$\partial_t A + \nabla\psi \times \widehat{\mathbf{z}} \cdot \nabla A = \eta\nabla^2 A + \widetilde{f}_{\mathrm{a}}. \quad (9.2)$$

In this model the magnetic flux function is of the form

$$A = B_0 y + \widetilde{A}, \quad (9.3)$$

so

$$\mathbf{B} = B_0\widehat{\mathbf{x}} + \widetilde{\mathbf{B}}. \quad (9.4)$$

As is the case in β-plane models, β corresponds to the horizontal gradient of the Coriolis parameter, i.e. $\beta = (2\Omega/r_0)\cos\theta_0$, where r_0 is the radius of the shell, Ω is the rotation rate and θ_0 is the latitude at which the β-plane is tangent to the spherical surface.

In this work, $\widehat{\mathbf{x}}$ corresponds to the azimuthal direction, and is the direction of the mean toroidal field; $\widehat{\mathbf{y}}$ corresponds to the polar direction and $\widehat{\mathbf{z}}$ to the radial direction, in which the system is stably stratified. Thus, Rossby waves propagate in the $-\widehat{\mathbf{x}}$ direction, i.e. westward, along $\mathbf{B_0}$. For simplicity we take B_0 to be uniform. Note that in the unforced ($\widetilde{f} \to 0$, $\widetilde{f}_{\mathrm{a}} \to 0$), inviscid, ideal limit (i.e. $\nu, \eta \to 0$), β-plane MHD conserves:

(i) total energy, $E = \langle(\nabla\psi)^2 + (\nabla A)^2\rangle/2$;
(ii) total A-squared, $H = \langle A^2\rangle$;
(iii) total cross helicity, $H_{\mathrm{c}} = \langle\nabla\psi \cdot \nabla A\rangle$.

Of course, inclusion of the Lorentz force breaks enstrophy conservation, so two-dimensional MHD dynamics is quite different from that of two-dimensional hydrodynamics.

It is interesting to note that even a straightforward linearization of Equations (9.1) and (9.2) reveals certain fundamental trends in the system. Assuming plane wave solutions and neglecting forcing and dissipation gives the dispersion relation

$$\omega^2 + \omega_{\mathrm{R_k}}\omega - \omega_{\mathrm{A_k}}^2 = 0, \quad (9.5)$$

where

$$\omega_{\mathrm{R}_{\mathbf{k}}} = -\frac{\beta k_x}{k^2} \tag{9.6}$$

is the Rossby wave frequency, and

$$\omega_{\mathrm{A}_{\mathbf{k}}} = k_x V_{\mathrm{A}_0} \tag{9.7}$$

is the Alfvén wave frequency. In β-plane MHD, these two wave branches are coupled. Hence, we note that for $k^2 < \beta/V_{\mathrm{A}_0}$ (i.e. $\omega_{\mathrm{A}_{\mathbf{k}}} < \omega_{\mathrm{R}_{\mathbf{k}}}$), the Rossby wave character is dominant, while for $k^2 > \beta/V_{\mathrm{A}_0}$ (i.e. $\omega_{\mathrm{R}_{\mathbf{k}}} < \omega_{\mathrm{A}_{\mathbf{k}}}$) the Alfvénic character dominates. Thus, the wavenumber $k_{\mathrm{LR}} = (\beta/V_{\mathrm{A}_0})^{1/2}$ defines a scale which demarks the boundary between Alfvénic and Rossby dominated ranges. For $k < k_{\mathrm{LR}}$, the wave spectrum may be thought of as a gas or ensemble of strongly dispersive Rossby waves, while for $k > k_{\mathrm{LR}}$, the waves are Alfvén waves. It is well known that even in two dimensions, Alfvén wave turbulence supports a *forward*, rather than inverse cascade, so 'negative viscosity phenomena', such as zonal flow formation, must occur on scales $\ell > \ell_{\mathrm{LR}}$, and are thus likely driven by Rossby wave interaction, rather than by turbulence. Note that ℓ_{LR} constitutes a scale which is somewhat reminiscent of the Rhines scale, familiar from discussions of geostrophic turbulence, and so is called the 'linearized Rhines scale' ℓ_{LR} here.

At this point, the critical reader is no doubt motivated to ask: *since B_0 flips every 11 years while tachocline formation proceeds over* 10^6 *years, why doesn't B_0 simply 'average out' on dynamically interesting time scales, allowing us to ignore B_0 in consideration of the formation of the tachocline?* This sentiment was more eloquently espoused in the 1998 paper by Gough & McIntyre, who suggested the following.

> Any field from a putative dynamo in the convection zone could be "dredged" into the tachocline by the meridional flow and thereby influence the dynamics, but it seems unlikely that the rapidly oscillating field associated with the solar cycle would contribute significantly to the dynamics in the radiative zone, particularly in view of the 10^6 year tachocline ventilation time τ_v.

Here we argue that while this timescale separation may justify ignoring the effects of B_0, *it does not justify the neglect of MHD effects*! The reason is simple – *in high-Rm two-dimensional MHD turbulence, $\langle \widetilde{B}^2 \rangle \gg \langle B \rangle^2$, so mean square magnetic fluctuation levels in the system are large, even if the mean field is weak.* This is a direct consequence of the Zeldovich theorem for 2D MHD, which is directly applicable to β-plane MHD (Zeldovich 1957; Diamond *et al.* 2005a; Tobias 2005). Below, we discuss the Zeldovich theorem and its implications.

It is now well known that in two-dimensional high-*Rm* MHD turbulence, the magnetic fluctuation intensity usually exceeds the mean field intensity by a large factor. This is a consequence of the stretching of magnetic fields by turbulence, and is encapsulated by the Zeldovich theorem, which follows from the conservation (up to resistive diffusion) of magnetic potential along fluid trajectories in two-dimensional incompressible flow. Here we generalize the Zeldovich theorem to account for direct forcing of the magnetic potential. Note that the presence or absence of β has no explicit impact upon the $\langle A^2\rangle$ budget. Though we usually do not think of the magnetic field as being stirred, such forcing is quite relevant to tachocline physics, since overshooting plumes naturally entrain convection zone magnetic fields while they plunge into the tachocline from above. During the course of this discussion, we also address and clarify some basic aspects of the Zeldovich theorem.

In two-dimensional incompressible β-plane MHD, the magnetic potential fluctuation satisfies

$$\frac{\partial \widetilde{A}}{\partial t} + \nabla\psi \times \widehat{\mathbf{z}} \cdot \nabla\widetilde{A} = -\widetilde{V}_y\langle B\rangle + \eta\nabla^2\widetilde{A} + \widetilde{f}_{\mathrm{a}}, \tag{9.8}$$

where we have assumed $\langle A\rangle = \langle A(y)\rangle$. Here $\langle B\rangle$ is the mean field and $\widetilde{V}_y = -\partial_x\psi$. Multiplying by $\widetilde{A}$, taking the fluctuation correlation length to be smaller than the scale of $\langle B\rangle$ variation, and integrating over space (denoted by $\langle\ \rangle$) we find

$$\frac{\partial\langle\widetilde{A}^2\rangle}{\partial t} + \langle\nabla\cdot\mathbf{V}\widetilde{A}^2\rangle = 2\big[-\langle\widetilde{V}_y\widetilde{A}\rangle\langle B\rangle + \eta\langle\widetilde{A}\nabla^2\widetilde{A}\rangle + \langle\widetilde{A}\widetilde{f}_{\mathrm{a}}\rangle\big]. \tag{9.9}$$

For stationary, homogeneous systems with periodic boundary conditions and no outflow (i.e. keep in mind that a spherical surface constitutes a closed system), the left-hand side of Equation (9.9) vanishes. Integrating once by parts on the right-hand side then gives the relation

$$\langle\widetilde{B}^2\rangle = \frac{-\langle\widetilde{V}_y\widetilde{A}\rangle\langle B\rangle + \langle\widetilde{A}\widetilde{f}_{\mathrm{a}}\rangle}{\eta}. \tag{9.10}$$

Note that either inhomogeneity or an in/out flow of magnetic potential can significantly alter the balance expressed by Equation (9.10). Finally, writing $\langle\widetilde{V}_y\widetilde{A}\rangle$ in the form of a Fick's law (i.e. $\langle\widehat{V}_y\widetilde{A}\rangle = -\eta_{\mathrm{T}}\partial\langle A\rangle/\partial y = -\eta_{\mathrm{T}}\langle B\rangle$, where η_{T} is a turbulent resistivity) and noting $\mathrm{d}\widetilde{A}/\mathrm{d}t = \widetilde{f}_{\mathrm{a}}$ on inertial scales yields

$$\langle\widetilde{B}^2\rangle = \frac{\eta_{\mathrm{T}}\langle B\rangle^2 + \langle\widetilde{f}_{\mathrm{a}}^2\rangle\tau_{\mathrm{a}}}{\eta}. \tag{9.11}$$

Here τ_{a} is the auto-correlation time of the (random) magnetic stirring force $\widetilde{f}_{\mathrm{a}}$.

Equation (9.11) extends the usual Zeldovich theorem balance ($\langle\widetilde{B}^2\rangle = (\eta_{\mathrm{T}}/\eta)\langle B\rangle^2$) to include the additional effect of random stirring of A. Note that writing $\langle B\rangle^2 = (\partial\langle A\rangle/\partial y)^2$ in Equation (9.11) suggests that the total magnetic fluctuation intensity is fed by both:

(i) turbulent mixing of gradients in $\langle A\rangle$ by ambient MHD turbulence, as parametrized by $\eta_T(\partial\langle A\rangle/\partial y)^2$;
(ii) external stirring by overshoot, as parametrized by $\langle\widetilde{f}_a^2\rangle\tau_a$.

These two stochastic processes are independent, so their contributions to $\langle\widetilde{B}^2\rangle$ are *additive*. As $\eta_T \gg \eta$, Equation (9.11) confirms that $\langle\widetilde{B}^2\rangle \gg \langle B\rangle^2$, even in the absence of direct stirring of $\widetilde{A}$. Strictly speaking, the Zeldovich theorem balance may be written as

$$\frac{\langle\widetilde{B}^2\rangle}{\langle B\rangle^2} = N_{u,m} + \langle\widetilde{f}_a^2\rangle\frac{\tau_a}{\eta}, \tag{9.12}$$

where $N_{u,m}$ is the 'magnetic Nusselt number', η_T/η, and τ_a is the forcing correlation time. In practice, for two-dimensional MHD, $N_{u,m}$ exhibits a strong scaling with magnetic Reynolds number Rm, consistent with both numerical calculations and theoretical expectations (Cattaneo & Vainshtein 1991; Diamond *et al.* 2005a). Note that while η_T is quenched, relative to kinematic based expectations, it still greatly exceeds the collisional resistivity η and so η_T quenching is indeed compatible with $\langle\widetilde{B}\rangle^2 \gg B_0^2$.

Frequently, $N_{u,m} \sim Rm$, though the universality of this putative scaling requires further study and documentation. However, it seems indisputable that in the turbulent tachocline, the β-plane MHD turbulence has $Rm \gg 1$ and so $\langle\widetilde{B}^2\rangle \gg \langle B\rangle^2$. Thus, the magnetic fluctuations and turbulence dominate the mean magnetic field, rendering the rapid reversals (on tachocline formation timescales) of the mean field direction a moot point. More succinctly put, while $\langle\mathbf{B}\rangle$ may 'average out' over long timescales on account of frequent reversals, $\langle\widetilde{B}^2\rangle$ will most certainly persist, albeit in different realizations, and in fact be the dominant repository of magnetic energy. Of course, $\widetilde{\mathbf{B}}$ has no systematic directionality. Hence, one should think of the tachocline as magnetized by layers of thin, quasi-two-dimensional stochastic magnetic networks (see Figure 9.2), which support the propagation of Alfvén waves and so 'elasticize' the tachocline layer.

9.3 The Rhines scale for MHD turbulence on a β-plane

A key element in our mental picture of hydrodynamic turbulence on a β-plane is the Rhines scale, which is of significance because it demarks the boundary between small-scale two-dimensional turbulence, composed of eddies and vortices, etc., and larger-scale Rossby wave turbulence. The physics of the Rhines scale is explained heuristically below. We then proceed to discuss the modifications of the Rhines scale introduced by coupling to stochastic magnetic fields in two-dimensional MHD on

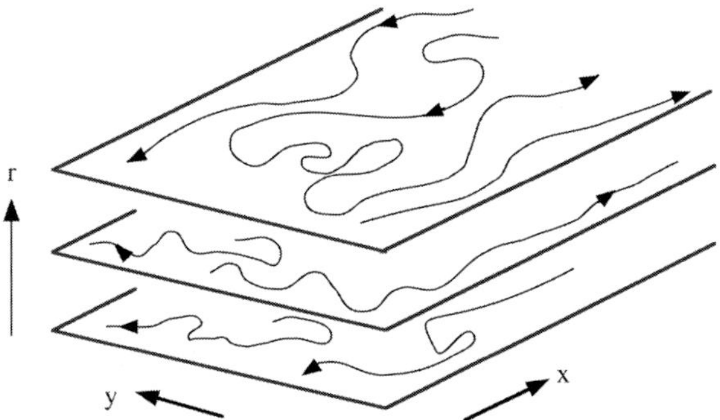

Figure 9.2. A sketch of turbulent magnetic field structure in the tachocline. The magnetic field is stochastic but organized into thin, quasi-two-dimensional layers or shells, on account of the strong, stable stratification. The *mean* magnetic field (not shown here) is primarily in the $\widehat{\mathbf{x}}$ direction.

the β-plane. We especially focus on possible *Rm* dependence of the Rhines scale and on its role in separating the region of forward MHD energy cascade from that of transfer of Rossby wave energy by nonlinear wave interaction.

In brief, the Rhines scale of quasi-geostrophic turbulence (Rhines 1975; Diamond *et al.* 2005b) is based on two facts, which are as follows.

(i) Each $\mathbf{k}$ or scale is characterized by a real frequency $\omega_{\mathbf{k}}$ and a self-decorrelation rate $\Delta\omega_{\mathbf{k}}$. Here $\Delta\omega_{\mathbf{k}}^{-1}$ may be thought of as an effective self-coherence time for the fluctuation with wavevector $\mathbf{k}$. In geostrophic turbulence, the real frequency is approximately the Rossby wave frequency $\omega_{\mathbf{k}} \cong \omega_{\mathrm{R}_{\mathbf{k}}} = -\beta k_x/k^2$.

(ii) At long wavelengths, Rossby waves are strongly dispersive, so it is extremely difficult to satisfy the three wave resonance condition for energy transfer unless either:

  (a) one member of the triad has $k_x = 0$ and so is a zonal flow,

  (b) $\Delta\omega_{\mathbf{k}} > \omega_{\mathbf{k}}$, so turbulence interaction effectively smears out wave resonance.

Thus, the scale at which $\omega_{\mathbf{k}} = \Delta\omega_{\mathbf{k}}$ naturally forms a boundary or dividing line between ranges of scales in which the nonlinear energy transfer is controlled by turbulent inverse cascade and resonant wave–wave interaction. This scale is referred to as the *Rhines scale*. Since for (strongly dispersive) Rossby waves, resonant wave interaction occurs only via scattering off an azimuthally symmetric zonal flow mode, the Rhines scale also sets the characteristic width of zonal flows. On dimensional grounds, the Rhines scale is usually estimated by taking $\Delta\omega_{\mathbf{k}} \sim k\widetilde{V}$, so

$$\omega_{\mathbf{k}} \approx k\widetilde{V} \tag{9.13}$$

implies that the Rhines wave number k_{R} is given by

$$k_{\mathrm{R}}^2 = \frac{\beta}{\widetilde{V}}, \tag{9.14}$$

and so the Rhines scale $\ell_R \sim (\widetilde{V}/\beta)^{1/2}$. Here $\widetilde{V}$ is a 'typical' eddy velocity, so ℓ_R exhibits some sensitivity to the structure of the spectrum. Moreover, concerns about Galilean invariance have motivated reconsideration of the definition of ℓ_R in terms of the local eddy strain rate (Vallis & Maltrud 1993). The resulting departures from the simple result of Equation (9.14) are, however, quite small (Nozawa & Yoden 1997). Thus, the Rhines scale ℓ_R is well established as a useful concept in, and as an element of, descriptions of geostrophic turbulence.

In two-dimensional β-plane MHD at large Rm, $\langle \widetilde{B}^2 \rangle \geq Rm \langle B \rangle^2$ (i.e. for simplicity we now take $N_{u,m} \sim Rm$) so the decorrelation rate $\Delta\omega_{\mathbf{k}}$ is simply $k\widetilde{V}_A$, where $\widetilde{V}_A^2 = \langle \widetilde{B}^2 \rangle / 4\pi\rho_0 \geq Rm V_{A_0}^2$. Here $\widetilde{V}_A$ may be thought of as an effective Alfvén or elastic velocity for propagation in the network of stochastic small-scale fields. Such stochastic fields are the principal agents of decorrelation here. Note that since the magnetic field allows large scales to damp small scales by Alfvénic coupling, concerns pertaining to Galilean invariance of the turbulence theory are moot in MHD (Moffatt 1978). Then, since $\widetilde{V}_A \gg V_{A_0}$, the effective boundary between turbulence and Rossby wave ranges in β-plane MHD is given by

$$\frac{\beta k_x}{k^2} \cong k\widetilde{V}_A, \tag{9.15}$$

so the MHD Rhines wavenumber k_{RM} is given by

$$k_{RM}^2 \cong \frac{\beta}{\widetilde{V}_A}, \tag{9.16}$$

and the effective Rhines scale for β-plane MHD is just $\ell_{RM} \sim (\widetilde{V}_A/\beta)^{1/2}$. As we will see, ℓ_{RM} is an important scale for the dynamics of β-plane MHD turbulence.

Several comments are appropriate here. First, note that ℓ_{RM} is similar to ℓ_{LR} from Section 9.2, the difference being that $\ell_{RM} \sim (\widetilde{V}_A/\beta)^{1/2}$ while $\ell_{LR} \sim (V_{A_0}/\beta)^{1/2}$, so $\ell_{RM}/\ell_{LR} \gtrsim Rm^{1/4}$. This once again reminds us that coupling to the stochastic small scale magnetic field $\widetilde{\mathbf{B}}$ is much stronger than the coupling to the mean field $\langle \mathbf{B} \rangle$, so the dynamically dominant Alfvén wave propagation is along the stochastic field. This appears in the theory as a decorrelation rate, rather than as a wave frequency, on account of the stochasticity of $\widetilde{\mathbf{B}}$. Second, note that ℓ_{RM} is manifestly Rm dependent, and $\sim Rm^{1/4}$ for the usual scaling of $\eta_T/\eta \sim Rm$. Thus, the range of MHD turbulence (i.e. all scales ℓ such that $\ell_d < \ell < \ell_{RM}$) *increases* with Rm. Third, it is *very important* to keep in mind that the presence of magnetic fields breaks enstrophy conservation, so that MHD turbulence cascades to small scales, even in two dimensions! In the case where the characteristic forcing scale $\ell_f < \ell_{RM}$, energy will forward cascade to small scale dissipation, while (non-scale-invariant) stochastic backscatter will gradually fill in $\ell_f < \ell < \ell_{RM}$. For $\ell_f > \ell_{RM}$, energy will be transferred forward, though wave interactions will play a role – see

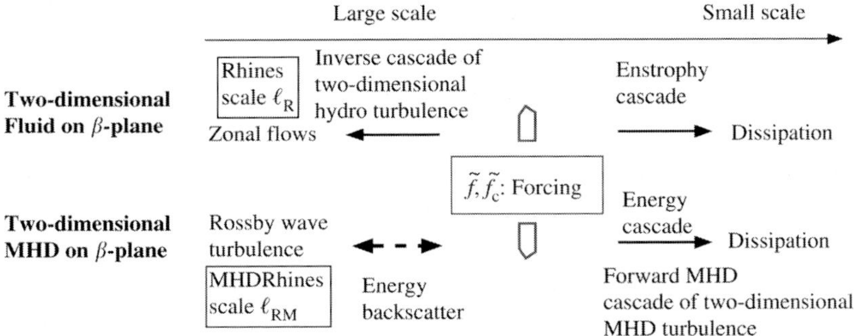

Figure 9.3. A cartoon contrasting the energy flow in two-dimensional hydrodynamic turbulence on a β-plane with that for two-dimensional MHD turbulence on a β-plane. A large-scale stochastic magnetic field lies in the plane.

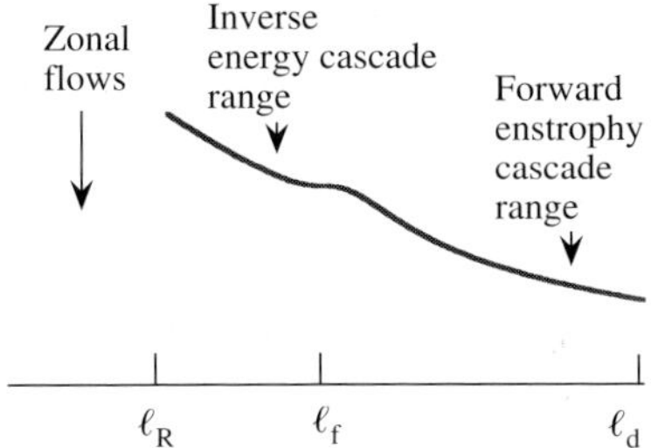

Figure 9.4. Cartoon of the energy spectrum for geostrophic turbulence. Note that ℓ decreases to the right.

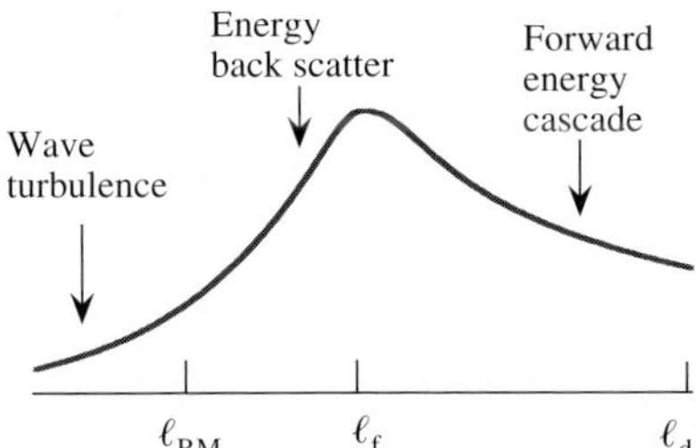

Figure 9.5. Cartoon of the energy spectrum for β-plane MHD turbulence. Note that ℓ decreases to the right.

Figures 9.3–9.5 for pertinent diagrams. Thus, any energy reaching ℓ_{RM} will eventually be coupled to small-scale dissipation. No inverse cascade of energy occurs. Thus, unlike the corresponding case in geostrophic turbulence, there is no reason to expect zonal flow formation at ℓ_{RM}, as *energy does not inverse cascade toward ℓ_{RM} in turbulent β-plane MHD*. Here, the effective Rhines scale ℓ_{RM} merely *separates* the range of forward cascading MHD turbulence from the range dominated by wave interaction.

Table 9.2. *Comparison of Rhines scales for HD and MHD β-Plane turbulence*

	β-Plane HD	β-Plane MHD
Rhines scale	$\ell_R = (\widetilde{V}/\beta)^{1/2}$	$\ell_{RM} = (\widetilde{V}_A/\beta)^{1/2}$ $\widetilde{V}_A^2 = \langle\widetilde{B}^2\rangle/4\pi\rho_0$
Fluctuation constituents	$\ell < \ell_R$ – eddies $\ell > \ell_R$ – Rossby waves, zonal flows	$\ell < \ell_{RM}$ – Alfvén waves, eddies $\ell > \ell_{RM}$ – Rossby waves, zonal flows, fields
Spectral energy flow	$\ell < \ell_R \rightarrow$ INVERSE cascade toward ℓ_R $\ell > \ell_R \rightarrow$ transfer by wave–zonal flow interaction	$\ell < \ell_{RM} \rightarrow$ FORWARD MHD cascade $\ell > \ell_{RM} \rightarrow$ transfer by wave–wave, wave–zonal structure interaction
Structure	Strong zonal flows on $\ell \sim \ell_R$ fed by inverse cascade from $\ell < \ell_R$	Weak zonal flows and fields on scales $\ell > \ell_{RM}$ fed by wave interaction from scales $\ell > \ell_{RM}$

To summarize this discussion and to understand and gain some perspective on the role of the MHD Rhines scale ℓ_{RM} in β-plane MHD, it is useful to systematically compare and contrast the physics of the Rhines scale ℓ_R familiar from neutral quasi-geostrophic turbulence with that of the MHD Rhines scale ℓ_{RM} for β-plane MHD. Table 9.2 summarizes this comparison. In the case of a neutral geostrophic fluid, the Rhines scale $\ell_R = (\widetilde{V}/\beta)^{1/2}$ separates eddy interaction dominated ($\ell < \ell_R$) and wave–zonal flow interaction dominated ranges ($\ell > \ell_R$). For β-plane MHD, the MHD Rhines scale $\ell_{RM} = (\widetilde{V}_A/\beta)^{1/2}$, where $\widetilde{V}_A^2 = \langle\widetilde{B}^2\rangle/4\pi\rho_0$, separates an MHD turbulence range ($\ell < \ell_{RM}$) composed of eddies and Alfvén waves from a range with $\ell > \ell_{RM}$, which is dominated by Rossby waves, with some zonal flows and fields present too. In the case of a geostrophic fluid, energy *inverse cascades* toward ℓ_R from all smaller scales, i.e. $\ell < \ell_R$. The strong dispersion of Rossby waves then forces further interaction to proceed primarily via wave–zonal flow scattering, thus generating zonal flows of characteristic scale ℓ_R. In this case, *nearly all* energy generated on scales with $\ell < \ell_R$ is ultimately fed into large-scale zonal flows. Thus, it is no surprise that zonal flows are a prominent feature of such systems. In the case of β-plane MHD, energy in the turbulent range on $\ell < \ell_{RM}$ *forward cascades*, toward small-scale dissipation. Energy contained on scales $\ell > \ell_{RM}$ participates in wave–wave and wave–zonal structure (i.e. flow and field) interaction. Note that in the case of β-plane MHD, most of the energy generated on scales $\ell < \ell_{RM}$ does *not* flow to large scales ($\ell > \ell_{RM}$). Such scales are fed only by (non-self-similar) stochastic back-scatter from smaller scales. Thus, we can expect zonal structures to be *significantly* less prominent features in β-plane MHD turbulence than in neutral geostrophic turbulence (Kim *et al.* 2001; Naulin *et al.* 2005).

9.4 Spectral transfer and turbulent dissipation in β-plane MHD

In this section, we examine the dynamics of interactions, spectral transfer and turbulent dissipation (η_{h} and ν_{h}) in β-plane MHD. Such a study necessarily builds upon existing understanding of two-dimensional MHD and Rossby wave turbulence, both of which have been extensively studied (Pouquet 1978; Horton 1999). This section should be construed only as an introduction to this large and complex subject, and thus should be viewed as a survey of tachocline-relevant issues in β-plane MHD. Some selected topics are pursued in depth here, but a complete discussion is far beyond the scope of this short article. Here, we shall discuss:

(i) spectral transfer of $\langle\widetilde{A}^2\rangle_{\mathbf{k}}$ and the turbulent diffusion of magnetic fields;
(ii) the turbulent transport of momentum and turbulent viscosity;
(iii) interactions of an ambient Rossby wave spectrum with a large-scale shear flow.

Other aspects of the problem are left for future publications. Throughout this section, we ignore any possible cross-correlation between fluid and magnetic forcing (i.e. we take $\langle\widetilde{f}\widetilde{f}_{\mathrm{a}}\rangle = 0$), so cross helicity may be zeroed *ab initio*. We emphasize, though, that this is only a crude approximation and that the relative coherence of $\widetilde{f}$ and $\widetilde{f}_{\mathrm{a}}$ is an important element of the physics of tachocline turbulence which should be considered carefully (Bracco *et al.* 1998).

Turbulent magnetic dissipation is best addressed by examining the spectral dynamics of $\langle\widetilde{A}^2\rangle$. The variance of the magnetic potential evolves according to

$$\frac{\partial}{\partial t}\frac{\langle\widetilde{A}^2\rangle}{2} + \frac{\langle\nabla\psi\times\widehat{\mathbf{z}}\cdot\nabla\widetilde{A}^2\rangle}{2} = -\eta\langle\widetilde{B}^2\rangle + \langle B\rangle\langle\widetilde{A}\partial_x\psi\rangle + \langle\widetilde{A}\widetilde{f}_{\mathrm{a}}\rangle, \tag{9.17}$$

or, equivalently, in k space

$$\frac{\partial}{\partial t}\langle\widetilde{A}^2\rangle_{\mathbf{k}} + T_{\mathbf{k}} = 2\left[\eta_{\mathrm{T}\mathbf{k}}\left(\frac{\partial\langle A\rangle}{\partial y}\right)^2 + \langle\widetilde{f}_{\mathrm{a}}^2\rangle_{\mathbf{k}}\tau_{\mathrm{a}\mathbf{k}} - \eta\langle\widetilde{B}^2\rangle_{\mathbf{k}}\right], \tag{9.18}$$

where the nonlinear transfer term $T_{\mathbf{k}}$ is:

$$T_{\mathbf{k}} = \langle\nabla\psi\times\widehat{\mathbf{z}}\cdot\nabla\widetilde{A}^2\rangle_{\mathbf{k}}. \tag{9.19}$$

Note the terms on the right-hand side of Equation (9.18) are precisely those which determine the Zeldovich theorem relation. Hereafter we refer to the right-hand side of Equation (9.18) as $\mathcal{S}_{\mathbf{k}}$, the net *source* for the time evolution of $\langle\widetilde{A}^2\rangle_{\mathbf{k}}$. The Zeldovich theorem, then, is merely a statement that $\sum_{\mathbf{k}}\mathcal{S}_{\mathbf{k}} = 0$, so that there is a net balance of inflow and outflow of $\langle\widetilde{A}^2\rangle$.

We now calculate $T_{\mathbf{k}}$ via closure theory (Orszag 1970; Yoshizawa *et al.* 2003). Before proceeding to grind the crank, some general comments on $\langle \widetilde{A}^2 \rangle_{\mathbf{k}}$ transfer are in order. It is well known that, in two-dimensional MHD, magnetic potential isocontours:

(i) tend to be 'chopped up' by the turbulent flow until reaching small-scale dissipation, or until Lorentz force back-reaction becomes significant;
(ii) tend to coalesce and aggregate on large scales, on account of the mutual attraction of like-signed current filaments.

The competition between these two processes determines the net effective magnetic dissipation. Since 'chopping up' tends to win if $\langle \widetilde{V}^2 \rangle \gg \langle \widetilde{B}^2 \rangle$ while coalescence tends to win if $\langle \widetilde{B}^2 \rangle \gg \langle \widetilde{V}^2 \rangle$, the net turbulent magnetic dissipation tends to scale as

$$\eta_{\mathrm{T}} \sim (\langle \widetilde{V}^2 \rangle - \langle \widetilde{V}_{\mathrm{A}}^2 \rangle)\tau_{\mathrm{c}}, \tag{9.20}$$

where τ_{c} is a correlation time. Such a form for η_{T} has indeed been recovered from the results of renormalized closure theory (Pouquet 1978). Subsequent use of the Zeldovich theorem balance then gives the form of the 'quenched' magnetic diffusivity in two-dimensional MHD:

$$\eta_{\mathrm{T}} \sim \frac{\eta^k}{1 + Rm V_{\mathrm{A}_0}^2 / \langle \widetilde{V}^2 \rangle}, \tag{9.21}$$

where η^k is the familiar kinematic diffusivity $\eta^k \sim \langle \widetilde{V}^2 \rangle \tau_{\mathrm{c}}$ (Pouquet *et al.* 1976). In the interesting limit where $Rm V_{\mathrm{A}_0}^2 / \langle \widetilde{V}^2 \rangle \gg 1$, the corresponding limit of Equation (9.21) can be obtained directly from the Zeldovich theorem, with the additional reasonable assumption of approximate equipartition, so that $\eta_{\mathrm{T}} \sim \eta \langle \widetilde{V}^2 \rangle / \langle B \rangle^2$.

The main issues to be addressed here are:

(i) just what exactly is τ_{c} and how is it determined?
(ii) what are the effects of Rossby wave coupling on spectral transfer of $\langle \widetilde{A}^2 \rangle$?

Regarding (i), we have previously discussed the physics of τ_{c}, which is Alfvénic propagation along a fluctuating network of stochastic magnetic fields, so $1/\tau_{\mathrm{c}\mathbf{k}} = \Delta\omega_{\mathbf{k}} = k\widetilde{V}_{\mathrm{A}}$. Note that τ_{c} is itself necessarily slowly time-dependent. To assess the effects of Rossby coupling (i.e. item (ii)), a closure calculation is necessary.

Here, we present an eddy-damped quasi-normal Markovian (EDQNM) closure calculation of $T_{\mathbf{k}}$ for β-plane MHD. The EDQNM closure develops a set of coupled spectra equations from the assumption of weakly non-Gaussian mode amplitude statistics and from physically motivated choices (further details can be found in Orszag 1970; Pouquet 1978).

Writing $T_{\mathbf{k}}$ as

$$T_{\mathbf{k}} = \left\langle \sum_{\mathbf{k}'} (\mathbf{k} \cdot \mathbf{k}' \times \widehat{\mathbf{z}})(\widetilde{A}_{-\mathbf{k}}\widetilde{\psi}_{-\mathbf{k}'}\widetilde{A}^{(2)}_{\mathbf{k}+\mathbf{k}'} + \widetilde{A}_{-\mathbf{k}}\widetilde{A}_{-\mathbf{k}'}\widetilde{\psi}^{(2)}_{\mathbf{k}+\mathbf{k}'}) - \sum_{\substack{\mathbf{p},\mathbf{q} \\ \mathbf{p}+\mathbf{q}=\mathbf{k}}} (\mathbf{p} \cdot \mathbf{q} \times \widehat{\mathbf{z}})\widetilde{\psi}_{-\mathbf{p}}\widetilde{A}_{-\mathbf{q}}\widehat{A}^{(2)}_{\mathbf{p}+\mathbf{q}} \right\rangle, \tag{9.22}$$

we seek to calculate $\widetilde{A}^{(2)}_{\mathbf{k}+\mathbf{k}'}$ and $\widetilde{\psi}^{(2)}_{\mathbf{k}+\mathbf{k}'}$ such that $T_{\mathbf{k}}$ is independent of fluctuation phase. The calculation is simplified by taking $\langle \mathbf{V}\cdot\mathbf{B}\rangle = 0$ and ignoring $\langle \mathbf{B}\rangle$ relative to $\widetilde{B}_{\mathrm{rms}}$, which amounts to neglecting linear Alfvén wave propagation in comparison to stochastic Alfvénic decorrelation (i.e. since $kV_{A_0} < \Delta\omega_{\mathbf{k}}$). Thus, $\widetilde{A}^{(2)}_{\mathbf{k}+\mathbf{k}'}$ and $\widetilde{\psi}^{(2)}_{\mathbf{k}+\mathbf{k}'}$ are written as:

$$\widetilde{A}^{(2)}_{\mathbf{k}+\mathbf{k}'} = \int_{-\infty}^{t} \mathrm{d}t' \mathrm{e}^{-\Delta\omega_{\mathbf{k}+\mathbf{k}'}(t-t')}(\mathbf{k} \cdot \mathbf{k}' \times \widehat{\mathbf{z}})\widetilde{\psi}_{\mathbf{k}'}(t')\widetilde{A}_{\mathbf{k}}(t'), \tag{9.23}$$

$$\widetilde{\psi}^{(2)}_{\mathbf{k}+\mathbf{k}'} = \int_{-\infty}^{t} \mathrm{d}t' \mathrm{e}^{-(\mathrm{i}\omega_{\mathbf{k}+\mathbf{k}'}+\Delta\omega_{\mathbf{k}+\mathbf{k}'})(t-t')}(\mathbf{k} \cdot \mathbf{k}' \times \widehat{\mathbf{z}})\frac{(k^2-k'^2)}{k''^2}\widetilde{A}_{\mathbf{k}'}(t')\widetilde{A}_{\mathbf{k}}(t'). \tag{9.24}$$

Note that $\widetilde{A}^{(2)}_{\mathbf{p}+\mathbf{q}}$ is identical to $\widetilde{A}^{(2)}_{\mathbf{k}+\mathbf{k}'}$, up to a re-labelling. The two-time correlators for $\widetilde{\psi}$ and $\widetilde{A}$ (here written for some generic field F) are taken to have the approximate structure:

$$\langle F^*_{\mathbf{k}}(t)F_{\mathbf{k}}(t')\rangle = |F_{\mathbf{k}}(t)|^2 \mathrm{e}^{-(\mathrm{i}\omega_{\mathbf{k}}+\Delta\omega_{\mathbf{k}})|t-t'|}. \tag{9.25}$$

Here, rapid decay on the $(\Delta\omega_{\mathbf{k}})^{-1}$ timescale accounts for decorrelation of resonant triads due to nonlinear scrambling, while the slower envelope behaviour (i.e. $|F_{\mathbf{k}}(t)|^2$) accounts for evolution of the spectrum in time. Given all this, it follows that the renormalized $\langle \widetilde{A}^2\rangle_{\mathbf{k}}$ transfer rate $T_{\mathbf{k}}$ is given by:

$$T_{\mathbf{k}} = \sum_{\mathbf{k}'} (\mathbf{k} \cdot \mathbf{k}' \times \widehat{\mathbf{z}})^2 \theta^A_{\mathbf{k},\mathbf{k}',\mathbf{k}+\mathbf{k}'}|\widetilde{\psi}_{\mathbf{k}'}(t)|^2|\widetilde{A}_{\mathbf{k}}(t)|^2 + \sum_{\mathbf{k}'} (\mathbf{k} \cdot \mathbf{k}' \times \widehat{\mathbf{z}})^2 \theta^{\psi}_{\mathbf{k},\mathbf{k}',\mathbf{k}+\mathbf{k}'}\left(\frac{k^2-k'^2}{k''^2}\right)|\widetilde{A}_{\mathbf{k}'}(t)|^2|\widetilde{A}_{\mathbf{k}}(t)|^2 - \sum_{\substack{\mathbf{p},\mathbf{q} \\ \mathbf{p}+\mathbf{q}=\mathbf{k}}} (\mathbf{p} \cdot \mathbf{q} \times \widehat{\mathbf{z}})^2 \theta^A_{\mathbf{p},\mathbf{q},\mathbf{k}'}|\widetilde{\psi}_{\mathbf{p}}(t)|^2|\widetilde{A}_{\mathbf{q}}(t)|^2, \tag{9.26}$$

where $k''^2 = (\mathbf{k} + \mathbf{k}')^2$ and

$$\theta^A_{\mathbf{k},\mathbf{k}',\mathbf{k}+\mathbf{k}'} = \Re\left\{\frac{\mathrm{i}}{((\omega_{\mathbf{k}} + \omega_{\mathbf{k}'}) + \mathrm{i}(\Delta\omega_{\mathbf{k}} + \Delta\omega_{\mathbf{k}'} + \Delta\omega_{\mathbf{k}+\mathbf{k}'}))}\right\}, \tag{9.27}$$

$$\theta^\psi_{\mathbf{k},\mathbf{k}',\mathbf{k}+\mathbf{k}'} = \Re\left\{\frac{\mathrm{i}}{((\omega_{\mathbf{k}} + \omega_{\mathbf{k}'} - \omega_{\mathbf{k}+\mathbf{k}'}) + \mathrm{i}(\Delta\omega_{\mathbf{k}} + \Delta\omega_{\mathbf{k}'} + \Delta\omega_{\mathbf{k}+\mathbf{k}'}))}\right\}. \tag{9.28}$$

Here θ^A represents the triad coherence time for the first and third terms in the expression for $T_{\mathbf{k}}$ while θ^ψ represents that for the second. Equations (9.26), (9.27) and (9.28) then give the full result for the renormalized $\langle\widetilde{A}^2\rangle_{\mathbf{k}}$ transfer rate, $T_{\mathbf{k}}$.

Several aspects of Equations (9.26)–(9.28) merit discussion at this point. First, note that $T_{\mathbf{k}}$ has the usual structure of coherent damping terms (i.e. the first two) competing against incoherent emission (i.e. the third; Kraichnan 1959). The coherent damping terms determine the effective turbulent magnetic dissipation. Thus, we have the turbulent magnetic diffusivity:

$$\eta_{\mathrm{T}} \cong \sum_{\mathbf{k}'} k'^2 [\theta^A_{\mathbf{k},\mathbf{k}',\mathbf{k}+\mathbf{k}'} |\widetilde{\psi}_{\mathbf{k}'}(t)|^2 - \theta^\psi_{\mathbf{k},\mathbf{k}',\mathbf{k}+\mathbf{k}'} |\widetilde{A}_{\mathbf{k}'}(t)|^2]. \tag{9.29}$$

Here we have taken $|\mathbf{k}| \ll |\mathbf{k}'|$ (i.e. we consider the dissipation of larger scales than the scale on which the system is forced), so $(k^2 - k'^2)/k''^2 \to -1$. It is not surprising to see that η_{T} for β-plane MHD is quite similar to its counterpart for two-dimensional MHD, apart from the triad coherence factors θ^A and θ^ψ, which contain the wave frequency contributions. Note that there is a slight difference between the frequency dependencies of θ^A and θ^ψ. In particular, θ^ψ is considerably more sensitive to Rossby wave dispersion, in that $\omega_{\mathbf{k}} + \omega_{\mathbf{k}'} \simeq 0$ is easily satisfied but triad resonance, as in θ^ψ, is not. A detailed quantitative study of the implication of $\theta^A \neq \theta^\psi$ is beyond the scope of this chapter.

Several aspects of Equation (9.29) merit further discussion, as well. First, and most important, it is easy to see that for $\Delta\omega > \omega_{\mathbf{k}} + \omega_{\mathbf{k}'}$, $\Delta\omega > \omega_{\mathbf{k}} + \omega_{\mathbf{k}'} - \omega_{\mathbf{k}+\mathbf{k}'}$ (where $\Delta\omega$ refers to the sum of the three model decorrelation rates in θ^A and θ^ψ) Equation (9.29) passes smoothly to results previously obtained for two-dimensional MHD with $\beta = 0$ (Diamond *et al.* 2005a). This limit corresponds to length scales $\ell < \ell_{\mathrm{RM}}$. Thus, existing results from the theory of turbulent diffusion of magnetic fields in two-dimensional MHD predict that the horizontal turbulent resistivity η_{T} is (strongly) quenched, in comparison to kinematic turbulence predictions, i.e.

$$\eta_{\mathrm{T}} = \frac{\eta^k}{1 + Rm V^2_{\mathrm{A}_0}/\langle\widetilde{V}^2\rangle}. \tag{9.30}$$

For $Rm V^2_{\mathrm{A}_0}/\langle\widetilde{V}^2\rangle > 1$, η_{T} is well approximated by $\eta_{\mathrm{T}} \approx \eta\langle\widetilde{B}^2\rangle/\langle B\rangle^2$. For $\langle\widetilde{V}^2\rangle \sim \langle\widetilde{V}^2_{\mathrm{A}}\rangle$, these are equivalent to $\eta_{\mathrm{T}} \approx \eta\langle\widetilde{V}^2\rangle/V^2_{\mathrm{A}_0}$. *The reader should take* *note, however, that while η_h is quenched relative to the standard kinematic estimates, it still greatly exceeds the collisional magnetic diffusivity η, since $\langle \widetilde{B}^2 \rangle \gg \langle B \rangle^2$ etc.* Thus, *turbulent horizontal diffusion of magnetic fields* may still be a significant mechanism for dissipating magnetic energy in the solar tachocline. We will discuss this issue further in the concluding section.

Having addressed the question of turbulent resistivity, we now consider the physics of turbulent viscosity and turbulent momentum transport in β-plane MHD. Recall that, in the Spiegel–Zahn scenario of tachocline formation, the tachocline location in the solar core is determined by the balance of burrowing with horizontal transport (in latitude) of momentum. In β-plane MHD, the mean azimuthal flow evolves according to

$$\frac{\partial}{\partial t}\langle V_x \rangle = \frac{-\partial}{\partial y}\left\{ \langle \widetilde{V}_y \widetilde{V}_x \rangle - \frac{\langle \widetilde{B}_y \widetilde{B}_x \rangle}{4\pi\rho_0} \right\}, \tag{9.31}$$

where we have ignored molecular viscosity and the momentum source related to burrowing. Also, here ρ_0 is taken as a constant as we consider incompressible two-dimensional dynamics. Note that in MHD, the net flux transport is determined by the *difference* between fluid and magnetic stresses. Thus, in a perfectly Alfvénized state the total momentum flux *vanishes*. Of course, the traditionally invoked turbulent horizontal viscosity $\nu_{\rm h}$ is based upon a 'mixing length' approximation to the fluid stress $\langle \widetilde{V}_y \widetilde{V}_x \rangle$, which is constructed by asserting that fluctuations in $\widetilde{V}_y$ occur via mixing of $\langle V_x \rangle$, i.e.

$$\begin{aligned} \widetilde{V}_y &= \langle V_x(y-\ell) \rangle - \langle V_x(y) \rangle \\ &\cong -\ell \frac{\partial \langle V_x \rangle}{\partial y}, \end{aligned} \tag{9.32}$$

so

$$\begin{aligned} \langle \widetilde{V}_y \widetilde{V}_x \rangle &= -(\widetilde{V}_x \ell) \frac{\partial \langle V_x \rangle}{\partial y} \\ &\equiv -\nu_{\rm h} \frac{\partial \langle V_x \rangle}{\partial y}. \end{aligned} \tag{9.33}$$

Here ℓ is the mixing length. Note that the main novel feature in the case of MHD is the competition between the two stresses. Thus, we focus our attention on this competition. It is useful to split the integration over scales (implicit in the averages in Equation (9.31)) into ranges of $k_<$ and $k_>$, where the $k_<$-range includes all $\mathbf{k}$ such that $|\mathbf{k}| < k_{\rm RM} = 2\pi/\ell_{\rm RM}$ and the $k_>$-range includes all $\mathbf{k}$ such that $|\mathbf{k}| > k_{\rm RM}$. Thus, the $k_>$-range corresponds to the range of *forward* cascading of energy in two-dimensional MHD turbulence for which wave interactions are subdominant, while the $k_<$-range is the range where nonlinear transfer etc. are controlled by *Rossby–Alfvén wave interactions* (see Figure 9.5). Denoting the total momentum flux by

Γ_v, we thus can write:

$$\frac{\partial \langle V_x \rangle}{\partial t} = -\frac{\partial}{\partial y}\Gamma_v$$
$$= -\frac{\partial}{\partial y}\left(\sum_{k_<}\Gamma_{v\mathbf{k}} + \sum_{k_>}\Gamma_{v\mathbf{k}}\right), \tag{9.34}$$

where

$$\Gamma_{v_\mathbf{k}} = -k_x k_y (|\psi_\mathbf{k}|^2 - |A_\mathbf{k}|^2) \tag{9.35}$$

and the y-dependence of Γ_v is, by definition, 'slow', as it corresponds to variation on scales larger than those typical of the turbulence.

We now discuss the contributions to the momentum flux coming from the $k_>$ and $k_<$-ranges. The $k_>$-range exhibits two-dimensional MHD-like turbulence dynamics. One of the most robust features of two-dimensional MHD is the trend toward approximate equipartition between hydrodynamic and magnetic energy on inertial range scales, i.e. $|\widetilde{\mathbf{V}}_\mathbf{k}|^2 \simeq |\widetilde{\mathbf{B}}_\mathbf{k}|^2$ (Pouquet 1978). This is yet another manifestation of 'Alfvénization', a ubiquitous feature of MHD turbulence. Apart from a small 'residual energy', significant departure from equipartition is due only to those effects which force an *imbalance* between fluid and field, such as deviation of the magnetic Prandtl number from unity (i.e. $P_\mathrm{m} \neq 1$), differences between $\langle \widetilde{f}^2 \rangle_\mathbf{k}$, $\langle \widetilde{f}_\mathrm{a}^2 \rangle_\mathbf{k}$, etc. In the tachocline $P_\mathrm{m} \leq 1$ but not drastically so, and $\langle \widetilde{f}^2 \rangle_\mathbf{k}$, $\langle \widetilde{f}_\mathrm{a}^2 \rangle_\mathbf{k}$ must have finite correlation, as both are due to convective overshoot and its consequent entrainment of convection zone magnetic fields. Thus, it seems eminently reasonable to expect significant competition and cancellation between fluid and magnetic stresses in the $k_>$-range, resulting in a substantial shortfall in turbulent momentum transport, relative to expectations. This tendency toward cancellation is a trivial consequence of the fluid and magnetic stresses tending toward equality and entering Γ_v with opposite sign. Thus, we can write $\Gamma_{v_>}$, the momentum flux due to fluctuations in the $k_>$-range, as

$$\Gamma_{v_>} = \sum_{k_>} -k_x k_y r_\mathbf{k} |\psi_\mathbf{k}|^2 \simeq r \langle \widetilde{V}_y \widetilde{V}_x \rangle, \tag{9.36}$$

where $r_\mathbf{k} \ll 1$, and is the 'residual' factor, dependent upon the quantities causing imbalances, so that,

$$r_\mathbf{k} = r_\mathbf{k}(P_\mathrm{m}, \langle \widetilde{f}^2 \rangle, \langle \widetilde{f}_\mathrm{a}^2 \rangle, \ldots) \ll 1, \tag{9.37}$$

means that the mean flow is effectively 'laminarized', and the turbulent viscosity (due to $|\mathbf{k}| > k_{RM}$) is effectively 'quenched' (note that the sign of $r_\mathbf{k}$ may vary with $\mathbf{k}$). The reduction of momentum transport due to Alfvénization of turbulence is well known in the context of the theory of 'ω-quenching' (Craddock & Diamond

1991; Küker *et al.* 1993; Kim & Dubrulle 2001; Kim *et al.* 2001) and in relation to the reduction of the rate of zonal flow generation as drift-Alfvén turbulence becomes more Alfvénic in character (Naulin *et al.* 2005). The upshot of this trend is that the contribution of the $k_>$-range to ν_h will likely be feeble in the absence of some mechanism which feeds the imbalance between fluid and magnetic energies, such as the magneto-rotational instability.

On scales $|\mathbf{k}| < k_{RM}$ (i.e. in the $k_<$-range), the fluctuation characteristics are predominantly those of Rossby waves, with quite modest magnetic perturbations. The wave turbulence nature of the $k_<$-range fluctuations precludes direct application of the 'conventional wisdom' of strong hydrodynamic turbulence in two dimensions. Here, we explore the interaction of an ambient Rossby wave spectrum with a weak, large scale 'test' shear spectrum as a means to ascertain the nature of the effective viscosity of a Rossby wave gas. Specifically, should the waves *gain* energy from the shear, the effective viscosity is *positive*, while if the waves *lose* energy to the shear, the effective viscosity is *negative*. Therefore, we proceed by examining the *modulational stability* of an ambient spectrum or gas of Rossby waves to a large-scale shear-flow perturbation (Diamond *et al.* 2005b).

Noting that large scale magnetic fluctuations are weak, that the Rossby wave energy density $\mathcal{E}_\mathbf{k} = k^2|\psi_\mathbf{k}|^2$, and considering a weak, large scale test flow δV_x, we have, from Equation (9.31),

$$\frac{\partial}{\partial t}\delta V_x = \frac{\partial}{\partial y}\sum_{\mathbf{k}_<}\frac{k_x k_y}{k^2}\widetilde{\mathcal{E}}_\mathbf{k}, \tag{9.38}$$

where $\widetilde{\mathcal{E}}_\mathbf{k}$ indicates the *modulation* in the wave energy induced by δV_x. Since the Rossby wave population density N is simply the enstrophy density (Dubrulle & Nazarenko 1997), we can re-write Equation (9.38) as

$$\frac{\partial}{\partial t}\delta V_x = \frac{\partial}{\partial y}\sum_{\mathbf{k}_<}\frac{k_x k_y}{k^4}\widetilde{N}_\mathbf{k}, \tag{9.39}$$

where

$$\frac{\partial \widetilde{N}}{\partial t} + \mathbf{v}_g\cdot\nabla\widetilde{N} + \delta\omega_\mathbf{k}\widetilde{N} = \frac{\partial(k_x\delta V_x)}{\partial y}\frac{\partial\langle N\rangle}{\partial k_y} \tag{9.40}$$

is the linearized wave kinetic equation for $\widetilde{N}$. The wave kinetic equation simply states that wave population density is an adiabatic invariant for slowly varying, large scale shear flows and is conserved along rays, up to scrambling. The population perturbation varies adiabatically with the large-scale flow perturbation. Here $\delta\omega_\mathbf{k}$ accounts for the finite lifetime of a wave packet induced by nonlinear scrambling, $\mathbf{v}_g$ is the packet group velocity, and $\langle N\rangle$ is the mean Rossby wave enstrophy spectrum.

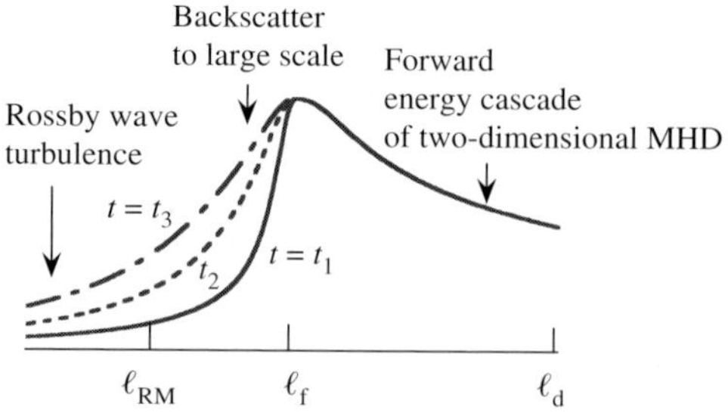

Figure 9.6. A cartoon showing the development in time of the large-scale portion of the energy spectrum for β-plane MHD turbulence. Note that since the Rossby wave scales satisfy $\ell_{\mathrm{R}} < \ell_{\mathrm{RM}} < \ell_{\mathrm{f}}$ and since the MHD energy cascade is *forward* from ℓ_{f}, the spectral slope is non-negative in the Rossby wave dominated range of scales. Note that scale decreases towards the right.

Note that Equation (9.40) determines the modulation in N induced by the weak 'test' shear flow. Writing

$$\delta V_x = \sum_{\mathbf{q},\Omega} \widetilde{V}_{\mathbf{q},\Omega} \mathrm{e}^{\mathrm{i}(\mathbf{q}\cdot\mathbf{x}-\Omega t)}, \tag{9.41}$$

a straightforward calculation gives

$$\Im\{\Omega_{\mathbf{q}}\} = -q_y^2 \sum_{\mathbf{k}} \left(\frac{k_x^2 k_y}{k^4}\right)\left(\frac{\delta\omega_{\mathbf{k}}}{(\Omega - \mathbf{q}.\mathbf{v}_{\mathrm{gr}})^2 + \delta\omega_{\mathbf{k}}^2}\right)\frac{\partial\langle N\rangle}{\partial k_y}, \tag{9.42}$$

as the rate of growth of the test shear. Thus, for $\partial\langle N\rangle/\partial k_y < 0$ the shear is *amplified*, hence indicating a *negative viscosity*. Note this is the case for the forward enstrophy cascade range, for which $\partial\langle N\rangle/\partial k < 0$. However, if $\partial\langle N\rangle/\partial k_y > 0$, the *shear gives energy to the turbulence* (i.e. *the shear is damped*), so the effective viscosity is *positive*. In the likely event that the forcing spectrum $\langle\widetilde{f}^2\rangle_{\mathbf{k}}$ peaks on small scales, i.e. on $\ell < \ell_{\mathrm{RM}}$, the $k_<$-range is energized by the *backscatter* of energy toward large scale Rossby waves, as shown in Figure 9.6. The enstrophy is thus necessarily *increasing* with $\mathbf{k}$ in the $k_<$-range, so $\partial\langle N\rangle/\partial k_y > 0$ is possible there and the effective viscosity would therefore be *positive*. The value of ν_{h} departs considerably from simplistic expectations and is smaller than standard estimates by the factor $(\delta\omega/\mathbf{q}\cdot\mathbf{v}_{\mathrm{gr}})^2 < 1$.

At this point it seems fair to say that the nature of the effective turbulent viscosity of β-plane MHD turbulence is a subtle question, indeed! Having divided the set of scales available to excitation into two classes, the $k_<$-range and the $k_>$-range, we have seen that the (directly excited) $k_>$-range contributes little to the turbulent viscosity, on account of the process of Alfvénization and the consequent near-cancellation of fluid and magnetic stresses. Rossby wave turbulence in the $k_<$-range generates a positive viscosity, but one which is small in $(\delta\omega/\mathbf{q}\cdot\mathbf{v}_{\mathrm{gr}})^2$. Thus, the effective turbulent viscosity is substantially reduced or quenched, in comparison to expectations. It is also amusing to note that our predictions concerning ν_h disagree with those of *both* Spiegel & Zahn and Gough & McIntyre! Recall that Spiegel & Zahn assume a *positive* viscosity, linked to tachocline excitation by plumes and to various large scale flow profile-driven instabilities, which were subsequently investigated in some detail (Chaboyer & Zahn 1992; Zahn 1992). Building upon the conventional intuition for two-dimensional quasi-geostrophic turbulence, Gough & McIntyre argue that hydrodynamic turbulence will produce a *negative* viscosity, and will drive potential vorticity homogenization (Rhines & Young 1982). Gough & McIntyre then use this argument to further claim that hydrodynamic turbulence cannot sustain a stationary tachocline against spin-down driven 'burrowing', thus bolstering their argument that a magnetic field must exist in the radiative core of the Sun in order to restrict tachocline penetration. We argue, however, that most scales ($\ell \lesssim \ell_{\mathrm{RM}}$) contribute *essentially nothing* to turbulent viscosity, since Alfvénization of the β-plane MHD turbulence results in near-cancellation of fluid and magnetic stresses, as in the case of ω-quenching. We do suggest the possibility that energy exchange between the large scale Rossby wave spectrum and the mean flow may persist. This interaction may be (loosely) thought of as a 'viscosity'. However, in contrast to the standard simple mixing models, such wave-flow interaction is quite sensitive to the structure of the Rossby wave spectrum at large ($\ell > \ell_{\mathrm{RM}}$) scales. The wave spectrum structure in this range emerges from stochastic backscatter from smaller scales.

9.5 Discussion and conclusions

This chapter has discussed the physics of turbulent dissipation in the solar tachocline. We have identified β-plane MHD as the 'minimal model' of tachocline turbulence, and have investigated the mechanisms of energy transfer, turbulent transport and dissipation of mechanical and magnetic energy according to this model. The principal results are as follows.

(i) A key scale, ℓ_{RM}, that demarks the boundary between two-dimensional MHD dynamics and Rossby wave dynamics was identified. This scale is somewhat analogous to the Rhines scale, familiar from hydrodynamic geostrophic turbulence. However, for $\ell < \ell_{\mathrm{RM}}$, a forward cascade of energy occurs in β-plane MHD. Moreover, ℓ_{RM} depends upon magnetic Reynolds number *Rm*.

(ii) Turbulence on scales $\ell < \ell_{\mathrm{RM}}$ tends to 'Alfvénize', and thus will not substantially mix and transport momentum. These scales do *not* contribute to ν_h. This is a consequence of close competition between fluid and magnetic stresses. Some momentum transport due to the nonlinear interaction of large scale Rossby waves may occur.

(iii) Turbulent resistivity is quenched in β-plane MHD, as in two-dimensional MHD. However, even the 'quenched' η_h greatly exceeds the collisional resistivity. Thus, turbulent *horizontal* diffusive dissipation (η_h) of magnetic fields in the tachocline may pose a significant limitation on tachocline penetration.

The implications of these results for tachocline formation models require some discussion. Indeed, the results and ideas presented here may not be a welcome addition to the theory of the solar tachocline! The 'turbulent horizontal viscosity' invoked in the Spiegel–Zahn scenario is *problematic*. 'Generic' turbulence on scales $\ell < \ell_{RM}$ will not significantly mix or transport momentum to a large extent. Rossby waves on scales $\ell > \ell_{RM}$ may drive some transport, but this process is *very* sensitive to the structure of the Rossby wave spectrum and depends upon the large-scale, low-energy tail of the spectrum, about which very little is known. Alternatively, some flow profile-driven instability may produce a ν_h, but despite extensive study, the specific mechanism involved has yet to be identified. Also, proponents of this type of viscosity mechanism must explain why such an instability will not simply hover near marginality, producing a state of 'self-organized criticality' rather than steady viscous dissipation (Diamond & Hahm 1995). Work on simple systems has shown that transport and relaxation in a continuum SOC are *not* well modelled by simple diffusion (Hwa & Kardar 1992). Thus, considerable clarification of the dynamics underlying the 'horizontal viscosity' invoked in the Spiegel–Zahn scenario is necessary in order to solidify the foundations of that model.

In the Gough–McIntyre scenario, the principal effect of turbulence is to introduce turbulent horizontal diffusion of magnetic fields, so that the balance of shearing of poloidal field B_0 with dissipation of toroidal field B_ψ now becomes

$$-B_0 \sin\theta \frac{\partial \bar{\Omega}}{\partial \theta} = \left(\eta \partial_r^2 + \eta_T \partial_h^2\right) B_\psi, \tag{9.43}$$

where ∂_h refers to a horizontal derivative. Here $\eta_T/\eta \approx \langle \widetilde{B}^2 \rangle / \langle B \rangle^2 \sim Rm$, so even the 'quenched' turbulent resistivity greatly exceeds the collisional value. The 'bottom line' here is that for $\eta_T/\eta \sim Rm > r_0^2/\Delta_T^2$, where r_0 is the tachocline radius and Δ_T its thickness, *turbulent horizontal diffusion of magnetic fields will dissipate magnetic energy faster than radial collisional resistive diffusion does, as assumed in the Gough–McIntyre scenario*. Since $r_0 \sim 0.7 R_\odot$ and $\Delta_T \sim 0.03 R_\odot$, in practice this means that for $Rm \gtrsim 500$ (probably satisfied in the solar tachocline!) horizontal turbulent diffusion is *the* dissipation process that limits tachocline burrowing. Here Rm is, of course, the Reynolds number for the horizontal motion, i.e. $Rm = v_h l/\eta$. Hence, the scaling of the tachocline thickness and its dependence on B_0 (predicted by Gough & McIntyre) should be reconsidered in the light of this observation.

This chapter poses more questions than it answers. Indeed, it should be viewed only as an introduction to the problem of β-plane MHD turbulence in the tachocline. Several future investigations are strongly suggested. These include, but are not limited to, the following.

(i) Completion of a rigorous analysis in the vein begun here, and accompanied by related numerical simulations which test the theory.
(ii) Examining the effects of $\langle \widetilde{f}\,\widetilde{f}_{\mathrm{a}} \rangle \neq 0$ correlations and finite cross helicity on the turbulent dissipation processes (Bracco *et al.* 1998).
(iii) Study of a two-layer β-plane MHD model, in which only the upper layer is forced by convective overshoot.
(iv) Extension of this study to geostrophic MHD turbulence on a sphere and in a spherical shell. In this regard, we note that the existing large scale numerical calculations of tachocline dynamics are purely hydrodynamic (Miesch 2001, 2003 and Chapter 5 of this book).
(v) Consideration of large scale flow structure and its coupling to tachocline turbulence dynamics (Gilman & Fox 1997).
(vi) Study of the types and physics of coherent magnetic structures formed in tachocline turbulence. Such structures may have the form of magnetic vortices (Kinney *et al.* 1995; Gruzinov *et al.* 2002) or zonal magnetic fields (Gruzinov *et al.* 2002). It has been suggested that magnetic structures formed in the tachocline may leave an 'imprint' on the magnetic fields ultimately observed in the photosphere (E. A. Spiegel, private communication).
(vii) Consideration of the effects of tachocline turbulence and flow structure on the solar differential rotation (Itoh *et al.* 2005).
(viii) Study of the dissipation of magnetic fields by *vertical mixing* in the stably stratified tachocline. Internal wave interaction is a possible agent of such mixing.

Of course, more consideration should also be given to the physics of vertical mixing in the strong, stable stratification environment of the tachocline. In particular, the effects of wave interactions on vertical transport of magnetic fields and on vertical momentum transport (including possible 'negative viscosity effects') merit further investigation. We hope that such studies will help elucidate the structure of the solar tachocline.

Acknowledgments

We thank (in alphabetical order) P. H. Haynes, D. W. Hughes, M. E. McIntyre, E. A. Spiegel, S. M. Tobias and A. Yoshizawa for stimulating discussions on these and related topics. We also thank O. Gurcan, D. W. Hughes, S. Keating and S. M. Tobias for critical readings of the manuscript. P.H.D. and L.J.S. would like to thank the Isaac Newton Institute for hospitality during the Tachocline

Workshop and during the performance of part of this work. P.H.D. and K.I. also thank Kyushu University for hospitality during the performance of part of this work. This work was supported by DoE Grant No. DE-FG02-04ER54738, NASA Grant No. NNG04GK96G, and by the Grant-in-Aid for Specially-Promoted Research of MEXT (16002005).

References

Bracco, A., Provenzale, A., Spiegel, E. A. & Yecko, P. (1998). *'Spotted Disks'*, astro-ph/9802298 23 Feb.

Bretherton, F. & Spiegel, E. A. (1968). *Astrophys. J.*, **153**, L77.

Cattaneo, F. & Vainshtein, S. I. (1991). *Astrophys. J.*, **376**, L21.

Chaboyer, B. & Zahn, J. P. (1992). *Astron. Astrophys.*, **253**, 173.

Craddock, G. & Diamond, P. H. (1991). *Phys. Rev. Lett.*, **67**, 1535.

Diamond, P. H. & Hahm, T. S. (1995). *Phys. Plasmas*, **2**, 3640.

Diamond, P. H., Hughes, D. W. & Kim, E.-J. (2005a). In *Fluid Dynamics and Dynamos in Astrophysics and Geophysics*, ed. A. M. Soward, C. A. Jones, D. W. Hughes & N. O. Weiss. London: CRC Press, p. 145.

Diamond, P. H., Itoh, S.-I., Itoh, K. & Hahm, T. S. (2005b). *Plasma Phys. Cont. Fusion*, **47**, R35.

Dubrulle, B. & Nazarenko, S. (1997). *Physica D*, **110**, 123.

Eddington, A. S. (1926). *The Internal Constitution of the Stars*. Cambridge: Cambridge University Press.

Gilman, P. A. & Fox, P. A. (1997). *Astrophys. J.*, **484**, 439.

Gough, D. O. & McIntyre, M. E. (1998). *Nature*, **394**, 755.

Gruzinov, A. V. & Diamond, P. H. (1994). *Phys. Rev. Lett.*, **72**, 1651.

Gruzinov, I., Das, A. & Diamond, P. H. (2002). *Phys. Let. A*, **302**, 119.

Horton, C. W. (1999). *Rev. Mod. Phys.*, **71**, 735.

Hughes, D. W. (1991). In *Advances in Solar System Magnetohydrodynamics*, ed. E. R. Priest & A. W. Hood. Cambridge University Press, Cambridge, p. 77.

Hwa, T. & Kardar, M. (1992). *Phys. Rev. A*, **45**, 7002.

Itoh, S.-I., Itoh, K., Yoshizawa, A. & Yokoi, N. (2005). *Astrophys. J.*, **618**, 1044.

Kim, E.-J. & Dubrulle, B. (2001). *Phys. Plasmas*, **8**, 813.

Kim, E.-J., Hahm, T. S. & Diamond, P. H. (2001). *Phys. Plasmas*, **8**, 3576.

Kinney, R., McWilliams, J. C. & Tajima, T. (1995). *Phys. Plasmas*, **2**, 3623.

Kraichnan, R. H. (1959). *J. Fluid Mech.*, **5**, 497.

Kraichnan, R. H. (1965). *Phys. Fluids*, **8**, 1385.

Küker, M., Rüdiger, G. & Kitchatinov, L. L. (1993). *Astron. Astrophys.*, **279**, 1.

McIntyre, M. E. (2000). In *Perspectives in Fluid Dynamics*, ed. G. K. Batchelor, H. K. Moffatt & M. G. Worster. Cambridge: Cambridge University Press, p. 557.

McIntyre, M. E. (2003). In *Stellar Astrophysical Fluid Dynamics*, ed. M. J. Thompson & J. Christensen-Dalsgaard. Cambridge: Cambridge University Press, p. 111.

Mestel, L. (1999). *Stellar Magnetism*. Oxford: Clarendon Press.

Miesch, M. S. (2001). *Astrophys. J.*, **562**, 1058.

Miesch, M. S. (2003). *Astrophys. J.*, **586**, 663.

Miesch, M. S. (2005). *Living Rev. Solar Phys.*, **2**, 1 (www.livingreviews.org/lrsp-2005-1).

Moffatt, H. K. (1978). *Magnetic Field Generation in Electrically Conducting Fluids*. Cambridge: Cambridge University Press.

Naulin, V., Kendl, A., Garcia, O. E., Nielsen, A. H. & Rasmussen, J. J. (2005). *Phys. Plasmas*, **12**, 052515.

Nozawa, T. & Yoden, S. (1997). *Phys. Fluids*, **9**, 2081.

Orszag, S. A. (1970). *J. Fluid Mech.*, **41**, 363.

Parker, E. N. (1966). *Astrophys. J.*, **145**, 811.

Parker, E. N. (1993). *Astrophys. J.*, **408**, 707.

Pedlosky, J. (1987). *Geophysical Fluid Dynamics*. New York: Springer-Verlag.

Pouquet, A. (1978). *J. Fluid Mech.*, **88**, 1.

Pouquet, A, Frisch, U. & Léorat, J. (1976). *J. Fluid Mech.*, **77**, 321.

Rhines, P. B. (1975). *J. Fluid Mech.*, **69**, 417.

Rhines, P. B. & Young, W. R. (1982). *J. Fluid Mech.*, **122**, 347.

Rüdiger, G. & Kitchatinov, L. L. (1977). *Astron. Nachr.*, **318**, 273.

Schou, J., Antia, H. M., Basu, S. *et al.* (1998). *Astrophys. J.*, **505**, 390.

Spiegel, E. A. & Zahn, J.-P. (1992). *Astron. Astrophys.*, **265**, 106.

Sweet, P. A. (1950). *Mon. Not. Roy. Astron. Soc.*, **110**, 548.

Tobias, S. M. (2005). In *Fluid Dynamics and Dynamos in Astrophysics and Geophysics*, ed. A. M. Soward, C. A. Jones, D. W. Hughes & N. O. Weiss. London: CRC Press, p. 193.

Tobias, S. M., Brummell, N. H., Clune, T. L. & Toomre, J. (2001). *Astrophys. J.*, **549**, 1183.

Vallis, G. K. & Maltrud, M. E. (1993). *J. Phys. Ocean.*, **23**, 1346.

Yoshizawa, A., Itoh, S.-I. & Itoh, K. (2003). *Plasma and Fluid Turbulence: Theory and Modelling*. Bristol: IoP Publishing.

Zeldovich, Ya. B. (1957). *Sov. Phys. J.E.T.P.*, **4**, 460.

Zahn, J.-P. (1992). *Astron. Astrophys.*, **265**, 115.

Part V

Instabilities

10

Global MHD instabilities of the tachocline

Peter A. Gilman & Paul S. Cally

The combination of differential rotation and toroidal fields believed to exist in the solar tachocline should be unstable to global MHD modes, typically dominated by longitudinal wavenumber $m = 1$ modes for toroidal fields of peak value 30 kG and higher, and a broader range of low m values for weaker fields. For toroidal field bands, the high field instability takes the form of a 'tipping' of the band away from coincidence with circles of latitude. For a wide range of toroidal fields and differential rotations, and in both the overshoot and radiative parts of the tachocline, the unstable modes grow in a time short compared to a solar cycle, and are therefore of interest for the solar dynamo problem, as well as for creation of longitude-dependent magnetic patterns seen at the solar surface. The latitudinal momentum transport by Reynolds and Maxwell stresses associated with unstable modes provides a way to mix angular momentum in latitude, and help limit the thickness of the tachocline.

10.1 Introduction

The study of global MHD instabilities of differential rotation and toroidal fields that might be present in the solar tachocline began with Gilman & Fox (1997). Their original motivation was to see whether the magnetic field could destabilize the differential rotation of the tachocline, estimated to be stable to hydrodynamical disturbances by itself. Spiegel & Zahn (1992) had argued that anisotropic, quasi-two-dimensional turbulence was needed in the tachocline to transport angular momentum from low latitudes to high, and thereby prevent the spread of the tachocline into the deep interior of the Sun during its lifetime. They speculated that this anisotropic turbulence could come from a hydrodynamic instability of the latitudinal differential rotation there. However, it is well established that the helioseismically determined latitudinal rotation gradient is in fact hydrodynamically stable, at least to global two-dimensional disturbances (Charbonneau *et al.* 1999b), or at most very weakly unstable (Garaud 2001), in the sense of saturating nonlinearly at very

The Solar Tachocline, D. W. Hughes, R. Rosner and N. O. Weiss.
Published by Cambridge University Press.

low levels. Gilman & Fox's basic result is that latitudinal differential rotation is virtually always unstable to two-dimensional global MHD disturbances, even for very weak toroidal fields. The unstable modes transmit angular momentum to high latitudes from low, but mainly via the Maxwell stress rather than the Reynolds stress. We explore and summarize the details of this instability in the sections that follow.

A global MHD instability of the tachocline is of much interest for the Sun beyond explaining the existence and thickness of the solar tachocline. The Sun has a 'solar cycle', with complex magnetic field patterns that show field reversals about every 11 years. These patterns are almost certainly maintained by dynamo action, and there are strong reasons for assuming that as part of this dynamo the solar tachocline contains strong ($\sim$100 kG) toroidal magnetic fields. Thus a global MHD instability in the tachocline should contribute to the solar dynamo. We discuss properties of the instability that could be important for the solar dynamo in later sections.

Beyond the differential rotation of the tachocline (discussed in detail in Chapter 3 by Christensen-Dalsgaard & Thompson) its most important property for considering global instabilities is its stratification and, in particular, whether the stratification is subadiabatic, and if so, by how much. Helioseismology tells us that the bottom of the convection zone, defined as the place where the temperature gradient changes from being nearly adiabatic to substantially subadiabatic in radiative equilibrium, occurs at a radius $r = 0.713R_{\odot}$ (Christensen-Dalsgaard *et al.* 1991). Charbonneau *et al.* (1999a) show that the tachocline straddles this radius, with roughly one third above it and two-thirds below it. More refined helioseismic measurements may change this result, but it indicates that both the overshoot layer at the bottom of the convection zone, and the much more subadiabatic radiative core, are contained within the tachocline.

Helioseismology cannot tell us, but within the overshoot layer we expect the stratification to be slightly subadiabatic, of order 10^{-4} to 10^{-6} of the adiabatic gradient (Rempel 2004, and references therein), while within the radiative layer below, the subadiabaticity rapidly approaches 10^{-1} with increasing depth. This subadiabatic stratification favours motions that are nearly horizontal in spherical shells, since vertical motions, particularly of global scale, have to work against the subadiabatic gradient. This property justifies the study of two-dimensional global MHD instabilities for the tachocline (see also the discussion of the Miles–Howard theorem in Section 10.2), though we will see in Section 10.4 that even small departures from two dimensions can be quite important in numerous ways. In our description below, we first focus on the strictly two-dimensional case, into which the subadiabatic stratification does not explicitly enter, and then consider its simplest generalization to three dimensions, namely an MHD generalization (Gilman 2000a) of the so-called 'shallow water' equations, widely used in geophysical fluid dynamics.

In order to do a rigorous calculation of an instability, say as an eigenvalue problem, one must start from a steady equilibrium of forces that is to be perturbed. In the solar tachocline, such equilibria should be possible with both toroidal fields and differential rotation present. The radial force balance should be magnetohydrostatic. In latitude in such a spherical shell, there is a magnetic curvature stress from the toroidal field that, if not opposed, will pull the field and attached material toward the poles. For toroidal fields of broad latitudinal profile, the curvature stress is most easily balanced by an equatorward hydrostatic pressure gradient. For narrow toroidal bands, this still works, but may be supplemented or replaced by an equatorward directed Coriolis force from a prograde fluid jet inside the toroidal ring (Rempel *et al.* 2000; Dikpati *et al.* 2003). In our analysis below, we consider both kinds of equilibria.

The instability we study here comes from a general class of ideal fluid shear flow instabilities for which so-called singular points of the system are important for determining properties of the instability and the structure of unstable modes. In the hydrodynamic system, the singular point occurs where the longitudinal phase velocity of an unstable mode equals the local rotation rate of the system. There is a large literature on this type of instability (see for example the discussion of the inviscid Orr–Sommerfeld equation in Drazin & Reid 1981, chapter 4). In the two-dimensional MHD case, this point is no longer singular, but it is replaced by other points where the difference between the mode phase speed and the local rotation rate equals the local Alfvén speed of the toroidal field. There is a much smaller literature for this case. This singular point remains the critical one in the MHD shallow water system, but the hydrodynamic singular point reappears, as well as a third one, involving the gravity wave speed as well as the phase velocity, rotation and Alfvén speeds.

Another point to keep in mind for this class of instability is that there is a significant difference between the unstable modes allowed in, say, an infinite channel, a very common configuration studied in both the HD and MHD cases, and in a spherical shell such as the solar tachocline. The difference is that in the channel all disturbance wavelengths along the channel are, in principle, excitable, while in the spherical shell the wavelengths are discrete, defined by the integer longitudinal wavenumber m. In the two-dimensional case, the lowest longitudinal wavenumber allowed is $m = 1$, so shear flow configurations that might be unstable to long-wave modes in a channel may be stable to discrete modes in the spherical shell.

The Alfvénic singular points in the shell defined above separate latitudinal domains where energy conversions, from the background state to the perturbations, are occurring, from other domains where the perturbations have amplitude but are energetically neutral, that is, no energy conversions are taking place. This means that in all specific cases, the energy conversions in part are maintaining neutral

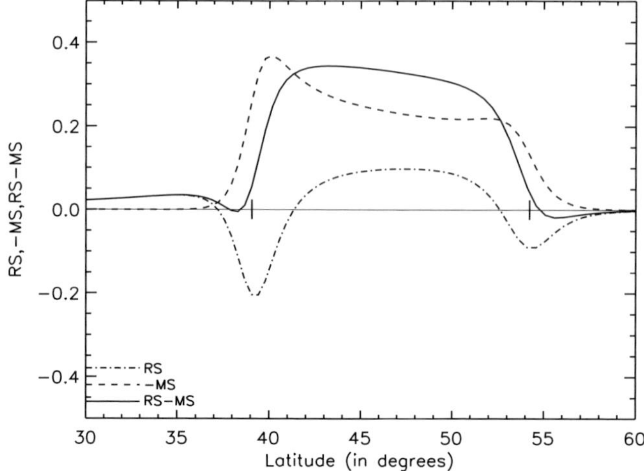

Figure 10.1. Example of profile of Reynolds (RS) and Maxwell (MS) stresses from a typical global MHD unstable mode, showing the transport of angular momentum from the equatorward side to the poleward side of a toroidal band of full width at half maximum 16° in latitude, which extracts kinetic energy from the differential rotation (taken from Dikpati & Gilman 1999). There is also energy extracted from the toroidal field itself by the 'mixed' stress associated with the same disturbances. Vertical marks on the horizontal axis show the location of the two singular points in the domain, illustrating how the stresses are organized about these points. Both stresses are essentially zero outside the latitude band defined by these points, even though there is substantial flow amplitude there (not shown). These flows are therefore energetically inactive (no stresses or transports), while the flow and magnetic perturbations between the singular points are energetically active.

wavelike structures by latitudinal work through pressure forces. For broad toroidal profiles, examples are given in Gilman & Fox (1999a,b). For narrow toroidal bands, the picture is simpler, because the two Alfvénic singular points in each hemisphere are always found on the shoulders of the toroidal profile. All of the energy conversions to drive the instability are going on between these two points, or inside the toroidal band. Everything outside is energetically neutral, and of course the perturbation magnetic fields are confined to the neighbourhood of the band, since there is a virtually field free domain away from the band.

An example of the organization of the energy conversion processes according to the singular points is given in Figure 10.1, which depicts the Reynolds and Maxwell stresses for an unstable disturbance of a toroidal band of 16° latitudinal half width, placed at 45° latitude. The vertical marks on the horizontal axis at about 39° and 54° are the location of the two singular points for this case. Clearly both Maxwell and Reynolds stress profiles change abruptly in their neighbourhood, and together lead to a smooth profile of transport of angular momentum from low latitudes to high across the toroidal band. Outside these latitudes both stresses are essentially

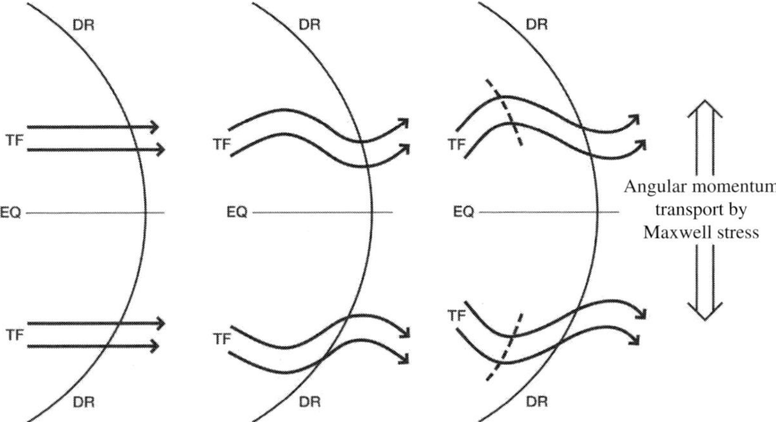

Figure 10.2. Schematic of how a longitudinally periodic disturbance in the toroidal magnetic field, interacting with the ambient differential rotation, creates a Maxwell stress by tilting the perturbation field lines, which extracts kinetic energy from the differential rotation and transports angular momentum from low latitudes to high (from Gilman 2005).

zero, in the domain where the disturbances are energetically neutral. Dikpati & Gilman (1999) give numerous other examples, including the structure of the mixed stress, determined also by the singular points, that extracts magnetic energy from the toroidal band itself.

Without even doing a formal perturbation analysis, we can demonstrate heuristically that combinations of differential rotation and toroidal field should be unstable when perturbed by a longitudinally periodic disturbance. Figure 10.2 shows schematically how this works. If the toroidal band shown in the left-hand schematic is perturbed, as in the centre schematic, the differential rotation will tilt the field lines as shown in the right-hand schematic. This tilt immediately implies a Maxwell stress that transports angular momentum down the rotation gradient, in this case toward higher latitudes. This extracts kinetic energy out of the differential rotation to drive the instability. By contrast, a velocity perturbation shaped like the wavy arrows in the right-hand schematic, which represents the hydrodynamic case, would yield angular momentum transport up the gradient, which would suck energy out of the disturbance and render it stable. This picture is qualitatively correct so long as the toroidal field is not so strong as to resist being tilted and deformed.

Before going into the formal instability analysis, it is helpful to consider qualitatively the type of disturbances to the toroidal field, particularly toroidal bands, we can expect. Figure 10.3 shows three different types of disturbance of a toroidal band, for longitudinal wave numbers $m = 0$, 1, and 2. The $m = 0$ mode is impossible to realize in two dimensions, because there is no place for the mass on the poleward side to go. It can be realized in the MHD shallow water case, but generally only

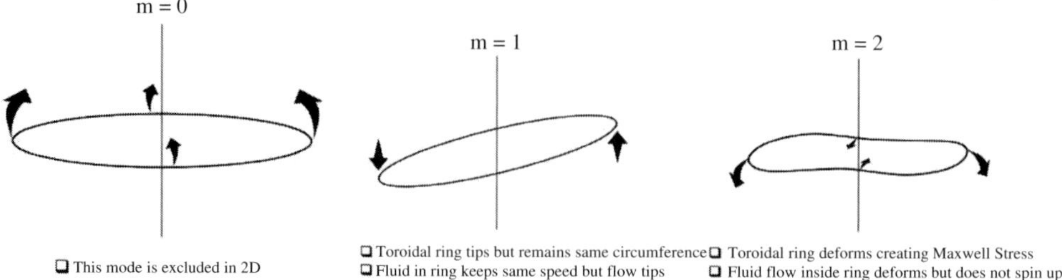

Figure 10.3. Schematic of displacement and deformation of a toroidal ring by periodic disturbances of differing longitudinal wavenumber m. Left-hand panel: $m = 0$ modes do not exist in the two-dimensional (2D) case, but do in shallow water and three-dimensional (3D) cases; ring can be displaced and shrunk or expanded. Centre: for $m = 1$ modes, which are most common for strong magnetic field, the ring tips. Right: for $m = 2$ (and higher) modes are deformed but not tipped. (Adapted from Dikpati *et al.* 2004a.)

for toroidal fields orders of magnitude larger than for $m > 0$ modes, and so is of relatively little interest.

We will find that the $m = 1$ mode is the one most commonly unstable, and the only one unstable for high toroidal fields (cf. Tayler 1973). As seen in Figure 10.2, it represents a 'tipping' of the toroidal field, with no deformation (or stretching or shrinking, as it turns out). Modes with $m = 2$ or higher all represent deformations of the toroidal ring about its original latitude, with more lobes for higher m. These modes are generally unstable only for small and moderate toroidal field. When the field is stronger, it resists such deformations, and therefore resists being unstable. The $m = 1$ tipping mode is not deforming, so strong field is no barrier to becoming unstable by tipping.

Differential rotation and toroidal fields present in the equilibrium state provide two reservoirs of energy to drive the instability; these will produce two forms of perturbation energy – kinetic and magnetic. Therefore we can expect fairly intricate flows of energy when the system is unstable. As usual, the unstable modes alter the differential rotation and toroidal fields to draw energy from them, thereby eventually bounding their own growth. The connections among the various energy reservoirs of this MHD system are illustrated schematically in Figure 10.4 below.

10.2 Classical shear and magneto-shear instabilities

Before embarking upon a discussion of possible shear instabilities in the solar tachocline, we briefly review the now classical literature on shear and magneto-shear instabilities, beginning with the simplest case of incompressible hydrodynamic

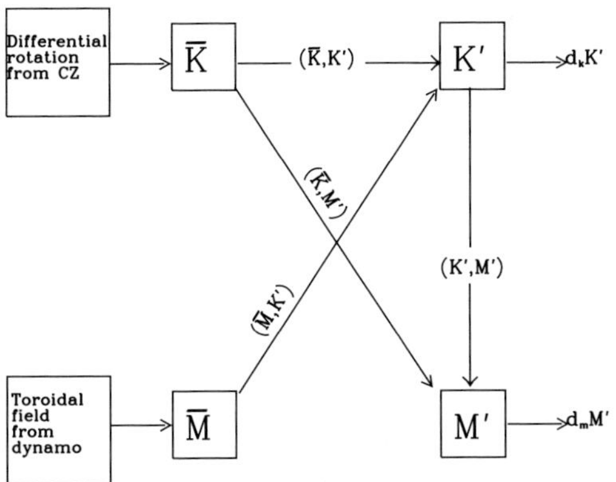

Figure 10.4. Schematic energy diagram for all the possible energy flows from growing unstable MHD modes. K represents kinetic energy and M magnetic. Energy flow directions shown are commonly found, but reversed flows are possible in some cases, so long as at least one is from a reference state reservoir. Overbars denote unperturbed state energies, primes perturbation energies; d_{k} and d_{m} are respectively kinetic and magnetic drag (see Section 10.3.4). Square boxes on the left denote processes that restore energy to the reference state. (From Dikpati *et al.* 2004a.)

shear flow, before adding magnetic fields. Ideal (dissipationless) systems are assumed throughout.

Two notable general results on inviscid hydrodynamic shear instabilities for flows of the form $u(z)\hat{\mathbf{x}}$ are (i) the Miles–Howard theorem (Miles 1961; Howard 1961): *a necessary condition for instability is that the Richardson number* $Ri = N^2/u'^2$ *is somewhere less than* $\frac{1}{4}$*, where N is the Brunt–Väisälä frequency, and furthermore, the growth rate* ω_{i} *of any instability is limited by* $\omega_{\mathrm{i}}^2 \leqslant \sup_z(\frac{1}{4}u'^2 - N^2)$; and (ii) Howard's semicircle theorem (Howard 1961): *the complex phase speeds* ω/k *associated with instability in a parallel shear flow with flow speed range* $\Delta u = \sup_z u - \inf_z u$ *lie in the semicircle of radius* $\frac{1}{2}\Delta u$ *centred on* $\bar{u} = \frac{1}{2}(\sup_z u + \inf_z u)$*, i.e.* $|\omega/k - \bar{u}|^2 < \frac{1}{4}(\Delta u)^2$. These are discussed at length by Drazin & Reid (1981). In the radiative part of the tachocline, the Brunt–Väisälä frequency is easily large enough for the Miles–Howard theorem to guarantee stability of the radial shear profile, at least in the absence of magnetic field. The Miles–Howard and semicircle theorems have been generalized to compressible shear flows by Chimonas (1970) and Eckart (1963), respectively.

The Miles–Howard theorem depends on shear profile, and supplies both a necessary instability criterion and a growth rate bound. The hydrodynamic semicircle theorem on the other hand relates to the total shear Δu, and gives information only on unstable eigenvalues. Various attempts have been made to generalize each to shear flows with velocity-aligned magnetic field added (e.g. Adam 1980), though most have not been very practical. One easily applicable sufficient criterion for stability is (Cally 2000): *an incompressible parallel shear flow with aligned magnetic field is stable if there exists a Galilean frame in which the flow is nowhere super-Alfvénic.*[1] Hughes & Tobias (2001) extend this further to yield the very satisfying generalization of Howard's semicircle theorem to MHD flow: *unstable phase speeds lie in the semicircle defined by* $|\omega/k-\bar{u}|^2 < \frac{1}{4}(\Delta u)^2 - \inf_z a^2$, *where* $a(z)$ *is the Alfvén speed.* The Hughes–Tobias theorem is easily shown to subsume Cally's result. On the basis of these results, Cally (2000) argues that a toroidal magnetic field of several kilogauss in the solar tachocline is sufficient to stabilize the *radial* shear against *local* shear instabilities, even without the stabilizing influence of stratification.

However, the topic of this chapter is *global* instabilities, not local. By this we mean instabilities at low longitudinal wavenumber m that are dependent on the spherical geometry. As discussed in Section 10.1, the $m = 1$ 'tipping mode' plays a particularly important role. Tayler (1973) found that $m = 1$ instabilities in *non*-rotating stars with toroidal magnetic field may be dominant, and are preferentially located near the axis of symmetry. This is apparently equivalent to the 'polar kink' instability of Cally (2003) (see Section 10.4.2), which is stabilized by rotation speeds in excess of the Alfvén speed. In any case, it is certainly not *driven* by rotation.

On the other hand, Acheson (1978) notes that the $m = 1$ instability is also most readily excited in rapidly rotating stars with toroidal magnetic field in the regime $a^2 \ll \Omega^2\varpi^2 \ll c^2 \lesssim g\varpi$, where Ω is angular velocity and $\varpi = r\sin\theta$ is the cylindrical radius. In the convection zone, this occurs if $-\varpi^3\,\partial\Omega^2/\partial\varpi > a^2$. In the core though, where the stratification is strongly subadiabatic, the instability is associated with almost horizontal motions, and sets in (again most strongly at $m = 1$) if $-\partial\Omega^2/\partial\theta > 2a^2\varpi^{-2}\csc 2\theta$. However, with anticipated field strengths of up to 10^5 G, it follows that $a \lesssim \Omega\varpi$ in the solar tachocline, and so this analysis, though suggestive, is not strictly relevant to strong-field latitudes.

The most famous magneto-rotational instability in astrophysics is the MRI of Balbus & Hawley (1991) (see also Ogilvie & Pringle 1996 and the discussion by Ogilvie in Chapter 12 of this book; Kitchatinov & Rüdiger 1997) – first applied to weakly magnetized accretion discs. The MRI is local in nature, and depends on an interplay between centrifugal force and the magnetic tension of poloidal magnetic field lines which thread the disc. It is unrelated to any stellar instabilities involving toroidal field.

[1] This also applies to compressible flow with gravity neglected, but with the Alfvén speed replaced by the cusp speed $c_T = ac/\sqrt{a^2+c^2}$.

10.3 Two-dimensional shell models

10.3.1 Basic equations: nonlinear and linear

The nonlinear equations for incompressible two-dimensional magneto-shell oscillations are most conveniently written in terms of the streamfunction ψ and magnetic streamfunction χ, from which the fluid velocity $\mathbf{v} = v\,\hat{\mathbf{e}}_\theta + w\,\hat{\mathbf{e}}_\phi$ and vector Alfvén velocity $\mathbf{a} = \mathbf{B}/\sqrt{\mu\rho} = a_\theta\,\hat{\mathbf{e}}_\theta + a_\phi\,\hat{\mathbf{e}}_\phi$ may be derived (spherical coordinates (r, θ, ϕ) are used here, where θ is colatitude and ϕ is longitude):

$$(-w, v) = \nabla\psi, \qquad (-a_\phi, a_\theta) = \nabla\chi. \tag{10.1}$$

In terms of the radial vorticity $\Omega = -\nabla^2\psi$ and the scaled electric current density $J = -\nabla^2\chi$, the evolution equations may be written in a concise form derived in Cally (2001):

$$\frac{\mathrm{D}\Omega}{\mathrm{D}t} = \mathbf{a}\cdot\nabla J + \eta_\mathrm{k}\left(\nabla^2\Omega + \frac{2}{r^2}\Omega\right), \tag{10.2}$$

$$\frac{\mathrm{D}\chi}{\mathrm{D}t} = \eta_\mathrm{m}\nabla^2\chi, \tag{10.3}$$

where kinetic and magnetic diffusivities η_k and η_m have been retained. This system exactly conserves angular momentum. In the absence of the kinetic and magnetic diffusivities, it also conserves total energy $\frac{1}{2}\int\int_S \left(|\nabla\psi|^2 + |\nabla\chi|^2\right)\mathrm{d}S$, cross helicity $\int\int_S \Omega\chi\,\mathrm{d}S = \int\int_S \mathbf{v}\cdot\mathbf{a}\,\mathrm{d}S$, and mean square magnetic potential $\frac{1}{2}\int\int_S \chi^2\,\mathrm{d}S$, where the integrals are over the whole spherical surface S (Cally 2001).

The most convenient way to derive the equations governing linear instabilities in the two-dimensional shell is to directly linearize Equations (10.2) and (10.3). Setting $\psi(\mu,\phi,t) = \psi_0(\mu) + \psi_1(\mu,\phi,t)$ and $\chi(\mu,\phi,t) = \chi_0(\mu) + \chi_1(\mu,\phi,t)$, the equations may be linearized in ψ_1 and χ_1. If furthermore, a ϕ and t dependence of the form $\exp[im(\phi - ct)]$ is assumed, where $c = c_\mathrm{r} + ic_\mathrm{i}$ is the (generally complex) longitudinal phase speed and m is an integer, then it is easily found that

$$(\omega_0 - c)\mathcal{L}\psi_1 - \psi_1\frac{\mathrm{d}^2}{\mathrm{d}\mu^2}[(1-\mu^2)\omega_0] - \alpha_0\mathcal{L}\chi_1 + \chi_1\frac{\mathrm{d}^2}{\mathrm{d}\mu^2}[(1-\mu^2)\alpha_0] = 0, \tag{10.4}$$

and

$$(\omega_0 - c)\chi_1 = \alpha_0\psi_1. \tag{10.5}$$

Here

$$\mathcal{L} = \frac{\mathrm{d}}{\mathrm{d}\mu}\left[(1-\mu^2)\frac{\mathrm{d}}{\mathrm{d}\mu}\right] - \frac{m^2}{1-\mu^2} \tag{10.6}$$

is the Legendre operator, $\omega_0 = \mathrm{d}\psi_0/\mathrm{d}\mu$ is the rotational angular frequency, and $\alpha_0 = \mathrm{d}\chi_0/\mathrm{d}\mu$ is the analogous Alfvén frequency. Since ψ_1 and χ_1 must vanish at the poles, Equations (10.4) and (10.5) may be solved numerically for specific rotational and Alfvén profiles to obtain eigenvalues c and eigenfunctions ψ_1 and χ_1. This has been done in several papers, starting with Gilman & Fox (1997).

When the underlying rotational and magnetic profiles display the expected symmetry about the equator, namely $\mathbf{v}$ symmetric (i.e. ψ_0 antisymmetric) and $\mathbf{a}$ antisymmetric (χ_0 symmetric), the eigenfunctions may be either symmetric (ψ_1 even in μ, χ_1 odd) or antisymmetric (ψ_1 odd, χ_1 even). Depending on the specific model, either one or the other may dominate as regards linear growth rate. This carries over to the nonlinear regime only partially though. Equations (10.2) and (10.3) clearly maintain antisymmetric structure if it is imposed as an initial condition, but a symmetric perturbation ψ_1 breaks the antisymmetry of ψ, and conversely for χ. Consequently, an initially antisymmetric perturbation will remain forever antisymmetric, but a symmetric perturbation ψ_1 to the antisymmetric ψ_0 (antisymmetric χ_1, symmetric χ_0) will see both the symmetric and antisymmetric parts evolve, to first and second order in perturbations respectively.

The schematic diagram Figure 10.4 shows how kinetic and magnetic mean energies $\bar{K}$ and $\bar{M}$, which are supplied from the convection zone and dynamo respectively, in turn supply energy to the kinetic and magnetic perturbations K' and M' through various stresses, and then possibly back to their original sources through drag (see Section 10.3.4). Details will be discussed in following sections.

10.3.2 Ideal fluid linear MHD instabilities

Based on Equations (10.4) and (10.5), Gilman & Fox (1997) deduced several general properties of unstable modes for the two-dimensional MHD case. Perhaps the most important of these is that the eigenvalue of unstable modes satisfies the semicircle theorem, originally proved for HD channel flow by Howard (1961). This theorem puts limits on both the growth rates and longitudinal phase velocities of unstable modes. Most remarkably, these limits are independent of the toroidal field strength or profile. Unstable modes must have phase velocities essentially between the minimum and maximum rotation rates of the system, and the growth rates are also bounded by the amount of differential rotation present.

We interpret this result as saying that the unstable modes have no properties resembling an Alfvén wave. Such waves are present in the system, but they are neutral waves with much different phase velocities than the unstable modes. It should follow that in the limit of vanishing differential rotation, there should be no instability, no matter what the toroidal field amplitude or profile is, and that is indeed what is found. Having a phase velocity between the minimum and maximum rotation of the system ensures that the magnetic patterns can be sheared as shown in Figure 10.1, and thereby extract energy from the differential rotation for the instability. The much faster Alfvén waves will not be sheared in this way, because the displacement of fluid particles and associated field lines oscillates so fast that little shearing can occur.

A second general result found by Gilman & Fox (1997) is that the necessary condition for instability, namely that the latitude gradient of total vorticity of the unperturbed state must change sign in the domain, no longer applies as soon as a magnetic field is added. The reason is simply that with any magnetic field present, vorticity is no longer conserved in the system. This result suggests the possibility of instability occurring even for weak magnetic fields, and that is what has been found.

To focus on MHD instabilities of the tachocline in detail, we start with differential rotation profiles that Charbonneau *et al.* (1999b) showed were stable there, namely profiles of the form $\omega_0 = s_0 - s_2\mu^2$, the same as used by Watson (1981). In that case, to be hydrodynamically unstable s_2 must exceed about 0.29, measured in units of the equatorial rotation, far above tachocline values. They also showed that if a term of the form $-s_4\mu^4$ is added, the profile is unstable for considerably smaller total differential rotations, but for amplitudes of s_2 and s_4 deduced from helioseismic measurements, the tachocline differential rotation is still stable, at least marginally, to two-dimensional hydrodynamic disturbances.

There is very little observational or theoretical evidence to guide us in a choice of toroidal field profiles to assume for the instability calculations. Current solar dynamo models (Dikpati *et al.* 2004b, and references therein) rely upon assuming a low magnetic diffusivity below the solar convection zone in order to generate toroidal fields as large as 50–100 kG, needed to produce rising loops that lead to emergence of sunspots at low latitudes. These models generally generate broad poloidal fields, which differential rotation in the tachocline and above shear into latitudinally broad toroidal fields.

On the other hand, at any given time in a sunspot cycle, spots emerge only in a rather narrow latitudinal range (the width of one wing in the classical 'butterfly diagram'), which can be interpreted as caused by a tachocline toroidal field band no greater in latitudinal extent than 10°–15°. Given these uncertainties, we have looked at the instability for both broad and narrow profiles. Broad profiles are covered in Gilman & Fox (1997, 1999a,b), narrow profiles in Dikpati & Gilman (1999) and Gilman & Dikpati (2000). Broad profiles include those that have no node within a hemisphere, and those that change sign at one latitude in each hemisphere. Only toroidal fields that are antisymmetric about the equator have been studied in detail, since it is known that the Sun's toroidal field, by Hale's polarity laws, is predominantly antisymmetric. But there is no doubt this assumption is not a requirement for instability.

In addition to the general properties of the MHD instability described above, the instability results for broad and narrow toroidal profiles have certain other characteristics in common. First, the instability occurs for virtually all magnetic field amplitudes and profiles, and for all differential rotations considered, which includes total differential rotation from above tachocline values, down to essentially zero. The only exception is that the instability ceases if the toroidal band is narrow enough, generally only 3° half width in latitude. For differential rotation typical of the tachocline, unstable mode e-folding growth times for both broad and narrow profiles range from several months to a few years, short enough to be of interest in considering the evolution of a solar cycle.

Consistent with the semicircle theorem, phase velocities for longitudinal propagation depend rather little on the strength of the toroidal field (confirming again their lack of Alfvén wave-like character), but within the range of speeds defined by the range of differential rotation, they depend sensitively on the profile of toroidal field with latitude. For narrow toroidal bands, the speed is that of the rotation of the latitude where the toroidal band is centred. The picture is more complex for broad toroidal fields, especially those with a node or change of sign within each hemisphere. Details are found in Gilman & Fox (1999a,b).

Which symmetry of disturbance about the equator is unstable, or more unstable, depends sensitively on the placement in latitude of a toroidal band (Gilman & Dikpati 2000) or the location in latitude of the node (if any) in a broad toroidal field (Gilman & Fox 1997, 1999a,b). Since as a solar cycle advances, the toroidal field in the tachocline, broad or narrow, advances toward the equator, we should expect the dominant symmetry of the instability about the equator to change, perhaps abruptly. If patterns of field associated with this instability have any effect on magnetic fields seen in the photosphere, we might expect to see evidence of such sudden shifts in symmetry. So far as we know, this has not been tried systematically.

In general, the lower the differential rotation, the lower the growth rate, even when, at high assumed toroidal field, most of the energy is coming from the toroidal field rather than the differential rotation. Lower differential rotation leads to less shearing of the perturbation magnetic field, so the Maxwell stress is smaller and less energy is extracted from the differential rotation (and smaller differential rotation means there is less energy available to extract). These observations are illustrated in Figure 10.5 for the simplest differential rotation $\omega_0 = 1 - s_2\mu^2$ and broad magnetic profile, $\alpha_0 = a\mu$, where $a = 1$ corresponds to a peak field of 100 kG. On the other hand, sufficiently narrow toroidal bands are stable because, for a given differential rotation, they feel less shear across the latitude range where there is any field.

Whether they contain a node in each hemisphere or not, for broad toroidal fields generally only $m = 1$ or tipping modes are unstable, for all field strengths. But as the toroidal field is narrowed, higher m's start to become unstable, unless the field

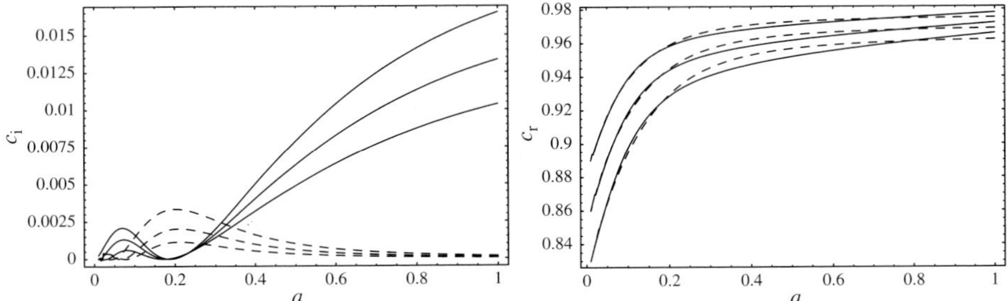

Figure 10.5. Linear growth rates c_i and phase speeds c_r as a function of magnetic field strength parameter a, where $\alpha_0 = a\mu$ and $\omega_0 = 1 - s_2\mu^2$, for the $m = 1$ instabilities. Full curves: antisymmetric modes; dashed curves: symmetric modes. Left-hand panel: $s_2 = 0.12, 0.15, 0.18$, from bottom to top. Right-hand panel: $s_2 = 0.12, 0.15, 0.18$, from top to bottom.

is very strong (Dikpati & Gilman 1999). For example, for toroidal bands placed at 45° latitude, $m = 2$ is excited for bands less than 29° full width, half maximum, $m = 3$ below 17°, and $m = 4$ below 7°, with a toroidal field of maximum value of 10 kG in the tachocline. The $m = 2$ mode is also the most unstable for bands between 15° and 5° latitude width.

But all these higher modes stabilize for peak toroidal fields of 20 kG and above, as the toroidal ring becomes too rigid to deform, for differential rotations found in the tachocline.

In terms of energetics and the energy flow diagram shown in Figure 10.4, for toroidal fields up to about equipartition with the differential rotation, the dominant energy flow is from the differential rotation to perturbation magnetic energy by means of the Maxwell stress, thence to perturbation kinetic energy by means of the perturbation $\mathbf{j} \times \mathbf{B}$ force, and sometimes even back into the differential rotation via the Reynolds stress. For higher toroidal fields, the flow is from toroidal field energy into perturbation kinetic energy via the 'mixed stress' or cross correlation between perturbation magnetic and velocity fields, whence by the perturbation $\mathbf{j} \times \mathbf{B}$ force into perturbation magnetic energy. For low toroidal fields, perturbation kinetic and magnetic energies are in near equipartition, while for high toroidal fields the perturbation magnetic energy dominates. When only $m = 1$ modes are unstable with high toroidal field, the tipping simply converts $m = 0$ magnetic energy to $m = 1$ magnetic energy, with no change in the total magnetic energy of the system. This is discussed in more detail in the next section on nonlinear effects.

10.3.3 Ideal fluid nonlinear MHD instabilities

Linear eigenvalue analyses reveal the presence and initial growth rates of instabilities, and something of their physical nature. However, in view of the requirement

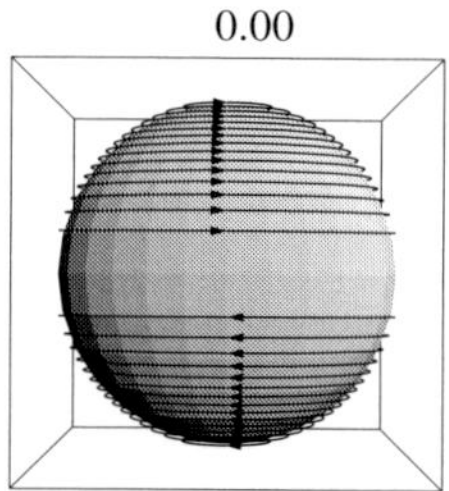

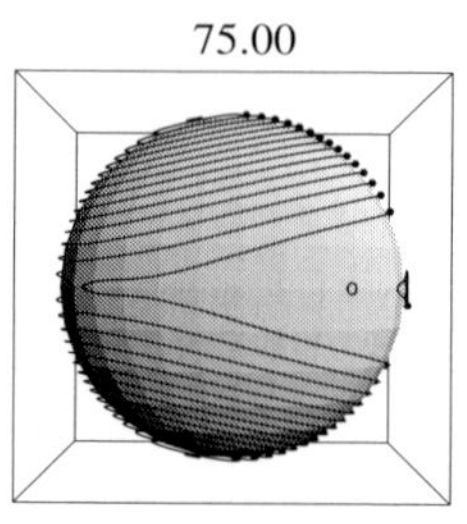

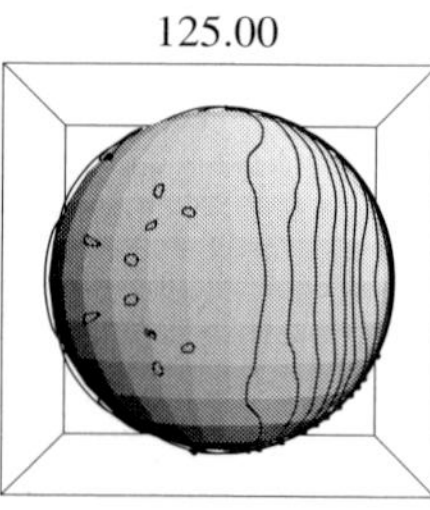

Figure 10.6. Snapshots of the development of the clam instability for shear profile $\omega_0 = 1 - 0.18\mu^2$ and magnetic profile $\alpha_0 = \mu$ (Figure 1 from Cally *et al.* 2003). In traditional dimensionless units, angular velocity ω_0 is set to unity at the equator. The labels indicate dimensionless time, where $t = 90$ corresponds to about 1 year. The initial broad magnetic field has a maximum strength of about 10^5 G at $\pm 45°$ latitude.

to test the Spiegel & Zahn (1992) hypothesis that the latitudinal differential shear profile can give rise to two-dimensional turbulence, it is necessary to extend modelling to the nonlinear regime. This was first carried out by Garaud (2001) in the hydrodynamic weakly nonlinear case, finding that the instabilities saturate at a low level, and do not appear to generate turbulence.

The first fully nonlinear simulations were presented by Cally (2001), and extended by Cally *et al.* (2003), who used a two-dimensional spherical harmonic spectral MHD code to re-examine many of the cases treated linearly before. Specifically, they examined solar-like differential rotation profiles coupled with both broad toroidal magnetic field patterns and banded patterns.

The results for strong broad magnetic profiles were spectacular and surprising. The dominant antisymmetric $m = 1$ instabilities discovered in the linear regime by Gilman & Fox (1997) are found to 'open up' the toroidal field like a clam shell (Figure 10.6), typically in little over a year. This is very characteristic of $\sim$100 kG broad fields combined with solar-type differential rotation. Such radical behaviour, for which there is no observational evidence, argues against such magnetic profiles existing in the solar tachocline. It may instead be that toroidal field is (i) much weaker than 100 kG in general; (ii) concentrated in bands, as suggested by sunspot emergence latitudes, or (iii) that the field has significant three-dimensional structure not accounted for in the shell model.[2]

In further support of option (i), broad toroidal field of 100 kG may be a problem for solar dynamo theory when $\mathbf{j} \times \mathbf{B}$ feedbacks are included, because they would tend to suppress the differential rotation; however, there is no helioseismic evidence that this happens in the Sun. Although rising flux tube considerations suggest that the

[2] However, the 3D 'polar kink instability' (Section 10.4.2) may also beset broad magnetic profiles and is even faster than the clam. It only exists for strong magnetic fields though.

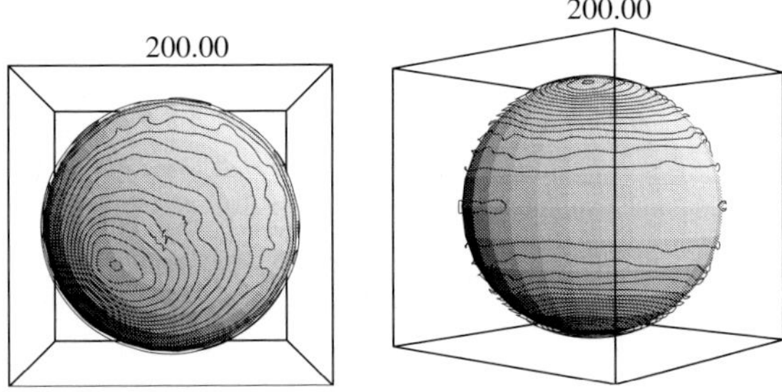

Figure 10.7. Two views of the nonlinear development of the axisymmetric $m = 1$ instability for $\omega_0 = 1 - 0.18\mu$, $\alpha_0 = 0.07\mu$ after dimensionless time 200. Left: from directly above the north pole. Right: from directly above the equator, which is horizontal on the image.

progenitor flux elements of active regions must have strengths around 100 kG when they detach from the tachocline, it is difficult to understand how such a strongly superequipartition field could be generated broadly in the tachocline. Field strengths of 10 kG are more plausible. Figure 10.5 indicates that a ~10 kG broad toroidal field is far less unstable. As suggested by c_r (right frame), the instability at low field strengths a is associated predominantly with the polar regions, where the rotation speed is a minimum, $1 - s_2$. Nonlinear simulations at $a = 0.07$ (7 kG), $s_2 = 0.18$, show that in fact the toroidal field 'slips off' the poles at high latitude during the $m = 1$ antisymmetric instability, leaving low latitude field relatively unchanged (Figure 10.7). There is no 'clam' runaway. The maximally unstable symmetric mode ($a \approx 0.1$) behaves similarly.

Option (ii) has led to detailed consideration of banded toroidal fields. Once again, $m = 1$ instabilities dominate at high (100 kG) field strengths, because of the difficulty in bending such strong field lines. The instabilities, as for the clam, are essentially *tipping instabilities*. Figures 10.8, 10.9 and 10.10 illustrate a variety of behaviours. In all cases, tipping settles towards a limiting value dependent predominantly on latitude and band width: wide bands tip further than narrow bands, and mid-latitude bands tip further than those at low latitude. Field strength affects the approach to the new equilibrium, but seemingly not its ultimate tip. Weaker bands of around 10 kG do not develop significant tip, but instead are significantly distorted by, and distort, the differential rotation. This happens because $m > 1$ modes tend to dominate, allowed because for weaker toroidal fields the ring deforms more easily.

The mechanism by which strong magnetic bands limit their tip is to develop prograde velocity jets, or at least shoulders, at the band latitudes (Figure 10.9). The added angular momentum resists further tipping.

Norton & Gilman (2005) have now found observational evidence of tipped toroidal fields from the location patterns of sunspots.

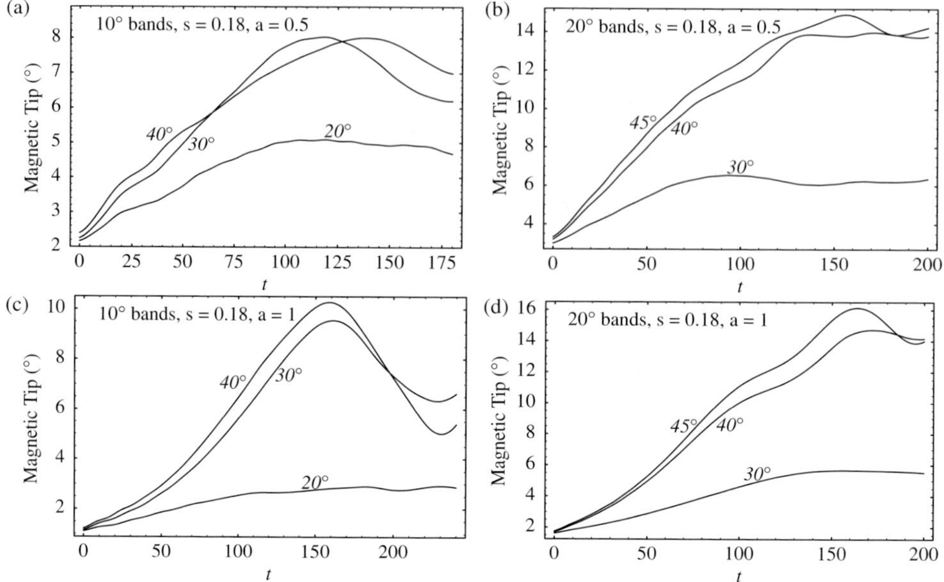

Figure 10.8. Magnetic tip angles for toroidal bands of various latitudes, widths and strengths (as labelled) combined with differential rotation $\omega_0 = 1 - 0.18\mu^2$ (Figure 7 of Cally *et al.* 2003). The Gaussian magnetic field profiles of Dikpati & Gilman (1999) are used throughout.

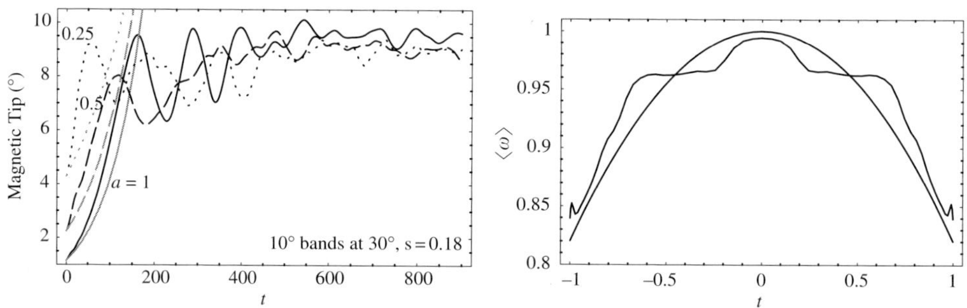

Figure 10.9. Left-hand panel: magnetic tip angles for 10° bands at ±30° latitude and various field strengths (labelled). The lighter curves indicate linear growth. Right-hand panel: longitudinally averaged angular velocity for the $a = 1$ (100 kG) case at dimensionless times $t = 0$ (parabola) and 500 (Figures 8 and 9 of Cally *et al.* 2003).

10.3.4 Effects of drag

A rudimentary way to account for the possible tendency of neighbouring shells to diffusively limit the development of two-dimensional instabilities is to incorporate a 'Newton's cooling' drag into both the vorticity and induction equations

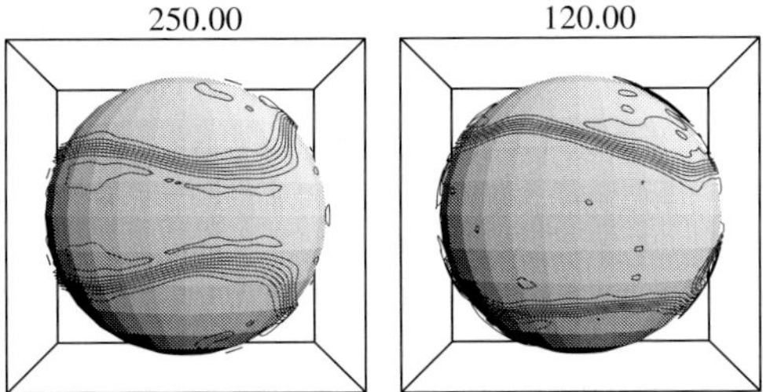

Figure 10.10. Nonlinear development of a 10° 'weak' band ($a = 0.1$, i.e. 10 kG) at ±30° initial latitude. Left: antisymmetric initial perturbation. Right: symmetric initial perturbation. These much weaker bands cannot resist the bending influences of the flow. As expected, the initial symmetry of the perturbation in the right-hand frame does not survive nonlinear evolution.

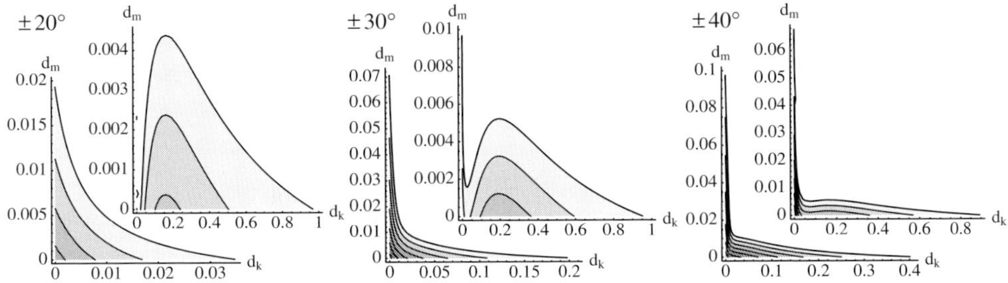

Figure 10.11. Linear growth rates for $a = 1$ bands of 10° width at three latitudes (labelled). The lower row corresponds to the symmetric modes, and the upper row corresponds to the antisymmetric. The white regions are stable, and successively darker shading corresponds to faster growth rates. The contours of c_i are 0.0, 0.002, 0.004, etc.

(Dikpati *et al.* 2004a):

$$\frac{D\Omega}{Dt} = \mathbf{a} \cdot \nabla J + \eta_k \left(\nabla^2 \Omega + \frac{2}{r^2} \Omega \right) - d_k \left(\Omega - \Omega_{eq} \right), \tag{10.7}$$

$$\frac{D\chi}{Dt} = \eta_m \nabla^2 \chi - d_m \left(\chi - \chi_{eq} \right), \tag{10.8}$$

where Ω_{eq} and χ_{eq} are the steady toroidal equilibrium solutions, and d_k and d_m are the kinetic and magnetic drag coefficients respectively. Figure 10.11 indicates schematically how kinetic and magnetic perturbation energies are ultimately removed from the system by drag.

The linear stability of this system is remarkably straightforward in the case $d_k = d_m = d$. It is easily seen that then the complex eigenvalues (phase speed) c are simply replaced by $c - id$. Consequently, any d greater than the maximum c_i stabilizes the system. Not surprisingly, d less than this leaves the system unstable, but with a reduced growth rate. However, this special case gives no clue to the much richer range of behaviours when $d_\mathrm{k} \neq d_\mathrm{m}$, both linearly and nonlinearly. As an example, Figure 10.11 depicts the instability regions and growth rates in the d_k–d_m plane for 10° banded profiles of 100 kG strength at three different latitudes. For the most part, it is clear that magnetic drag is far more potent than kinetic drag in slowing or eliminating instability. The reason for this is that in vorticity terms the MHD instability is driven by the $\mathbf{j} \times \mathbf{B}$ force acting as a source of vorticity. Magnetic damping of the perturbation fields directly reduces this source and weakens the instability. Conversely, particularly for strong unperturbed toroidal fields, the kinetic drag that damps vorticity must be quite large to compete with the strong magnetic source.

In the nonlinear regime, Dikpati *et al.* (2004a) find that drag can suppress the clam instability in broad magnetic profiles. It can also greatly slow the development of tip in banded profiles, but generally does not greatly reduce the ultimate degree of tip.

10.4 Towards three dimensions

Since the tachocline straddles the base of the convection zone, its stiffness to radial motions varies over many orders of magnitude. Within the convection zone it is effectively neutrally stable, whilst in the radiative core, buoyancy periods of only a few hours are much faster than any of the instabilities discussed in this chapter. The nature and thickness of the intervening overshoot layer is very much an open question, so we can only suppose that there *may* be a layer of non-negligible thickness in which buoyancy and the instabilities operate on similar timescales, and may therefore interact.

Several questions arise when considering the three-dimensional tachocline. One consequence of the two-dimensional shell model is that plasma cannot cross field lines (in the absence of magnetic diffusion), leading to significant constraints on allowable evolution. For example, the polar slip instability is ruled out. On the other hand, in three dimensions, unmagnetized plasma may of course pass over or under a flux tube or field concentration, or even through it by splitting it apart. One task for three-dimensional modelling therefore is to determine the extent to which two-dimensional-like plasma isolation actually occurs. If the field actually exists as flux tubes of $\sim$100 kG strength, their magnetic energy density is well in excess of the available kinetic energy associated with shear, so the tubes should be able to maintain their integrity, and indeed be quite rigid in the face of flows. In that case, if plasma is to cross from one side of the tube to the other, it must go over or under. In the radiative tachocline, the relevant criterion then involves the Froude number $F = V/hN$, where h is the tube height, N is the Brunt–Väisälä frequency, and V is the typical two-dimensional flow speed relative to the tube. If F is less than about 1, the flow has insufficient available kinetic energy to supply the required potential energy jump, and plasma isolation is maintained (assuming no siphon mechanism is operating). On the other hand, a 10 kG flux tube is unlikely to be able to maintain its identity, perhaps unless it is significantly twisted (which may be the case), and the isolating property of two-dimensional shell models will not be replicated in three dimensions. In the convective tachocline, there is little reason to believe that plasma cannot cross from one side of a tube to the other, whatever its strength.

As yet, there is no fully nonlinear theory of magneto-shear instabilities in the three-dimensional tachocline. However, there are at least two partial treatments. The first adopts the 'shallow water' formalism used to great effect in geophysical fluid dynamics, in which vertical structure is assumed linear and hydrostatic. The other is in a sense exactly opposite to this: it looks for solutions with fine scale oscillatory structure radially. Both are reviewed in this section. Since each currently assumes that the equilibrium field is uniform throughout the layer, neither addresses the issue of plasma isolation. Nevertheless, each predicts novel behaviour, which a full three-dimensional treatment should address.

10.4.1 Shallow 'water' theory

10.4.1.1 Equations

In its hydrodynamic form, 'shallow water' theory dates from the late nineteenth century. It was first formulated to deal with global ocean tides, and is described in Hough (1898). In its simplest form for the sphere, it applies to a single incompressible shell of fluid whose inner boundary is fixed, and whose outer boundary is allowed to deform. Gravity is included, and the fluid disturbances are assumed to be large in longitude and latitude compared to the thickness of the shell. In that case the pressure field is hydrostatic, determined at every point by the weight and therefore the thickness of the fluid above it. This hydrodynamic version has been applied to the problem of the instability of latitudinal differential rotation of the tachocline by Dikpati & Gilman (2001a).

Unlike the strictly two-dimensional system we have discussed above, the shallow water generalization allows for vertical motions that are tied to the deformation of the 'free surface' at the outer boundary of the shell. The horizontal motions remain independent of height, but they can have horizontal divergence, consistent with vertical motions that are linear functions of height, zero at the bottom, maximum at the top.

Gilman (2000a) generalized these classical shallow water equations to the MHD case, to apply to the solar tachocline. The key additional assumption he made is that the upper surface is a material surface, with no magnetic flux crossing it. This leads to a modified continuity equation for the magnetic field that states that the horizontal divergence of the total magnetic flux (as opposed to magnetic field, as in the two-dimensional case) vanishes. In this system, horizontal flows and magnetic fields are both independent of height, while both vertical flows and fields are linear functions of height. The modified magnetic field continuity equation is

$$\nabla \cdot [(1+h)\mathbf{B}] = 0. \tag{10.9}$$

In Equation (10.9), $1+h$ is the local shell thickness, measured relative to the average thickness, and $\mathbf{B}$ is the (scaled) horizontal vector magnetic field.

In this approximation, the induction equation becomes

$$\frac{\partial \mathbf{B}}{\partial t} = \nabla \times (\mathbf{v} \times \mathbf{B}) + (\nabla \cdot \mathbf{v})\mathbf{B} - (\nabla \cdot \mathbf{B})\mathbf{v}, \tag{10.10}$$

in which we see the added horizontal divergence terms, absent in the strictly two-dimensional case.

In the MHD case, the hydrodynamic shallow water relation between the pressure and the thickness is modified to

$$\nabla p = gH\nabla(1+h) - \nabla\left(\tfrac{1}{2}\mathbf{B}\cdot\mathbf{B}\right), \tag{10.11}$$

in which g is the dimensional gravity. We see that the magnetic field presence reduces the hydrostatic pressure from the non-magnetic case, since across the top boundary the total pressure must be continuous. Equation (10.11) then leads to a modified horizontal vector equation of motion, given by

$$\frac{\partial \mathbf{v}}{\partial t} = \nabla\left(\tfrac{1}{2}\mathbf{B}\cdot\mathbf{B} - \tfrac{1}{2}\mathbf{v}\cdot\mathbf{v}\right) - (\hat{\mathbf{k}} \times \mathbf{v})\,\hat{\mathbf{k}}\cdot\nabla\times\mathbf{v} + (\hat{\mathbf{k}}\times\mathbf{B})\,\hat{\mathbf{k}}\cdot\nabla\times\mathbf{B} - gH\nabla(1+h), \tag{10.12}$$

in which $\hat{\mathbf{k}}$ is the local vertical unit vector.

The MHD shallow water system is completed by the inclusion of the usual shallow water equation for continuity of mass:

$$\frac{\partial(1+h)}{\partial t} + \nabla \cdot [(1+h)\mathbf{v}] = 0. \tag{10.13}$$

Equations (10.9)–(10.13) have many interesting solutions, for equilibrium states, linear and nonlinear waves, and HD and MHD instabilities. MHD shocks and solitary waves are possible in this system. These equations have also recently received more theoretical attention (Dellar 2002, and references therein).

The original hydrodynamic shallow water equations assumed an air–water interface at the outer boundary, so the gravity felt is essentially the full value, given the large density difference. But it is possible to relate this system to a continuously stratified shell by means of a 'reduced' gravity, proportional to the fractional difference between the actual radial temperature gradient and the adiabatic gradient. This correspondence is demonstrated in detail in the Appendix of Dikpati *et al.* (2003). This reduced gravity is conveniently defined as a dimensionless parameter G, given by

$$G = \frac{1}{2} \frac{g_{\odot} |\nabla - \nabla_{\mathrm{ad}}|}{r_0^2 \omega_c^2} \frac{H^2}{H_{\mathrm{p}}} = \frac{H^2 |N^2|}{2 r_0^2 \omega_c^2}, \tag{10.14}$$

in which $g_{\odot}$ is the dimensional gravity at tachocline depths, $|\nabla - \nabla_{\mathrm{ad}}|$ is the fractional departure from the adiabatic gradient, H is the shell thickness, H_{p} is the local pressure scale height, r_0 is the shell radius, ω_{c} is the rotation rate of the shell, and N is the Brunt–Väisälä or buoyancy frequency. If we evaluate G for the tachocline, we find $10 < G < 10^3$ for the radiative part, and $10^{-2} < G < 1$ for the overshoot layer. If we apply the shallow water model to the radiative part of the tachocline, then its top surface corresponds to the overshoot layer, which is certainly easier to deform than the even more subadiabatic domain below the radiative part of the tachocline. Similarly, if we apply the shallow water model to the overshoot layer, its top is the unstable convection zone, which offers no resistance to deformation, compared to the radiative part of the tachocline below. Thus the simple shallow water system we have chosen can plausibly apply to both parts of the tachocline. In addition, of course, both layers could be included in the same model by generalizing it to be a two-layer shallow water system, with different effective gravities in the two layers. That more complex system has yet to be studied in the MHD case.

10.4.1.2 Reference states

There are a variety of ways to define unperturbed equilibrium states for the MHD shallow water system. If we specify the differential rotation and toroidal field as we have done in the two-dimensional system, then we must solve for the equilibrium thickness they imply, by integrating the latitudinal equation of motion, and applying appropriate constraints, such as that the total mass of the shell remains constant whatever the configuration of its outer boundary. Solutions for the hydrodynamic case are shown in Dikpati & Gilman (2001a). These indicate that for tachocline amplitude differential rotations, the shell thickness is a minimum in mid-latitudes (relative to the oblate figure from the constant interior rotation of the Sun) increasing toward both the poles and the equator (see Figure 2 of Dikpati & Gilman 2001a).

Not surprisingly, the amplitude of this deformation is inversely proportional to G. For smaller G, a larger deformation is needed to create a large enough latitudinal pressure gradient to balance the Coriolis force associated with the differential rotation. For a G of 0.1, typical of the overshoot layer, the shell thickness variations with latitude can be as large as a few tens of percent of the average thickness. By contrast, in the radiative layer the deformations are very small.

Equilibrium states for the MHD case are computed in Dikpati & Gilman (2001b), Gilman & Dikpati (2002) and Dikpati *et al.* (2003). Just as in the hydrodynamic case, for a given toroidal field profile assumed, the thickness variations are again proportional to G^{-1}, and so are much larger in the overshoot layer. For both broad and narrow toroidal field profiles, the shell must be thicker on the poleward side of the peak fields than on the equatorward side. The narrower is the profile the steeper is the height gradient in the neighbourhood of the peak field. For peak fields as large as 100 kG, the deformation of the overshoot layer can be such as to make the shell thickness nearly vanish in low latitudes. Thus there is a limit to the strength of field that can be kept in equilibrium.

One way the system can equilibrate when there are strong, narrow toroidal fields without having such large thickness variations is to have a prograde fluid jet inside the toroidal band, so that the resulting equatorward Coriolis force takes the place of the hydrostatic pressure gradient in achieving balance (see Rempel *et al.* 2000 and Dikpati *et al.* 2003 for examples). The jet required can be large: for a peak field of 100 kG at 45° latitude, the linear velocity of the jet relative to the interior rotation is $\sim$200 m s^{-1}. Clearly, the Sun can choose any combination of latitudinal pressure gradient and jet amplitude the physics allows to achieve this balance.

10.4.1.3 Hydrodynamic instabilities

In the shallow water system, both hydrodynamic and MHD instabilities are possible, and may occur together, so we examine each in turn. For the Sun, hydrodynamic instabilities have been considered almost exclusively in Dikpati & Gilman (2001a). They show that, for high G, the instability is virtually identical to that of the two-dimensional case. Not surprisingly, with high effective gravity the vertical displacements and velocities are particularly small compared to their horizontal counterparts, leading to essentially two-dimensional conditions.

By contrast, for low G, an additional regime of instability appears, occurring for lower differential rotation than is unstable with two dimensions, actually lower than observed tachocline values. Thus, in the overshoot layer of the tachocline, the theory says we should find this instability, to longitudinal wavenumbers $m = 1$ and 2. But with still lower G, the instability vanishes.

The approximate explanation for these results is as follows. The domain of intermediate G with stronger instability arises because, in the shallow water case, instability occurs when the latitudinal gradient of potential vorticity (vorticity divided by thickness) changes sign in the domain, or where the potential vorticity has an inflection point. For small enough G, the shell thickness vanishes where this inflection point is found, dividing the fluid shell into polar and equatorial domains, neither one of which has an inflection point. Then the instability stops. When the system is unstable, the dynamics, in terms of transport of angular momentum by Reynolds stresses, the formation of high latitude jets, etc., is very similar to the two-dimensional case.

Unstable modes in the hydrodynamic shallow water case have some particularly interesting properties from the perspective of solar dynamo theory. They generally contain kinetic helicity, which in dynamo models gives rise to an 'alpha-effect' which lifts and twists toroidal fields into poloidal fields. Since this instability would occur within the solar tachocline, it would provide an additional source of generation of poloidal fields there from the strong toroidal fields there. But to determine the range of toroidal fields over which this instability can be expected to occur requires solution of the MHD instability problem. Strong toroidal fields may radically alter or suppress this hydrodynamic instability. On the other hand, if the tachocline is host to narrow isolated toroidal bands, there is no reason why the hydrodynamic instability cannot occur away from the band. Both types of behaviour have been found.

10.4.1.4 Magnetohydrodynamic instabilities

Global instabilities of the MHD shallow water system are analysed principally in Gilman & Dikpati (2002) for broad toroidal field profiles, and in Dikpati *et al.* (2003) for narrow toroidal bands. In both cases, for high G the MHD instabilities are virtually the same as their two-dimensional counterparts. For broad toroidal fields, this similarity holds for all G greater than unity, greatly extending the applicability of the two-dimensional results. For banded profiles the situation is more complex, depending on the strength of the band and its placement in latitude.

We show representative domains of instability and growth rates for MHD disturbances in Figures 10.12 and 10.13. In both we have chosen to display results that could best apply respectively to the overshoot and radiative tachoclines, for which we take $G = 0.2$ (broad) or 0.1 (banded) for the overshoot case, and $G = 100$ for the radiative case.

In Figure 10.12 (left-hand panel) we see that the hydrodynamic shallow water instability found by Dikpati & Gilman (2001a) carries over into the MHD case for broad toroidal fields up to several kilogauss. Above that, these modes are damped and eventually suppressed with increasing field amplitude, but replaced by purely MHD instability modes of similar growth rates, which peak in the neighbourhood of peak toroidal fields of about 100 kG. At these high fields, only the tipping mode is excited, and only the mode that is antisymmetric about the equator. From the right-hand panel of Figure 10.12, we see that in the radiative tachocline there is no hydrodynamic instability, only MHD unstable modes. These occur only for peak

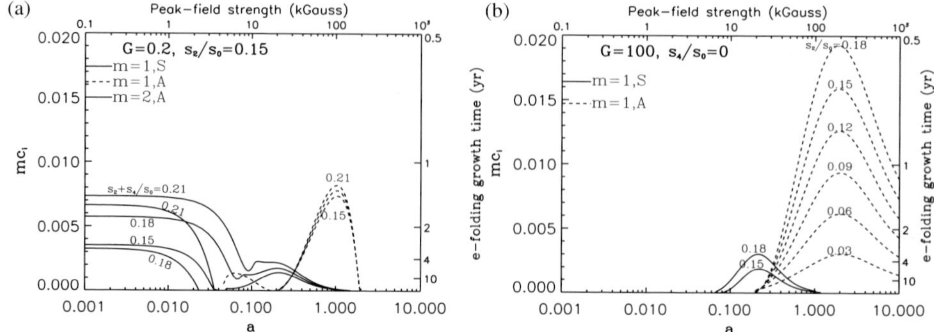

Figure 10.12. Disturbance growth rates for broad toroidal fields of peak amplitude shown on the horizontal axis, for overshoot and radiative parts of the tachocline; e-folding growth times are shown in years on the right-hand vertical axes. (Adapted from Gilman & Dikpati 2002.)

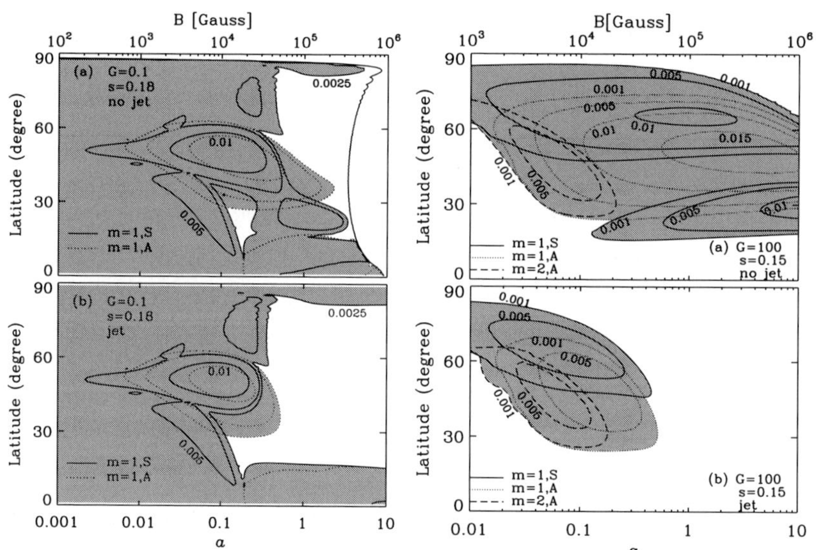

Figure 10.13. Domains of instability and growth rate contours for toroidal bands of width 10° placed at the latitudes shown on the vertical axes. Left-hand panels are for overshoot tachocline stratification, right-hand panels are for radiative tachocline stratification. Upper panels show results when the reference state contains no prograde fluid jet inside the toroidal band, lower panels are for the case for which the whole poleward magnetic curvature stress is balanced by the equatorward Coriolis force of such a jet. (Adapted from Dikpati *et al.* 2003.)

fields near 10 kG and above, with by far the strongest instability for peak fields of order 100 kG. Here again, only $m = 1$, the tipping mode, is unstable.

In Figure 10.13, we show growth rates and domains of instability as functions of latitude placement of bands of 10° half width, and peak field strength.

Overshoot layer results are on the left, radiative layer results on the right. Upper panels are for reference states with no prograde jet inside the band, lower panels include a 'full' jet, full in the sense that the Coriolis force completely replaces the latitudinal hydrostatic pressure gradient in the latitudinal force balance. In the overshoot layer, without jets bands are unstable at virtually all latitudes, for peak fields up to several tens of kilogauss, and in sunspot latitudes for fields up to several hundred kilogauss. Which mode-symmetry dominates depends on the latitude of band placement, as in the two-dimensional case described earlier. Comparing upper and lower panels, we see that a jet makes the system much less unstable, cutting off instability well below 100 kG at all latitudes except the very lowest and highest. This is because the jet resists tipping, which amounts to a change of axis of the angular momentum contained in the jet, much as happens in a gyroscope.

In the radiative layer, the range of latitudes of toroidal band placement that lead to instability is somewhat narrower, but without jets instability is found up to much higher field strengths. This is because in the radiative layer case, the shell thickness varies very little with latitude, while in the corresponding overshoot case, it goes to zero along the nearly vertical curve shown near the right-hand edge of the upper left-hand panel. The radiative tachocline also shows that which mode-symmetry is preferred depends on latitude placement of the band. In addition, in the radiative tachocline, $m = 2$ is also unstable for peak fields up to $\sim$10 kG. Notice also that while the jet suppresses $m = 1$ instability above several tens of kilogauss peak fields, it has almost no effect on $m = 2$. This is because at these field strengths the equilibrium jet is weaker, but also because $m = 1$ does not represent a change in the orientation of the angular momentum axis, so there is no gyroscopic effect.

Associated with each unstable MHD mode there is also a pattern of kinetic helicity that could contribute to dynamo action in the solar tachocline. The details are discussed in the references cited above. In the MHD case, these kinetic helicity patterns are particularly interesting because they are found in the neighbourhood of the toroidal band, and depend in part on the field strength. This would add a new nonlinearity to the dynamo problem. In the case of banded toroidal fields, examples have been found where kinetic helicity patterns of both hydrodynamic and MHD origin are found coexisting.

To sum up, in the MHD shallow water system, both broad and banded toroidal fields are commonly found to be unstable to longitudinal wavenumber $m = 1$, and sometimes $m = 2$, under both overshoot and radiative layer type subadiabatic stratifications. Thus if the shallow water model has some applicability to the solar tachocline, the results predict that under most conditions, and most phases of the solar cycle, the toroidal field should be tipped with respect to latitude, and the instability should contribute to the workings of the solar dynamo.

10.4.2 Fine radial structure instabilities

Cally (2003) examines the linear stability of a Boussinesq (thin layer) shell. The equilibrium state is assumed toroidal and independent of radius r, but both toroidal and poloidal perturbations are allowed in velocity and magnetic field. A uniform Brunt–Väisälä frequency N is assumed throughout, characterizing the departure from an adiabatic temperature gradient (see Equation (10.14) for the relationship between N and the shallow water reduced gravity G). It is supposed that the perturbed poloidal and toroidal kinetic and magnetic stream functions, and the entropy perturbation, may be Fourier expanded in radius r across the thin layer, with each proportional to $\sin k(r - r_0)$ or $\cos k(r - r_0)$ as appropriate, where the inner radius r_0 is taken to be rigid. The radial wavenumber k is a free parameter. Horizontal and time dependence are assumed to be of the form $\exp[\mathrm{i}m(\phi - c\,t)]$ as in the two-dimensional linear calculations of Gilman & Fox (1997). If a rigid or free upper boundary condition were imposed it would restrict k to certain discrete values, but this was left open and k unspecified.

The instabilities found through this analysis generalize both the two-dimensional results, and the three-dimensional instabilities discovered by Tayler (1973) in non-rotating stars. They appear in the opposite regime to 'shallow water', since they exhibit short-scale oscillatory radial behaviour rather than a simple linear profile. In that sense, this approach complements rather than overlaps the shallow water studies. Exploring the middle ground between the two will require more detailed numerical modelling.

As for the two-dimensional linear calculations, the aim is to calculate the eigenvalues $c = c_\mathrm{r} + \mathrm{i}c_\mathrm{i}$. The instability growth rate is then mc_i. With k set to zero the two-dimensional case is recovered, but with $k > 0$ it is possible to explore the growth of 'radially local' three-dimensional modes for various choices of magnetic field, latitudinal shear, and N. The question asked is: are the two-dimensional (or near-two-dimensional) instabilities dominant, or do distinct three-dimensional instabilities arise and take over? And in particular, what role does the Brunt–Väisälä frequency play?

As an example, consider the broad magnetic profile $\alpha_0 = \mu$ discussed earlier (see Figures 10.5 and 10.6). To maximize the potential for interaction with buoyancy, a Brunt–Väisälä frequency of $N = 1$ is adopted in Figure 10.14, measured in dimensionless units in which the equatorial rotational frequency $\omega_0(0)$ is also 1. The figure shows the two-dimensional growth rate $c_\mathrm{i} = 0.016$ is recovered at $k = 0$ as expected, and a slightly faster three-dimensional version at $k = 2$ (this may not be physically relevant if a top boundary condition restricts k to a discrete spectrum). However, most significantly, a new completely unrelated and very much faster three-dimensional instability sets in for $k > 7$, with maximal

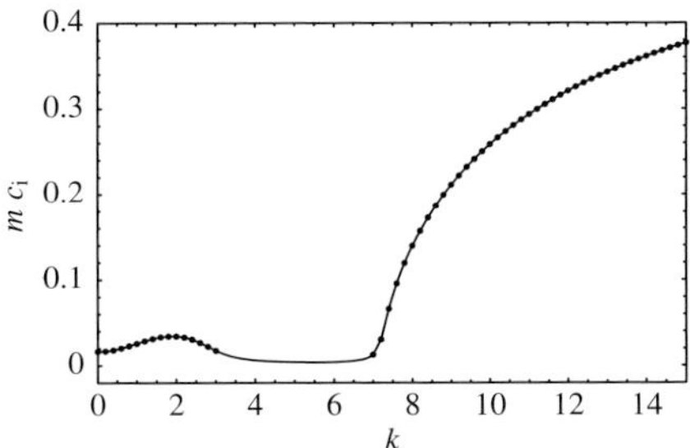

Figure 10.14. The growth rate as a function of radial wavenumber k for antisymmetric modes with $m = 1$, $N = 1$, $\omega_0 = 1 - 0.18\mu^2$, and $\alpha_0 = \mu$. The curve represents results obtained using the shooting method, whereas the points are from a Galerkin method (Figure 1 of Cally 2003).

growth rate

$$c_i \sim \sqrt{\alpha_p^2 - \omega_p^2} \quad \text{as } k \to \infty. \tag{10.15}$$

Here $\omega_p = \omega_0(1)$ is the rotational angular velocity at the pole $\mu = 1$, and α_p is the equivalent polar Alfvén angular speed. The growth rate (10.15) and the consequent instability criterion $\alpha_p^2 > \omega_p^2$ are rigorously derived (Cally 2003), and shown to only apply for $m = 1$. This 'polar kink instability', as its name suggests, corresponds to a kink instability in the magnetic loops about the poles first identified by Tayler (1973) in non-rotating stars. If the rotational plasma speed exceeds the Alfvén speed, the angular momentum is sufficient to stabilize the kink through the action of centrifugal force; the flow speeds up on the in-kink side of the tube owing to field line compression, and so tends to move the field lines back outward.

As the Brunt–Väisälä frequency N is increased, the polar kink instability is pushed to higher and higher onset wavenumbers k_{crit}; in fact, the critical wavenumber scales approximately with N, and the asymptotic growth rate (10.15) remains unchanged. Consequently, it is clear that a large subadiabaticity (large N) does not actually suppress three-dimensional instabilities on its own; rather, it pushes them to smaller radial scales (higher k), compressing them to a 'pancake'. This makes physical sense, since a three-dimensional instability is less able to move plasma radially against buoyancy as N increases. Conceivably, diffusive effects might be able to restrict instabilities at high k, though Cally (2003) finds that thermal diffusion does not destroy or even slow down the polar kink, which is predominantly magnetic in nature rather than thermal.

Although it is unlikely that strong fields are present near the poles in the tachocline, similar instabilities are found at lower latitudes, though their mathematical analysis is much more difficult. In particular, no simple stability criterion or growth rate is known.

Recently, Gilman *et al.* (2004) presented some preliminary results based on the Boussinesq thin layer hydrostatic primitive equations (HPE) of Miesch & Gilman (2004). This approach differs from the radially local analysis of Cally (2003) in that it is applied across a finite (though thin) layer with, in this case, rigid boundary conditions applied top and bottom. Nevertheless, qualitatively similar results were obtained. In particular, by examining fixed vertical wavenumbers n, very rapid, fine radial scale, tipping instabilities are found to develop at sufficiently high magnetic field strengths (of order 100 kG) and low reduced gravity G. As G (i.e. N) is increased, only higher n instabilities are excited, just as in Cally (2003). The correspondence between the two techniques has yet to be fully explored, but seems close. Three dimensional instabilities of toroidal fields in deep, uniformly rotating, spherical shells such as stellar interiors have been studied by Zhang *et al.* (2003). These may be related in a 'thin shell' limit to those discussed here, but this also has not been explored yet.

The lesson to be learned from these analyses is that the argument that the highly stable gravitational stratification of the lower tachocline will suppress three-dimensional instabilities, leading us to consider only the two-dimensional shell, is potentially misleading. In fact, three-dimensional instabilities that do not exist in strict two dimensions, such as the polar kink, are possible at arbitrarily large N. However, they will be very 'flat', thus appearing two-dimensional despite their intrinsic reliance on the radial direction. In the (many) cases, though, where the polar kink or similar instabilities do not exist, the two-dimensional shell model is a particularly useful indicator.

10.5 Conclusions: implications for the Sun

The results described in this chapter indicate clearly that any significant toroidal fields within the solar tachocline should render it unstable to global MHD disturbances. Even if toroidal fields there are concentrated into narrow bands, migration of the zone where sunspots are found toward the equator with the advance of each solar cycle ensures that a wide range of latitudes is affected by this instability. Indeed, toroidal bands may first occur at latitudes that are higher than the spot zones, with little signature of them seen at the visible surface. Since the instability efficiently transports angular momentum from low latitudes to high, its presence can easily prevent the tachocline thickness from growing into the deeper interior. The more vigorous is the transport, the thinner the tachocline can be. This transport occurs on a timescale of the solar cycle, many orders of magnitude shorter than other angular momentum transfer rates estimated for either the solar radiative interior or the solar wind.

All of the instability analysis discussed above was applied to the solar tachocline where there is known to be latitudinal differential rotation. Gilman (2000b) first made the explicit distinction between a 'fast' tachocline, where physical processes on time scales of months and years predominate, and a 'slow' tachocline, where processes act on vastly longer time scales. Gough & McIntyre (1998), among others, have argued that the radiative part of the tachocline, even though it is quite thin and immediately adjacent to the overshoot tachocline above (which everyone agrees is fast) is nevertheless a slow domain.

But this radiative part undeniably has latitudinal differential rotation, so to be slow, despite its latitudinal differential rotation, there must be virtually no toroidal field there. The instability results cited above show clearly that even toroidal fields of a few hundred gauss are enough to trigger the MHD instability, with growth rates that might be as long or longer than a solar cycle, but still very fast compared to the 'slow' times Gough & McIntyre assume to prevail there. It does not matter whether the toroidal field has its origins in the convection zone dynamo above, or comes from below by shearing of a global poloidal field there. Furthermore, weak toroidal fields lead not just to the simple tipping of a rigid toroidal ring, but to the excitation of other non-zero longitudinal wave numbers as well, and most likely to a more general mixing in latitude. We conclude that any part of the tachocline that has toroidal field in it must be 'fast'. Dikpati *et al.* (2006) have recently shown that flux transport dynamos that correctly simulate many features of the solar cycle also generate strong toroidal fields throughout the solar tachocline, if the magnetic diffusivity in the lower part of the tachocline is less than $\sim 10^6\,\mathrm{cm^2\,s^{-1}}$, much lower than the convection zone, but still much higher than molecular diffusivities characteristic of the radiative interior.

It is still possible, perhaps probable, that there is some domain of the solar interior below the tachocline (by definition, below the latitudinal and radial differential rotation) that is slow in the sense assumed by Gough & McIntyre (1998). But if there is any significant toroidal field in the radiative part of the tachocline, then the dynamics and MHD of this slow deep interior does not extend its influence significantly into the tachocline, and all of the tachocline is 'fast'. If this slow interior domain has no differential rotation, then the global MHD instability we have reviewed does not occur there, even if there are toroidal fields, so it can be slow. Gough & McIntyre's model is known generally to contain some toroidal field, and does not produce the global MHD instability. That is because their model is axisymmetric about the rotation axis, so all $m > 0$ modes, which are characteristic of the instability, are excluded by assumption.

In addition to limiting tachocline thickness, the global MHD instability may contribute several effects to the workings of the solar dynamo. It can create patterns of field with longitudinal wavenumber $m > 0$ that might be seen at the surface but also may make the whole global solar dynamo fundamentally non-axisymmetric. It may be possible to see traces of $m = 1$ 'tipping' modes in the pattern of locations of sunspots, and $m = 1$ and 2 may be related to the so-called sector structure seen in the heliosphere. The growth rates of unstable modes, with e-folding times of several months to a few years, are short enough for the dominant patterns produced to adjust to the dynamo advancing the tachocline toroidal field toward the equator. The longitudinal phase velocities of these modes, being intermediate between the minimum and maximum rotation of the system, and for toroidal bands at the rotation rate of the latitude of the band, ensure that the phase speeds are never very far from the commonly used 'Carrington rate' (rotation rate at $\sim 18°$ latitude at the solar surface), a benchmark for constructing synoptic maps of solar fields.

Since both unstable hydrodynamic and MHD modes contain kinetic helicity, this instability can also provide additional means of generating poloidal from toroidal fields at tachocline depths. Dikpati & Gilman (2001c) have shown that a source of poloidal field in the tachocline is very important for producing the correct symmetry of axisymmetric solar fields about the equator. In addition, the instability can also determine the equatorial symmetry of $m > 0$ fields, which can even switch at certain points in a solar cycle.

Both the hydrodynamic and MHD instabilities typically lead to the formation of fluid jets, particularly prograde jets, either at fixed latitudes or migrating with the toroidal field as a solar cycle progresses. Christensen-Dalsgaard *et al.* (2005) have found helioseismic evidence of persistent prograde jets near 60° latitude, that last through most of a solar cycle.

One of the most important solar properties left out of the analysis of this instability so far is the radial gradient of rotation, that of course defines the existence of the tachocline. Its inclusion into the instability problem makes the calculation much more difficult, turning it into a non-separable two-dimensional eigenvalue problem from a one-dimensional problem. But from Figure 10.2, we can surmise that with radial shear in the MHD case the instability might be enhanced, because a wavy magnetic pattern in the longitude-radius surface, sheared by the radial differential rotation, would extract kinetic energy and angular momentum from the radial gradient. The dominant longitudinal wavenumber for instability might change, but it is difficult to see how such a process would actually damp the instability. This effect of radial shear could damp the hydrodynamic instability however.

Similarly, the neglect of radial magnetic structure in the shallow water and radially local analyses means that the equilibrium field forms a 'wall' across the whole tachocline, thereby stopping flow of plasma from one side to the other. The questions raised in the prelude to Section 10.4 concerning the ability of magnetic bands to prevent flows across their latitudes are therefore unanswered. Future modelling must allow for both radially as well as latitudinally compact magnetic bands, lying wholly within the computational domain, to address these concerns.

The analyses described here provide insight into the kinds of MHD instabilities that should be expected to occur in global three-dimensional convection models for the solar convection zone and tachocline. A significant limitation of such models to date is their inability to deal with a tachocline with a realistic subadiabatic stratification. Until this limitation is overcome by much larger computer power, tractable models confined to the tachocline will remain useful.

Acknowledgments

The authors wish to acknowledge the assistance of Mausumi Dikpati in reviewing this chapter. NCAR is sponsored by the National Science Foundation.

References

Acheson, D. J. (1978). *Phil. Trans. R. Soc. Lond.*, A**289**, 459.
Adam, J. A. (1980). *J. Phys. A*, **13**, 3325.
Balbus, S. A. & Hawley, J. F. (1991). *Astrophys. J.*, **376**, 214.
Cally, P. S. (2000). *Sol. Phys.*, **194**, 189.
Cally, P. S. (2001). *Sol. Phys.*, **119**, 231.
Cally, P. S. (2003). *Mon. Not. Roy. Astron. Soc.*, **339**, 957.
Cally, P. S., Dikpati, M. & Gilman, P. A. (2003). *Astrophys. J.*, **582**, 1190.
Charbonneau, P., Christensen-Dalsgaard, J., Henning, R. *et al.* (1999a). *Astrophys. J.*, **527**, 445.
Charbonneau, P., Dikpati, M. & Gilman, P.A. (1999b). *Astrophys. J.*, **526**, 523.
Chimonas, G. (1970). *J. Fluid Mech.*, **43**, 833.
Christensen-Dalsgaard, J., Gough, D. O. & Thompson, M. J. (1991). *Astrophys. J.*, **378**, 413.
Christensen-Dalsgaard, J., Corbard, T., Dikpati, M., Gilman, P. & Thompson, M. J. (2005). In *Proc. 22nd International NSO/Sac Peak Workshop*, ed. K. Sankarasubramanian, M. Penn & A. Pevtsov (PASP Monograph Series) **346**, p. 115.
Dellar, P. J. (2002). *Phys. Plasmas*, **9**, 1130.
Dikpati, M. & Gilman, P. A. (1999). *Astrophys. J.*, **512**, 417.
Dikpati, M. & Gilman, P. A. (2001a). *Astrophys. J.*, **551**, 536.
Dikpati, M. & Gilman, P. A. (2001b). *Astrophys. J.*, **552**, 348.
Dikpati, M. & Gilman, P. A. (2001c). *Astrophys. J.*, **559**, 428.
Dikpati, M., Gilman, P. A. & Rempel, M. (2003). *Astrophys. J.*, **596**, 680.
Dikpati, M., Cally, P. S. & Gilman, P. A. (2004a). *Astrophys. J.*, **610**, 597.
Dikpati, M., de Toma, G., Gilman, P. A., Arge, C. N. & White, O. R. (2004b). *Astrophys. J.*, **601**, 1136.
Dikpati, M., Gilman, P. A. & MacGregor, K. B. (2006). *Astrophys. J.*, **638**, 564.
Drazin, P. G. & Reid, W. H. (1981). *Hydrodynamic Stability.* Cambridge: Cambridge University Press.
Eckart, C. (1963). *Phys. Fluids*, **6**, 1042.
Garaud, P. (2001). *Mon. Not. Roy. Astron. Soc.*, **324**, 68.
Gilman, P. A. (2000a). *Astrophys. J.*, **544**, L79.
Gilman, P. A. (2000b). *Sol. Phys.*, **192**, 27.
Gilman, P. A. (2005). *Astron. Nachr.*, **326**, 208.

Gilman, P. A. & Dikpati, M. (2000). *Astrophys. J.*, **528**, 552.
Gilman, P. A. & Dikpati, M. (2002). *Astrophys. J.*, **576**, 1031.
Gilman, P. A. & Fox, P. A. (1997). *Astrophys. J.*, **484**, 439.
Gilman, P. A. & Fox, P. A. (1999a). *Astrophys. J.*, **510**, 1018. Erratum: **534**, 1020 (2000).
Gilman, P. A. & Fox, P. A. (1999b). *Astrophys. J.*, **522**, 1167.
Gilman, P. A., Dikpati, M. & Miesch, M. S. (2004). In *SOHO 14/GONG 2005 Workshop 'Helio- and Asteroseismology: Towards a Golden Future'*, ed. D. Danesy (ESA SP-559, Noordwijk, The Netherlands), p. 440.
Gough, D. O. & McIntyre, M. E. (1998). *Nature,* **394**, 755.
Hough, S. (1898). *Phil. Trans. R. Soc. Lond.*, A**191**, 139.
Howard, L. N. (1961). *J. Fluid Mech.*, **10**, 509.
Hughes, D. W. & Tobias, S. M. (2001). *Proc. R. Soc. Lond.*, A**457**, 1365.
Kitchatinov, L. L. & Rüdiger, G. (1997). *Mon. Not. Roy. Astron. Soc.*, **286**, 757.
Miesch, M. S. & Gilman, P. A. (2004). *Sol. Phys.*, **220**, 287.
Miles, J. W. (1961). *J. Fluid Mech.*, **10**, 496.
Norton, A. A. & Gilman, P. A. (2005). *Astrophys. J.*, **630**, 1194.
Ogilvie, G. I. & Pringle, J. E. (1996). *Mon. Not. Roy. Astron. Soc.*, **279**, 152.
Rempel, M. (2004). *Astrophys. J.*, **607**, 1046.
Rempel, M., Schüssler, M. & Tóth, G. (2000). *Astron. Astrophys.*, **363**, 789.
Spiegel, E. A. & Zahn, J.-P. (1992). *Astron. Astrophys.*, **265**, 106.
Tayler, R. J. (1973). *Mon. Not. Roy. Astron. Soc.*, **161**, 365.
Watson, M. (1981). *Geophys. Astrophys. Fluid Dyn.*, **16**, 285.
Zhang, K., Liao, X. & Schubert, G. (2003). *Astrophys. J.*, **585**, 1124.

11

Magnetic buoyancy instabilities in the tachocline

David W. Hughes

It is natural to associate the tachocline with the region of generation of a strong toroidal field by the winding-up of a weaker poloidal component. Here I discuss the break-up and subsequent escape of such a field via magnetic buoyancy instabilities. I consider the different modelling approaches that have been employed and discuss which have the most relevance in a solar context.

11.1 Introduction

For many years, a controversial issue of solar magnetism has been that of the location of the site (or sites) of the generation and storage of the Sun's predominantly toroidal magnetic field, which eventually escapes and rises to the surface, leading to active regions and, ultimately, to much of the exotic magnetic behaviour observed in the photosphere, chromosphere and corona. For two rather different reasons, the idea had been put forward that the bulk of the toroidal field must be stored either at the base of, or just beneath, the convection zone. From estimates of the rise times of magnetic flux tubes through the convection zone, Parker (1975) argued that the dynamo must operate only in the 'very lowest levels of the convective zone'. Golub *et al.* (1981) (see also Spiegel & Weiss 1980) proposed a similarly deep-seated layer of toroidal field, but from arguments based instead on the expulsion of magnetic fields by convective motions. The discovery of the tachocline by helioseismology provides probably the most compelling evidence for pinning down the location of the solar toroidal field. Although there is no consensus on how the solar dynamo operates (see, for example, the discussion in Chapter 13 by Tobias & Weiss), it *is* generally agreed that toroidal field is wound up from a relatively weak poloidal ingredient via strong differential rotation (the ω-effect of mean field dynamo theory). Consequently, the tachocline becomes the natural location for a deep-seated, predominantly toroidal magnetic field.

The Solar Tachocline, D. W. Hughes, R. Rosner and N. O. Weiss.
Published by Cambridge University Press.

The large-scale magnetic features observed at the solar surface, such as sunspots, which must originate from a global magnetic field, must therefore be manifestations of the instability and subsequent escape of magnetic field from the tachocline. The mechanism responsible is known as *magnetic buoyancy*. Although the tachocline may play host to a number of different types of hydrodynamic and hydromagnetic instabilities (some of which are reviewed by Gilman & Cally in Chapter 10), the only one with readily observable consequences is that due to magnetic buoyancy.

It is worth explaining, at this early stage, that the term 'magnetic buoyancy' is used within astrophysics (and, indeed, more pertinently, within solar physics) to refer to three related, but different, physical mechanisms. The idea of magnetic buoyancy was conceived by Parker (1955) and, coincidentally, by Jensen (1955), who considered the rise of isolated tubes of magnetic flux. The essential physics is captured by considering an isolated horizontal flux tube in pressure equilibrium with its non-magnetic surroundings; thus

$$p_{\rm i} + \frac{B^2}{2\mu_0} = p_{\rm e}, \tag{11.1}$$

where the tube has field strength B, and the internal and external gas pressures are denoted by $p_{\rm i}$ and $p_{\rm e}$. Hence $p_{\rm i} < p_{\rm e}$. If, for example, the tube is in thermal equilibrium with its surroundings then it follows from the gas law that $\rho_{\rm i} < \rho_{\rm e}$; i.e. that the tube is less dense than its surroundings and will thus rise under the influence of gravity. For what follows, it is worth stressing that this is not an instability but, rather, a lack of equilibrium.

The buoyant tendency of magnetic fields can though act as an instability mechanism of magnetized atmospheres in equilibrium, the simplest such case being that of an atmosphere with a horizontal field dependent only on height. This was first addressed by Newcomb (1961) and, in an astrophysical context, by Parker (1966), although Kruskal & Schwarzschild (1954) had previously considered the related instability of a discontinuous field.

The third usage of the term 'magnetic buoyancy' refers to the instability of isolated flux tubes. From the arguments advanced above, it can be seen that, in general, isolated flux tubes will not be in mechanical equilibrium. However, for one specific temperature difference between the tube and its surroundings, the internal and external densities will be equal. In such circumstances one may address the stability of such a tube (Spruit & van Ballegooijen 1982), any instability thereof also carrying the name 'magnetic buoyancy'.

These three magnetic buoyancy mechanisms, although clearly related through the principal role of the magnetic pressure, possess significant differences in their underlying physics and in their astrophysical implications. (Also, somewhat confusingly, in the astrophysical literature all three are, on occasion, referred to as the 'Parker instability'.) All three have been advanced as playing an important role in the instability of the magnetic field in the tachocline and its subsequent rise through the convection zone. Thus in this review I shall first consider each of these in turn, concentrating on the most important physical attributes of each, before discussing in Section 11.5 their role in the evolution of the Sun's large-scale magnetic field. I shall discuss them not, as above, in the chronological order of their formulation, but instead will first consider the two instability mechanisms, which are of potential relevance to the disruption of any field in the tachocline, and then consider the rise of magnetic flux tubes, the subsequent stage in the field's evolution.

11.2 The magnetic buoyancy instability of a large-scale field

Any astrophysical magnetic field will, in reality, vary in all directions. However, from the point of view of instabilities driven by magnetic buoyancy it is clearly the vertical variation that is most influential; therefore the simplest problem to consider is the instability of a magnetohydrostatic equilibrium in which all quantities vary only with height z. For the most part we shall consider equilibria with uni-directional fields; for such cases we shall suppose that the imposed magnetic field is in the x-direction. We shall denote the wavenumber vector by $\mathbf{k} = (k_x, k_y, k_z)$ and the (possibly complex) growth rate of linear perturbations by s.

11.2.1 Linear considerations

11.2.1.1 The instability mechanism

The essence of the magnetic buoyancy instability can be understood in terms of a simple 'parcel' argument (e.g. Tayler 1973; Moffatt 1978; Acheson 1979). Consider an atmosphere in equilibrium containing a horizontal magnetic field. Imagine that a parcel of gas is raised, with no bending of the magnetic field lines, from a height z to a height $z + \mathrm{d}z$; all diffusive effects are, for the moment, neglected. Suppose that the properties of the tube change from ϕ to $\phi + \delta\phi$ and that the variable ϕ takes the value $\phi + \mathrm{d}\phi$ at height $z + \mathrm{d}z$. Since the mass and magnetic flux of the parcel are conserved, and the parcel moves adiabatically, we have the following relations:

$$\frac{\delta B}{B} = \frac{\delta \rho}{\rho}, \qquad \frac{\delta p}{p} = \gamma \frac{\delta \rho}{\rho}. \tag{11.2}$$

Assuming that the parcel moves sufficiently slowly that it maintains total pressure equilibrium with its surroundings gives the relation

$$\delta p + \frac{B\delta B}{\mu_0} = \mathrm{d}p + \frac{B\mathrm{d}B}{\mu_0}. \tag{11.3}$$

The displaced parcel will be unstable and continue to rise if $\delta\rho < \mathrm{d}\rho$. Manipulation of expressions (11.2) and (11.3) then leads to the following criterion for instability:

$$\frac{-ga^2}{c^2}\frac{\mathrm{d}}{\mathrm{d}z}\ln\left(\frac{B}{\rho}\right) > N^2, \tag{11.4}$$

where a is the Alfvén speed, c is the adiabatic sound speed and N is the Brunt–Väisälä (or buoyancy) frequency, defined by:

$$a^2 = \frac{B^2}{\mu_0\rho}, \qquad c^2 = \frac{\gamma p}{\rho}, \qquad N^2 = \frac{g}{\gamma}\frac{\mathrm{d}}{\mathrm{d}z}\ln\left(p\rho^{-\gamma}\right). \tag{11.5}$$

Inequality (11.4) may be regarded as the modification by a stratified magnetic field of the Schwarzschild criterion. Of particular significance is that a horizontal magnetic field that decreases sufficiently rapidly with height can destabilize a convectively stable atmosphere (i.e. one with $N^2 > 0$). Clearly, magnetic buoyancy instability is of most significance when it is the sole instability mechanism available; it is therefore natural, in general, to consider sub-adiabatically stratified atmospheres. This is particularly true for the tachocline, where it is envisaged that magnetic buoyancy is responsible for the disruption of the magnetic field in a convectively stable region.

The problem of the ideal (diffusionless), linear instability of a stratified horizontal field was first solved by Newcomb (1961), in full generality, using the energy principle of Bernstein *et al.* (1958). Newcomb showed that instability to modes that do not bend the field lines (*interchange* modes) occurs if and only if (11.4) is satisfied *somewhere* in the fluid. More specifically, he expressed the criterion as

$$\frac{\mathrm{d}\rho}{\mathrm{d}z} > -\frac{\rho g}{a^2 + c^2}. \tag{11.6}$$

Expressions (11.4) and (11.6) are equivalent, using the equation of magnetohydrostatic equilibrium; whereas the latter resembles a slight modification to the non-magnetic criterion, it is the former that brings out most clearly the destabilizing influence of the field gradient.

Although (11.4) captures the essential physics of the magnetic buoyancy instability of a stratified magnetic field there are clearly many crucial effects that are neglected, particularly three-dimensional perturbations, diffusion and rotation. One of the most surprising aspects of magnetic buoyancy instability is that three-dimensional perturbations, despite having to do work against magnetic tension, can be more readily destabilized than interchange modes, which do not bend the field lines. Newcomb (1961) showed that the most readily destabilized three-dimensional perturbations are those with $k_x \to 0$, and that a necessary and sufficient condition for their instability is that the inequality

$$\frac{\mathrm{d}\rho}{\mathrm{d}z} > -\frac{\rho g}{c^2} \tag{11.7}$$

is satisfied somewhere in the plasma. At least in this guise, this expression is unchanged from its non-magnetic counterpart. The crucial role of the field was however elucidated by Thomas & Nye (1975), who showed (essentially) that criterion (11.7) can be expressed in the alternative form:

$$-\frac{ga^2}{c^2}\frac{\mathrm{d}}{\mathrm{d}z}\ln B > N^2. \tag{11.8}$$

Comparison of (11.4) with (11.8) shows that whereas the instability of interchange modes requires a sufficiently rapid decrease with height of B/ρ, that of three-dimensional modes requires the less stringent criterion of an equivalent decrease with height only of B. The physics underlying the instability mechanism was explored by Hughes & Cattaneo (1987). For interchange modes, the necessary work that must be done against gas pressure in order to create density perturbations is accompanied by unavoidable – but non-beneficial – work against magnetic pressure. Three-dimensional perturbations, however, with a long variation in the direction of the field, do work against gas pressure whilst minimizing that against magnetic pressure; the stabilizing effects of tension are negligible for such perturbations. It should though be noted that the beneficial feature of undulatory modes (i.e. modes with bent field lines) occurs only for three-dimensional motions; two-dimensional undulatory perturbations ($k_y = 0$) are unable to escape work against magnetic pressure. Although expression (11.7) is a necessary condition for the instability of such two-dimensional perturbations, it is by no means sufficient (see Hughes & Cattaneo 1987); in general much more severe field gradients are needed to drive any such instability.

The most extreme manifestation of magnetic buoyancy instability – and one whose nonlinear evolution we shall consider in Section 11.2.2 – is that arising from a discontinuity with height of the magnetic field. Such an equilibrium has a discontinuity in the gas pressure and, for continuous temperature profiles, a discontinuity in the density, with lighter gas supporting heavier gas. Such an equilibrium is thus susceptible to a Rayleigh–Taylor instability. This was first investigated by Kruskal & Schwarzschild (1954), who considered the stability of a plasma supported above a vacuum by a magnetic field. A local analysis (e.g. Parker 1979) yields the following dispersion relation governing the growth rate s:

$$s^2 = \frac{g\Delta\rho(k_x^2 + k_y^2)^{1/2} - \rho a^2 k_x^2}{2\rho + \Delta\rho}, \tag{11.9}$$

where $\Delta\rho$ denotes the jump in density at the interface. It is noteworthy that for this Rayleigh–Taylor type of instability – and in contrast to the instability of a smoothly varying field – the most readily destabilized modes are interchanges ($k_x = 0$), the effects of bending the field lines here being entirely penalizing.

11.2.1.2 The role of diffusion

So far we have considered only *ideal* (i.e. diffusionless) magnetic buoyancy instabilities. However, as with many instability mechanisms, the incorporation of diffusive effects can lead to significant qualitative and quantitative changes to the nature of the instability. From the simple parcel argument outlined above it is readily seen that it is beneficial to the instability if the magnetic diffusivity η is small (thus maintaining the destabilizing field gradient) and the thermal diffusivity κ is large (thereby eroding the stabilizing entropy gradient). A formal analysis (Gilman 1970; Acheson 1979) leads to the following instability criterion (neglecting viscosity and in the limit of $k_x \to 0$), which should be regarded as the diffusive modification to (11.8):

$$-\frac{ga^2}{c^2}\frac{\mathrm{d}}{\mathrm{d}z}\ln B > \frac{\eta}{\kappa}N^2. \tag{11.10}$$

(The full criterion, for non-zero viscosity and finite wavenumbers, can be found in Acheson (1979).) In stellar interiors, the laminar values of η and κ satisfy the inequality $\eta \ll \kappa$; with these values the stabilizing influence of the entropy gradient ($N^2 > 0$) is dramatically reduced. On the other hand, if one were to argue that laminar values are inappropriate and that all diffusivity ratios are $O(1)$ then the influence of any stabilizing entropy gradient is undiminished. This is possibly an important issue in the triggering of magnetic buoyancy instabilities in the tachocline, and one to which we shall return in Section 11.5.

In addition to important modifications to the instability criteria (11.4) and (11.8), which describe direct instabilities (i.e. $\mathrm{Re}(s) > 0$, $\mathrm{Im}(s) = 0$), diffusive effects can lead to the onset of oscillatory instabilities ($\mathrm{Re}(s) > 0$, $\mathrm{Im}(s) \neq 0$; sometimes referred to as *overstability*). Interestingly, the equations for interchange instabilities in the presence of diffusion can be transformed into those of thermosolutal convection (Spiegel & Weiss 1982; Hughes & Proctor 1988), the most extensively studied double-diffusive system (see, for example, Turner 1973). The transformation, although mathematically straightforward, is not what one might naïvely imagine on physical grounds (with the entropy gradient mapped to the thermal gradient, and the gradient of magnetic field mapped to that of salinity), but instead maps the thermal gradient of thermosolutal convection into a linear combination of the entropy and magnetic field gradients. This leads to the following criterion for the onset of overstability:

$$-\frac{ga^2}{c^2}\left(\eta+\nu-\kappa(\gamma-1)\right)\frac{\mathrm{d}}{\mathrm{d}z}\ln\left(\frac{B}{\rho}\right)$$
$$> (\kappa+\nu)(\kappa+\eta)(\nu+\eta)\frac{k^6}{k_x^2}+(\kappa+\nu)N^2, \qquad (11.11)$$

where $k^2 = k_x^2 + k_z^2$ (recall $k_y = 0$ here). Criterion (11.11) describes two rather different types of instability depending on the sign of $\eta + \nu - \kappa(\gamma - 1)$.

If $\eta + \nu > \kappa(\gamma - 1)$ (which is typically not the case in stellar interiors) then overstability can occur only for 'top-heavy' field gradients (i.e. B/ρ decreasing with height). The instability mechanism may be understood qualitatively in terms of a parcel argument of the type first proposed by Cowling (1957) in the context of magnetoconvection. For the sake of simplicity let us here assume that κ is extremely small such that we can neglect thermal considerations. Suppose a parcel (or flux tube) is displaced upwards and is denser than its surroundings. It will then fall and, as a consequence of mixing with adjacent gas of a weaker field strength during its voyage, will return to its original level with a weaker field – and hence a higher density – than it had initially. The parcel will then 'overshoot' on the downward side, repetition of this process leading to growing oscillations.

If, on the other hand, $\eta+\nu < \kappa(\gamma-1)$ then criterion (11.11) describes oscillatory instability for 'bottom-heavy' field gradients (i.e. B/ρ increasing with height); in other words, the rather surprising notion of instability when *both* gradients (field and entropy) are 'stabilizing' (Hughes 1985a) – and in contrast to thermosolutal convection where no instability can occur when both the thermal and salinity gradients are stabilizing. The instability mechanism can, however, again be understood in terms of a parcel argument, where here, for simplicity, we may neglect the effects of magnetic diffusion. The crucial feature of the instability is that when a flux tube is raised in an atmosphere in which B increases with height it is squashed by the stronger external field; if the squashing is sufficiently vigorous, the tube will be hotter than its surroundings. However, with B/ρ increasing with height the tube is guaranteed to be denser than its surroundings and hence will fall. Instability is then facilitated by thermal diffusion, which transmits heat away from the squashed (and heated) tube, causing it to return to its original level cooler, and hence denser, than it was initially. As explained above, this is precisely the recipe for overstability.

11.2.1.3 The influence of rotation

The simplest possible order of magnitude estimate for the growth time T of a direct magnetic buoyancy instability gives $T \sim (c/a)(d/g)^{1/2} \approx 30$ days, on adopting a plasma beta of $\beta = 10^8$ and taking d as the depth of the tachocline. Since this is comparable to the solar rotation period it suggests that rotation will be an important ingredient in the evolution of the instability. By far the most comprehensive study of magnetic buoyancy instability influenced by rotation is that of Acheson (1978), who considered the instability of a toroidal magnetic field in a cylindrical geometry, not only incorporating differential rotation $\Omega(r, z)$, but also the effects of latitudinal variation via radial and axial components of the acceleration due to gravity. His study covered not only instabilities driven by magnetic buoyancy but also those resulting from differential rotation (such as Goldreich–Schubert–Fricke instabilities) and from gradients in the magnetic field (even in the absence of gravity). His analysis – and indeed that of all studies of the role of rotation on magnetic buoyancy instabilities – was local, and hence could not capture the influence of velocity shear on magnetic buoyancy instabilities (or, indeed, shear flow instabilities themselves).

Interchange modes (the plane layer counterpart of axisymmetric modes) are the most affected by rotation through the angular momentum constraint (although this can be eased by viscosity); we shall therefore concentrate on undulatory modes. The full picture is extremely complicated (see Acheson 1978) and so here we shall focus on the simpler case of the *uniform* rotation of a plane layer, with the initial magnetic field, gravity and the rotation vector mutually orthogonal. Even for this simplified system analytical progress is, in general, not possible. However, for rapid rotation one may adopt the magnetostrophic approximation (neglecting inertial terms) and obtain the following criterion for the instability of low-frequency modes (the plane-layer equivalent of equation (7.11) of Acheson 1978):

$$-\frac{\gamma g}{c^2}\frac{\mathrm{d}}{\mathrm{d}z}\ln\left(\frac{B}{\rho}\right) > \frac{k_x^2 k^2}{k_y^2} + 4\frac{\eta^2\Omega^2}{a^4}\frac{k^4}{k_x^2} + \frac{(\gamma-1)^2 g^2}{4c^4} + \frac{\eta}{\kappa}\frac{N^2}{a^2} - \frac{(\gamma+1)^2 g^2}{4c^4\left(1 + 8\eta\kappa(\Omega^2/N^2a^2)(k^4/k_x^2)\right)^2}, \tag{11.12}$$

where $k^2 = k_x^2 + k_y^2 + k_z^2$, and where it has also been assumed that thermal diffusion is suitably fast. Using this expression, Acheson (1978) nicely tied earlier results of Gilman (1970), who considered a constant Alfvén speed atmosphere ($B \propto \rho^{1/2}$) and infinite thermal and electrical conductivities, together to subsequent extensions by Acheson & Gibbons (1978), who considered arbitrary field configurations, and Roberts & Stewartson (1977), who examined the regime of large but finite conductivities. In particular, Roberts & Stewartson demonstrated that for a constant Alfvén speed atmosphere, instability could be more easily facilitated for finite (as opposed

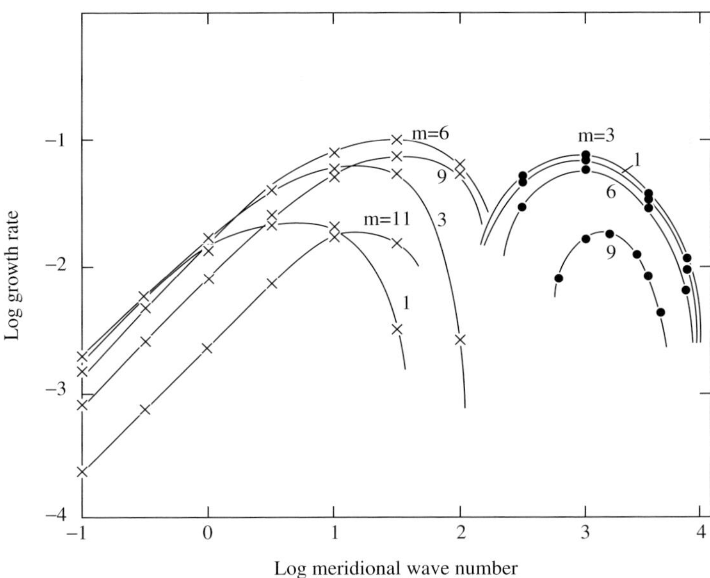

Figure 11.1. Growth rate versus meridional wavenumber for various values of the azimuthal wavenumber, here denoted by m. The parameters are chosen to represent conditions in the solar convective overshoot zone. Crosses denote a low-frequency instability; dots denote a high-frequency instability. (From Schmitt & Rosner 1983.)

to infinite) values of κ. Expression (11.12) makes clear, at least for the parameter regime of rapid rotation and fast diffusion, the identical behaviour of N^2 and $1/\kappa$. Thus, from the Roberts & Stewartson result, it follows, possibly somewhat surprisingly initially, that *in*creasing N^2 (i.e. increasing the stable stratification of the background state) can be *de*stabilizing – though ultimately is, of course, stabilizing. Simply speaking, there is an optimal frequency for oscillatory double-diffusive instabilities, such that diffusion acting on a displaced fluid element generates the maximal density shift for a returning parcel; increasing the stable stratification changes the oscillation frequency of fluid parcels and may act to shift it favourably for instability. Such behaviour has also been identified in other double-diffusive systems (e.g. Masuda 1978; Soward 1979; Pearlstein 1981).

The general dispersion relation, incorporating rotation, is a fifth-order complex equation, which typically can be solved only numerically. This has been performed by Schmitt & Rosner (1983) and by Hughes (1985b). Schmitt & Rosner focused their attention on the calculation of growth rates of unstable modes for parameter values appropriate to the Sun. They found that for a toroidal field whose strength decreased with radius in a slightly sub-adiabatic region, and assuming molecular diffusivities, both low and high frequency undulatory modes are unstable, with comparable growth rates, but very different meridional scales (see Figure 11.1).

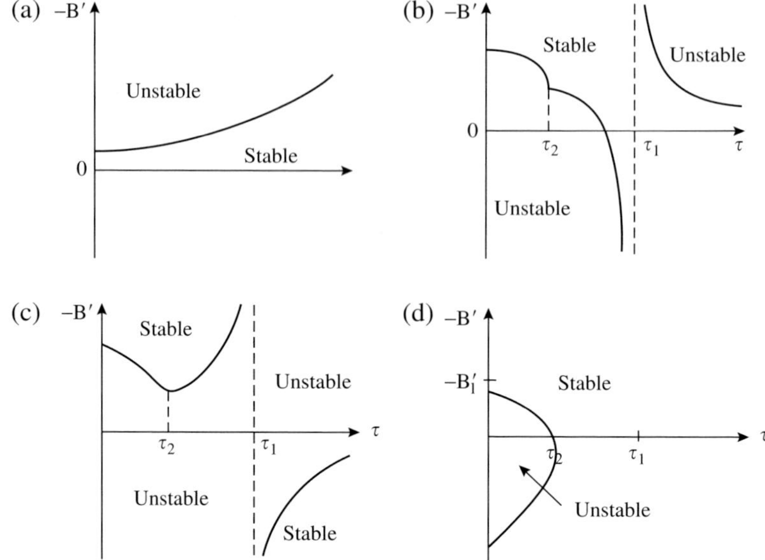

Figure 11.2. The stability boundaries of the various unstable modes plotted separately in the $B' - \tau$ plane, where $\tau = \eta/\kappa$. Plot (*a*) shows the boundary of the low-frequency (magnetostrophic), top-heavy mode; plots (*b*) and (*c*) depict the two possible configurations for the unstable regions for the high-frequency modes ($\tau_2 = (\gamma - 1)/2$). Plot (*d*) shows the boundary of the low-frequency, bottom-heavy modes. In all the plots $\tau_1 = \gamma - 1 - \nu/\kappa$. (After Hughes 1985b.)

Hughes, on the other hand, concentrated on determining the stability boundaries of the various modes present in the system. Expression (11.12) gives the stability boundary for the instability of low frequency modes for 'top-heavy' field gradients. Instability though is also possible for low frequency modes driven by bottom-heavy field gradients and also for high frequency modes with field gradients of either sign. The regions of instability are sketched in Figure 11.2.

11.2.1.4 The interaction with a shear flow

A key ingredient missing from any local analysis of the effect of differential rotation on magnetic buoyancy instability is that of velocity shear. Just as for classical (hydrodynamical) shear instabilities, these must be captured through an eigenvalue analysis. The identification of the solar tachocline, and its likely coincidence with the storage site for the Sun's toroidal magnetic field, highlights the importance of examining the interaction between shear flows and magnetic buoyancy instabilities.

This problem, though with a different motivation, was first investigated by Adam (1978), and has recently been extended by Tobias & Hughes (2004), both works examining the stability of equilibria with aligned field and flow, $\mathbf{B} = B(z)\hat{\mathbf{x}}$,

$\mathbf{U} = U(z)\hat{\mathbf{x}}$. The energy principle of Bernstein *et al.* (1958) may be employed to investigate ideal (diffusionless) instabilities, for which the profiles of field and flow can be arbitrary functions of z. However, the presence of a shear flow renders the linear operator non-self-adjoint (Frieman & Rotenberg 1960) and thus, in contrast to the case with no shear flow, it is only possible to derive sufficient conditions for the stability of the flow. Using the energy principle, Tobias & Hughes (2004) derived two stability criteria. The first is that stability is assured if there exists any constant U_0 such that everywhere both $(U - U_0)^2 = \widetilde{U}^2 < c_{\mathrm{T}}^2$ and

$$\widetilde{U}^2 \leq \frac{a^2(c^2\rho' - \rho g)}{(a^2 + c^2)\rho' - \rho g}, \tag{11.13}$$

where $c_{\mathrm{T}}(z)$ is the tube speed (or cusp speed), defined by

$$c_{\mathrm{T}}^2 = a^2c^2/(a^2 + c^2). \tag{11.14}$$

Criterion (11.13) extends the result of Adam (1978), who showed that inequality (11.13) guaranteed stability for the restricted class of two-dimensional undulatory modes. The second criterion of Tobias & Hughes (2004), which is somewhat more involved, involves the derivative of $U(z)$: stability is guaranteed if, everywhere, $\widetilde{U}^2 < c_{\mathrm{T}}^2$ and

$$-\frac{\mathrm{d}}{\mathrm{d}z}\left(\frac{\rho g k_x^2 \widetilde{U}^2(a^2 - \widetilde{U}^2)}{(k_x^2 + k_y^2)(a^2 + c^2)(c_{\mathrm{T}}^2 - \widetilde{U}^2) + k_x^2\widetilde{U}^4}\right)$$
$$\leq g\frac{\mathrm{d}\rho}{\mathrm{d}z} + \rho k_x^2(a^2 - \widetilde{U}^2) - \frac{\rho g^2(k_x^2 + k_y^2)(a^2 - \widetilde{U}^2)}{(k_x^2 + k_y^2)(a^2 + c^2)(c_{\mathrm{T}}^2 - \widetilde{U}^2) + k_x^2\widetilde{U}^4}. \tag{11.15}$$

The rather complicated expression on the left-hand side of (11.15) may, depending on the form of U, take either sign. Thus although the right-hand side of (11.15) is, for $\widetilde{U}^2 < c_{\mathrm{T}}^2$ (the range of validity of the criterion), a monotonically decreasing function of $\widetilde{U}^2$, there remains the interesting possibility that, for *fixed* values of k_x and k_y, an atmosphere that is unstable in the absence of a flow may be stabilized by the presence of a suitable shear. However, it should be noted that since this potentially stabilizing term tends to zero as $k_x \to 0$, criterion (11.15) can say nothing about whether overall stability – i.e. stability considering *all* values of k_x and k_y – can be attained by the incorporation of shear.

It is of interest to note that since the criterion (11.15) requires that $\widetilde{U}^2 < c_{\mathrm{T}}^2$ then, at least via the energy principle, we are unable to say anything (for non-zero U) about the stability of the non-magnetic case. Thus, unfortunately, expression (11.15) with $a \equiv 0$ cannot be interpreted as a Richardson-type criterion for compressible atmospheres.

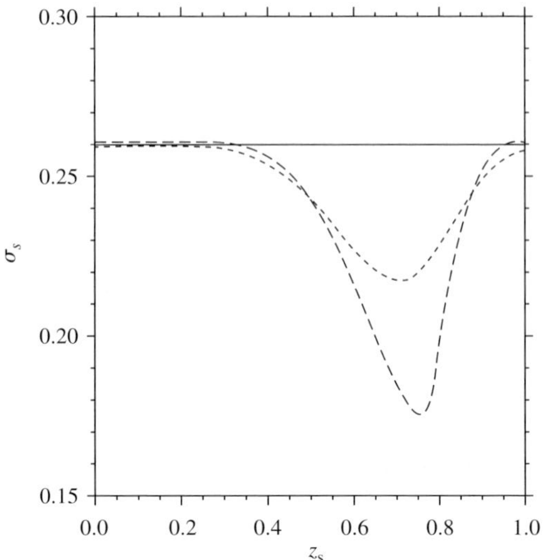

Figure 11.3. Growth rate as a function of the location of the shear (z_S) for the flow $U(z) = U_0 \tanh \alpha(z - z_S)$. Here $U_0 = 7$ (measured in terms of the Alfvén speed) and $\alpha = 1$ (solid line), 5 (dashed) and 10 (dot-dashed). (From Tobias & Hughes 2004.)

Expressions (11.13) and (11.15) yield important information on the nature of shear flows for which an atmosphere with a stratified magnetic field remains stable. However, the analysis reveals nothing about the influence of shear flows on unstable modes driven by magnetic buoyancy. Such information can be achieved only via a solution of the linear eigenvalue problem, which, typically, must be performed numerically. Attention has to be focused on specific choices of the profiles for $U(z)$ and $B(z)$ and thus it is not possible to obtain results of the same generality as the sufficient criteria for stability, expressions (11.13) and (11.15). Tobias & Hughes (2004) considered the influence of two different velocity shear profiles – one a cubic in z and the other varying as $\tanh(z)$ – on atmospheres unstable in the absence of shear and with a magnetic field decreasing linearly with height. They found that the shear ultimately has an 'axisymmetrizing' and stabilizing effect on the instability, although for certain modes (with fixed values of the wavenumbers) the initial effect is to destabilize the instability further. The stabilizing role of the shear depends crucially on the precise location of the region of strong shear; this is illustrated clearly in Figure 11.3, which depicts the growth rate for a shear flow of the form $U(z) = U_0 \tanh \alpha(z - z_S)$ for a range of z_S and three different values of α. It can be seen that the shear has its most stabilizing effect when it is localized about $z \approx 0.75$, the value of z about which the eigenfunction for the field in the absence of shear is peaked.

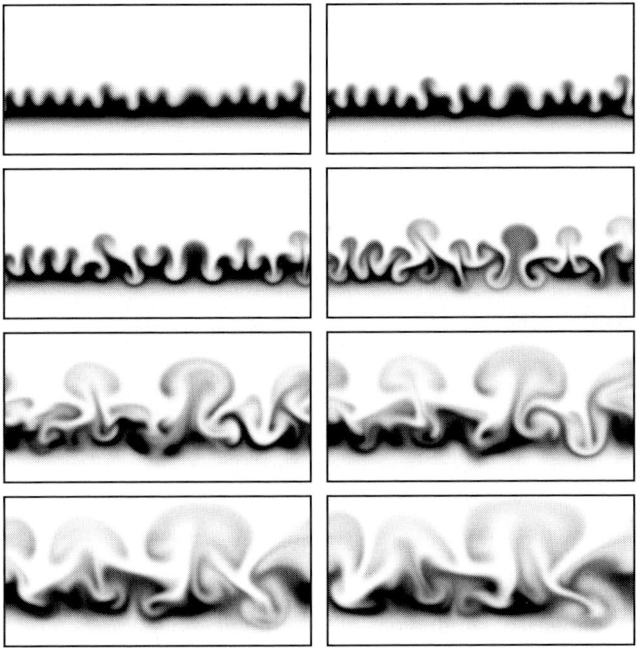

Figure 11.4. Snapshots of the evolution of the magnetic field for an interchange mode at eight different times; the plots are scaled independently. The field is directed into the page and is initially uniform, occupying the region $0.2 < z < 0.4$. The evolution is controlled by the interaction of vortex pairs. (After Cattaneo & Hughes 1988; courtesy Evy Kersalé.)

11.2.2 Nonlinear evolution

In terms of the deep-seated solar magnetic field, one of the crucial questions is whether the nonlinear evolution of the magnetic buoyancy instability of a smoothly varying field can give rise to concentrated clumps of field ('flux tubes') of the form that eventually protrude through the solar surface. This problem was first addressed by Cattaneo & Hughes (1988), who considered the nonlinear evolution of the two-dimensional (interchange) instability of a slab of uniform field embedded in an otherwise non-magnetic atmosphere. In this case the instability is driven solely by the density jump at the upper interface of the magnetic slab. As anticipated for a magnetically driven Rayleigh–Taylor instability, the rising field adopts the form of magnetic mushrooms. A strong shear along the magnetic interface results which, via a secondary Kelvin–Helmholtz instability, wraps the gas in the edges of the mushrooms into strong vortices. The subsequent evolution is then dominated by the pairwise interaction of neighbouring vortices from *different* mushrooms, which act in concert to drag down pockets of strong field despite their inherent buoyancy. The evolution of the field is portrayed in Figure 11.4. So although this very simple

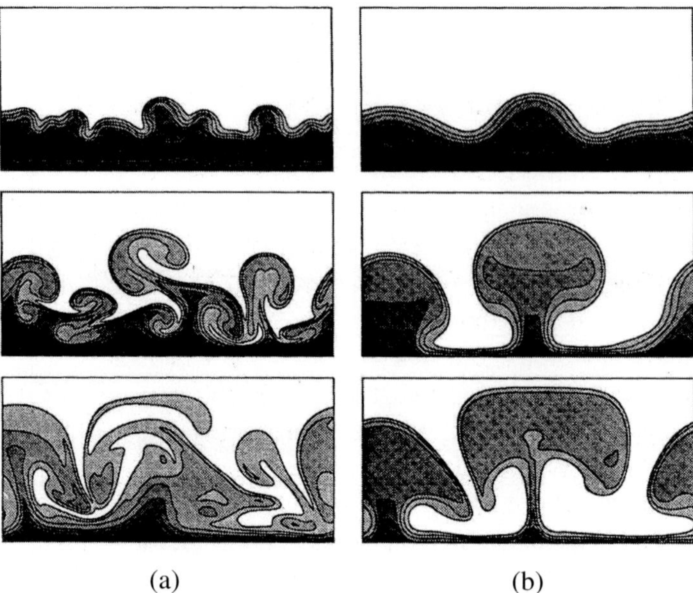

Figure 11.5. Numerical simulations of the nonlinear breakup of a magnetic layer. The strength of the dominant x-component (into the page) is shown, darker tones corresponding to stronger fields. In both cases the field is initially horizontal with a weak B_y component. In (a) the transverse (y) component vanishes at the upper interface and increases linearly with depth. In (b) it vanishes near the lower boundary and increases with height. The motions are independent of the x-direction into the page. In (a) the emerging flux is small scale and a significant fraction of the layer remains undisturbed, stabilized by the increasing transverse component. In (b) the whole layer is consumed by the instability; the emerging field retains coherence by virtue of its helical nature. (From Cattaneo *et al.* 1990b.)

model, in which the field lines remain straight throughout, does give rise to isolated regions of strong field, these 'tubes' do not rise but, surprisingly, are transported downwards.

Cattaneo *et al.* (1990a) extended the model to consider the instability of a weakly sheared magnetic field $\mathbf{B} = (B_x(z), B_y(z), 0)$ with $|B_y/B_x|$ small, but still restricted attention to x-independent motions, thereby distinguishing the x-direction – these modes may be thought of as the plane layer counterparts of the axisymmetric evolution of a mixed toroidal/poloidal field. The determining factor for the nonlinear evolution turns out to to be the location of the *resonant layer*, namely the height at which B_y vanishes. Here the magnetic field is locally untwisted with respect to x-independent modes, and is thus the place that offers the least resistance to the instability – which again is driven by a discontinuity in density at the upper interface of the field. Consequently, as illustrated in Figure 11.5, if the resonant layer is close to the upper interface then the instability adopts only a small vertical scale and the escaping field is essentially untwisted. By contrast, if the resonant layer is deeper-lying the instability adopts a longer vertical scale, it being energetically worthwhile to untwist all of the field above the resonant layer in order to feel the benefits of the untwisted field in the vicinity of the resonant layer. In this case the emerging field fragments are large scale and, as a result of their strong 'poloidal' component, maintain their coherence as they rise. Cattaneo *et al.* (1990b) speculated that a variation in the distribution of a weak poloidal ingredient of the field may account for the observed variation through the solar cycle in the scale and structure of emerging flux (e.g. Golub *et al.* 1981).

Clearly, however, a global magnetic field responsible for the sunspot belts has succumbed to a non-axisymmetric (undulatory) instability, and it is therefore important to investigate the unconstrained (three-dimensional) nonlinear evolution. As discussed above, the preferred mode of a Rayleigh–Taylor instability – i.e. one driven by a discontinuity in the field rather than by a smooth variation – is two-dimensional. Any three-dimensionality in this case must then arise either as a purely nonlinear phenomenon or as a result of a three-dimensional initial perturbation. Matthews *et al.* (1995) and Wissink *et al.* (2000) considered the same equilibrium state as Cattaneo & Hughes (1988), but allowed for a fully three-dimensional nonlinear development. The initial evolution is essentially two-dimensional, with the formation of strong, anti-parallel vortices; however, the crucial difference from the restricted two-dimensional evolution is that such vortex pairs are susceptible to a three-dimensional instability, which causes a strong arching of the vortices and hence of their associated magnetic field. The basic instability mechanism is that first studied by Crow (1970), and which has received considerable attention owing to its importance in the dynamics of trailing vortices from aircraft wings. The resulting structure of the field is illustrated in Figure 11.6a; it is pleasingly reminiscent of the notion of a tube of flux buckling and erupting through the solar surface, as illustrated in Parker's sketch in Figure 11.6b.

The most recent study to address the nonlinear development of magnetic buoyancy instabilities from the base of the convection zone is that of Fan (2001), who considered an initial state with a Gaussian profile for the magnetic field. The important difference, in comparison with the top hat profiles used in the work discussed above, is that the equilibrium quantities are such that the preferred modes are three-dimensional, with interchange modes stable. The evolution is therefore three-dimensional from the outset. In appearance it is rather similar to that found by Matthews *et al.* (1995) and Wissink *et al.* (2000), with the field adopting the form of arched structures. However, the three-dimensionality in this case is due to the initial instability, rather than to any subsequent nonlinear vortical interactions.

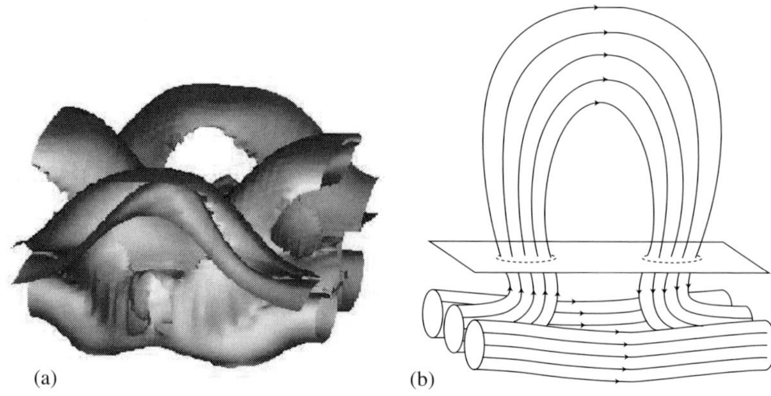

Figure 11.6. (a) Isosurface of the magnetic field following the three-dimensional evolution of an unstable magnetic layer. (From Matthews *et al.* 1995.) (b) Sketch of the field erupting through the solar surface to give rise to a bipolar sunspot pair. (After Parker 1979.)

11.3 The instability of isolated flux tubes

In Section 11.2 we discussed the instability of horizontally homogeneous magnetic fields, with a view to considering whether a large-scale, global solar field could give rise to the strong, isolated flux concentrations observed at the surface. A rather different viewpoint, based on observations of concentrated fields at the surface, is to suppose that the solar magnetic field exists *always* in the form of tubes, and to explore the consequences of this assumption. An isolated tube of flux must have a gas pressure deficit relative to its surroundings; its relative density will then depend on the temperature difference (if any) between the tube and its surroundings. As explained earlier, a tube that is always at the same temperature as its surroundings will have a density deficit and will thus be buoyant. If, however, heat transfer between a tube and its surroundings is weak then one may argue that a tube, even if initially in thermal equilibrium, will rise only until it attains mechanical equilibrium – of necessity at a lower temperature than its surroundings. Whether this occurs depends both on the heat transfer between the tube and its surroundings and also on the stratification of the atmosphere. For example, a tube rising adiabatically in an adiabatic atmosphere will always be buoyant, whereas in an isothermal atmosphere it will eventually attain mechanical equilibrium. In this section we investigate the nature of the instability of flux tubes in mechanical equilibrium.

The instability in its simplest form, in which the flux tube is perturbed bodily (without bending), may be addressed by a parcel argument rather similar to that employed in Section 11.2.1 for the interchange instability of a diffuse field. Formally the changes are minor, reflecting the fact that, even in the equilibrium state, the gas pressures of the tube and of the surroundings are different, and that external to the tube there is no magnetic field. Thus (cf. expressions (11.2) and (11.3)) we have the following conditions

$$\frac{\delta B}{B} = \frac{\delta \rho}{\rho}, \qquad \frac{\delta p}{p_{\mathrm{i}}} = \gamma \frac{\delta \rho}{\rho}, \qquad \delta p + \frac{B \delta B}{\mu_0} = \mathrm{d}p, \tag{11.16}$$

where p_{i} refers to the internal gas pressure, related to the external gas pressure p_{e} via (11.1). Manipulation of these expressions leads to the following criterion for instability, first obtained, in a more formal manner, by Spruit & van Ballegooijen (1982):

$$\delta > \frac{2 - \gamma}{\gamma(2 + \gamma\beta)}, \tag{11.17}$$

where $\delta = \mathrm{d}\ln T_{\mathrm{e}}/\mathrm{d}\ln p_{\mathrm{e}} - 1 + 1/\gamma$ is positive (negative) if the stratification of the external medium is superadiabatic (subadiabatic), and β denotes the ratio of the internal gas pressure to the magnetic pressure. There are two points of immediate note. One is that instability is possible only for convectively unstable atmospheres ($\delta > 0$), the other is that *in*creasing the field strength (decreasing β) is stabilizing. These two points highlight the stark difference between the instability of a magnetized atmosphere in lateral pressure equilibrium, and that of an isolated tube of magnetic flux. The rather counter-intuitive stabilizing effect of the magnetic field arises through the very assumption of an initial equilibrium; the stronger the magnetic field the cooler the tube must be, and consequently the more stable.

For more general motions of a flux tube one clearly has to progress beyond simple parcel arguments. Spruit (1981) derived the *thin flux tube equations*, a set of model equations that assumes the coherence of magnetic flux tubes, and models their evolution under the influence of buoyancy, magnetic tension and a drag force between the tube and the external medium. Spruit & van Ballegooijen (1982) found that instability to wavy modes occurs if

$$k_x^2 < \frac{1 + 1/\beta}{2H_{\mathrm{p}}^2}\left(1/\gamma + \beta\delta\right), \tag{11.18}$$

where H_{p} is the pressure scale height of the external medium. Thus instability will occur for sufficiently long wavelengths provided that $\beta\delta > -1/\gamma$, which should be contrasted with inequality (11.17), the instability criterion for modes that raise the entire tube without bending. Thus flux tubes in convectively stable atmospheres ($\delta < 0$) can be unstable, instability being facilitated by the flow along the tube, as envisaged by Parker (1955). It is though worth pointing out that in convectively unstable atmospheres ($\delta > 0$), increasing the strength of the field is a *stabilizing* influence, just as for the interchange modes. For a very weak field ($\beta \gg 1$) the growth rate of the instability is proportional to $\delta^{1/2}$, whereas for a very strong field ($\beta \ll 1$) it is reduced to $O(\beta^{1/2})$.

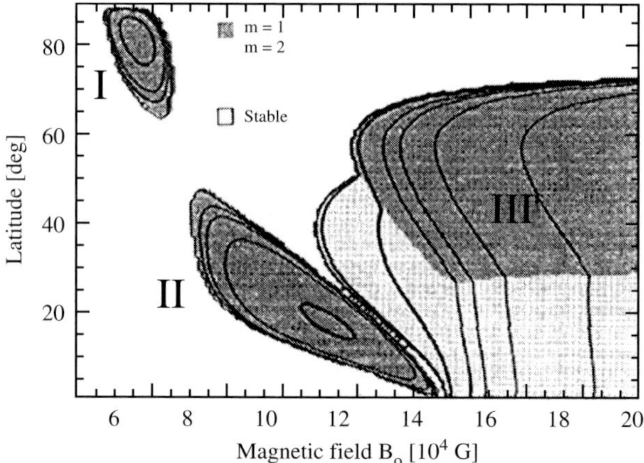

Figure 11.7. Regimes of instability for flux tubes stored in the overshoot region; *m* denotes the azimuthal wavenumber. (From Caligari *et al.* 1995.)

Caligari *et al.* (1995) have explored the consequences of this instability mechanism for flux tubes stored in the convective overshoot region. Their main result (Figure 11.7) is that fields up to $O(10^5)G$ (i.e. super-equipartition fields) can be stored before instability sets in, and that this therefore provides a natural explanation for the production of strong flux tubes that can subsequently rise unscathed through the convection zone.

11.4 The buoyant rise of isolated flux tubes

The appearance of active regions at the solar surface is certainly suggestive of tubes of magnetic flux erupting through the photosphere. The simplest model of a bipolar sunspot pair is of a 'magnetic sea serpent' protruding from the Sun. One of the most difficult problems in solar MHD is to relate the observed surface field to the posited deep-seated field. It is, in some sense, a 'post-tachocline' problem and one that, despite its importance, strictly lies outside the scope of this volume. That said, it is so closely linked to the instability mechanisms discussed above that it would be remiss not to include some discussion here.

The problem has received considerable attention, mostly based on models employing the thin flux tube approximation (for an extensive list of references see, for example, the reviews of Moreno-Insertis 1992; Fan 2004; Schüssler 2005). As discussed above, flux tubes can be unstable to undulatory modes. The nonlinear development of this instability leads to pronounced arching, with the top of the tube rising through the convection zone, and with the bottom trapped in the overshoot zone. There is, however, a serious problem in this description of the field. As the tube rises it expands dramatically, with a concomitant reduction in field strength, leading to what has been termed an 'explosion' of the flux tube. Various scenarios have been advanced for what may happen to the tube after the explosion. However, one must also consider the possibility that an explosion of the tube signifies the breakdown of the theory itself.

From surface observations we have a very good knowledge of the field strength at the solar surface, and how it varies depending on the size of the magnetic elements. However, we have an extremely limited knowledge of the field strength at depth. Within the thin flux tube theory a partial answer can be provided by considering the influence of rotation on rising tubes. Choudhuri & Gilman (1987) and Choudhuri (1989) have argued that the rise of flux tubes starting from the base of the convection zone with a field strength $\lesssim 10^4$ G is strongly constrained by rotation and, consequently, the field emerges in polar regions. Only if the initial field strength $\gtrsim 10^5$ G does the field emerge at low latitudes, as observed on the Sun. D'Silva & Choudhuri (1993) have used the same thin flux tube model to investigate the tilt of bipolar magnetic regions, and they conclude that for consistency with the observed tilt at the solar surface, the field strength at the base of the convection zone must be in the range 60–160 kG. For these reasons, together with the flux tube stability results discussed in Section 11.3, a field strength of 10^5 G at the base of the convection zone has become a widely quoted figure.

Notwithstanding the ability of the thin flux tube model to be able to deal with the dynamics of the global solar field, and indeed to make predictions concerning both flux emergence and the field strength at depth, it is still worthwhile casting a critical eye on the very roots of the approximation. In particular, even assuming for the moment that the field does assume the form of flux tubes – isolated concentrations of field surrounded by a flux surface – is its behaviour well-approximated by the thin flux tube approximation? This problem has been looked at in different guises by a number of authors. (e.g. Emonet & Moreno-Insertis 1998; Fan *et al.* 1998; Hughes *et al.* 1998; Hughes & Falle 1998). A common thread to emerge from all of these calculations is that the internal field structure, measured by its twist for example, is a crucial ingredient in determining the coherence of any rising tube. A tube needs to possess a sufficiently twisted field in order to prevent its annihilation through the incursion of vorticity. The internal dynamics, which is simply neglected or 'averaged over' in the thin flux tube approximation, probably plays a crucial role. Hughes & Falle (1998), via an adaptive grid, high Reynolds number simulation, showed also how the interactions between the tube and its surroundings can lead to a buoyant tube taking a path that is far removed from the vertical, and that, therefore, the interactions between the tube and its surroundings are far more complicated than can be modelled by a simple drag force.

11.5 Implications for the tachocline

Understanding the role of magnetic buoyancy in the evolution of the solar magnetic field entails two aspects. One is elucidating the various physical processes, such as instability mechanisms, that may be occurring. The other is deciding which of these are of relevance. The preceding sections have addressed the former issue; in this concluding section I shall address the latter. The first nettle to grasp concerns the very means by which the magnetic field should be treated. One possibility is to argue that the magnetic field exists only in the form of coherent tubes of flux, to derive equations for the dynamics of such tubes, and then to evaluate their consequences. The second approach is to stay more closely attached to the full MHD equations, though of course this renders progress rather more difficult.

The rationale for adopting the flux tube paradigm is, in the words of Schüssler (2005), 'simply because they are there'. Whereas it is undeniably true that observations of the solar surface reveal concentrations of magnetic field, one needs to be careful both in the description of these flux concentrations and also in extrapolating the structure of the magnetic field from the surface to the interior. Working down from the surface we must first enquire into the nature of the magnetic field in the convection zone. The flux tube picture essentially ignores the convective motions, on the grounds that the field is sufficiently strong to resist their influence. Even if this is so one must nonetheless ask whether the evolution of the field is governed by the thin flux tube equations. For a flux tube to maintain its coherence it must, of necessity, be twisted, otherwise it simply falls apart as it rises (Schüssler 1979). However, the internal dynamics – which is neglected in the thin flux tube approximation – then becomes an important issue. What is beyond doubt is that it is possible to construct a sequence of flux tubes with differing internal structures, that are nevertheless equally buoyant – and are thus all equal under the thin flux tube approximation – but that exhibit a wide variety in their rise as governed by the full MHD equations. This problem, even of itself, suggests that a certain degree of caution should be exercised, and raises the question of when, if ever, the thin flux tube equations are a rigorous approximation to the MHD equations. In reality the field, even if far from homogeneous, will not be in the form of tubes with a separation between field and field-free regions. As revealed in all numerical simulations (e.g. Cattaneo 1999) interactions between the fluid and field will be significant. A self-consistent model of the field in the convection zone that takes account of these interactions (but is still considerably simpler than the full MHD equations) would represent a significant breakthrough.

Describing the earlier stages of the field's evolution in terms of flux tubes is even more problematic, since it is much more difficult to argue that the solar field at depth exists as isolated tubes. Indeed, the global coherence of the solar field, manifested through Hale's polarity laws, is instead suggestive of an underlying *large-scale*, predominantly toroidal field. As discussed by Tobias & Weiss in Chapter 13, there is as yet no satisfactory answer to the problem of generating a sizeable global field. All the same, whatever the dynamo mechanism turns out to be, it seems almost inevitable that an important ingredient will be the dragging out of a toroidal field from a weaker poloidal component. This will immediately lead to a large-scale, mixed poloidal/toroidal field. Unless the motion is in the form of narrow coherent jets it will not form narrow tube-like structures in the magnetic field.

Consequently, despite the attractions arising from its simplicity, it seems that the thin flux tube approximation is not a valid description of the turbulent MHD processes that occur in stellar interiors. Thus we need to treat with some degree of caution any estimates of the interior field strength, for example, that arise from it. We are therefore forced back to the full MHD equations (or, preferably, to seek an approximation to them that retains the crucial physics). At present we have some limited knowledge of the separate parts that must eventually be put together to explain the workings of the solar interior magnetic field; we can construct a plausible picture of what might be happening, though it would be presumptuous to believe we can do more than this at the moment.

It seems inescapable, given the smooth meridional dependence of the differential rotation, that the field is fairly large-scale in the meridional direction (but possibly of narrow radial extent) and that the field we eventually observe at the surface in active regions is the result of an instability of this field. Clearly, in the very nature of a buoyancy-driven instability, motions in the vertical direction are significant. This obviously has a bearing on the structure of the tachocline and on the role of any predominantly two-dimensional (horizontal) instabilities, such as those discussed in Chapter 10; if these latter modes are to be of significance they must occur at a greater depth than instabilities due to magnetic buoyancy.

As explained in Section 11.2, provided the field decreases sufficiently rapidly with height it will be susceptible to magnetic buoyancy instability with a preferred undulatory mode. Indeed, less vigorous instabilities can occur even when the field increases with height. One criticism that has been levelled at the idea that the relevant instability is that of a layer of field, rather than that of a tube, is that there is no strong-field threshold for the instability; in particular, the Rayleigh–Taylor type modes will occur for *any* field strength, whereas the Sun certainly has the means of holding back at least some of its field until it has attained a certain strength. This point certainly needs addressing, but the answer will presumably lie in the physics so far neglected. A full explanation of the instability will require an understanding of the effect of a tangled, rather than unidirectional, field; of the role of convective overshoot; and of the influence of velocity shear. All of these are potentially stabilizing and may be responsible for holding fields down until they are of equipartition strength. A further topic of considerable interest concerns the role of thermal and magnetic diffusion on the instability, and their possible dependence on magnetic field strength. Schmitt & Rosner (1983) put forward the interesting idea that when the field strength is weak, the diffusivities assume their turbulent values (i.e. what we might think of as $O(1)$ Prandtl numbers) and instability is suppressed by, essentially, the stabilizing N^2 term on the right-hand side of inequality (11.10). Then, as a strong toroidal field is wound up by differential rotation, the turbulent diffusivities are suppressed, laminar values are assumed and the appropriate ordering becomes $\eta \ll \kappa$; the stabilizing entropy gradient is thus nullified and instability – of a strong field – can take place. This appealing idea indeed represents a specific aspect of a broader class of problems involving the suppression of turbulent transport by the dynamical feedback of a magnetic field or a shear flow – manifested not only by turbulent diffusivity but also, for example, by the α-effect of mean field electrodynamics. This somewhat controversial area is central to a complete understanding of the tachocline, as discussed by Diamond *et al.* in Chapter 9 of this book (see also the review by Diamond *et al.* 2005).

What about the field that breaks away from this layer? As discussed in detail in Chapter 13, an intriguing possibility is that, as it escapes, the field may contribute to its regeneration. Thelen (2000a,b), for example, has considered the nature of the α-effect driven by magnetic buoyancy and its role in a simplified dynamo model; Cline *et al.* (2003), with a rather different model, have shown how a combination of velocity shear and magnetic buoyancy is sufficient to drive a dynamo. As for the treatment of the field as it rises through the convection zone, I believe that modelling it in terms of isolated flux tubes lacks self-consistency, since it predicts a massive expansion and subsequent 'explosion'. The truth is probably that the field is not in the form of tubes with closed flux surfaces, but is instead a complicated tangled mess, albeit with a strong toroidal component, interacting inextricably with the fluid motions in the convection zone. To understand this, either via numerical simulations, or through a new theory that captures the essential physics but is simpler than the full MHD equations, remains one of the great challenges of solar MHD.

References

Acheson, D. J. (1978). *Phil. Trans. R. Soc. Lond.*, A**289**, 459.
Acheson, D. J. (1979). *Sol. Phys.*, **62**, 23.
Acheson, D. J. & Gibbons, M. P. (1978). *J. Fluid Mech.*, **85**, 743.
Adam, J. A. (1978). *J. Plasma Phys.*, **19**, 77.
Bernstein, I. B., Frieman, E. A., Kruskal, M. D. & Kulsrud, R. M. (1958). *Proc. R. Soc. Lond.*, A**244**, 17.
Caligari, P., Moreno-Insertis, F. & Schüssler, M. (1995). *Astrophys. J.*, **441**, 886.
Cattaneo, F. (1999). *Astrophys. J.*, **515**, L39.

Cattaneo, F. & Hughes, D. W. (1988). *J. Fluid Mech.*, **196**, 323.
Cattaneo, F., Chiueh, T. & Hughes, D. W. (1990a). *J. Fluid Mech.*, **219**, 1.
Cattaneo, F., Chiueh, T. & Hughes, D. W. (1990b). *Mon. Not. Roy. Astron. Soc.*, **247**, 6p.
Choudhuri, A. R. (1989). *Sol. Phys.*, **123**, 217.
Choudhuri, A. R. & Gilman, P. A. (1987). *Astrophys. J.*, **316**, 788.
Cline, K. S., Brummell, N. H. & Cattaneo, F. (2003). *Astrophys. J.*, **599**, 1449.
Cowling, T. G. (1957). *Magnetohydrodynamics*. Interscience.
Crow, S. C. (1970). *AIAA J.*, **8**, 2172.
Diamond, P. H., Hughes, D. W. & Kim, E.-J. (2005). In *Fluid Dynamics and Dynamos in Astrophysics and Geophysics*, ed. A. M. Soward, C. A. Jones, D. W. Hughes & N. O. Weiss. London: CRC Press, p. 145.
D'Silva, S. & Choudhuri, A. R. (1993). *Astron. Astrophys.*, **272**, 621.
Emonet, T. & Moreno-Insertis, F. (1998). *Astrophys. J.*, **492**, 804.
Fan, Y. (2001). *Astrophys. J.*, **546**, 509.
Fan, Y. (2004). *Living Rev. Solar Phys.*, **1**, 1 (http://www.livingreviews.org/lrsp-2004-1).
Fan, Y., Zweibel, E. G. & Lantz, S. R. (1998). *Astrophys. J.*, **493**, 480.
Frieman, E. & Rotenberg, M. (1960). *Rev. Mod. Phys.*, **32**, 898.
Gilman, P. A. (1970). *Astrophys. J.*, **162**, 1019.
Golub, L., Rosner, R., Vaiana, G. S. & Weiss, N. O. (1981). *Astrophys. J.*, **243**, 309.
Hughes, D. W. (1985a). *Geophys. Astrophys. Fluid Dyn.*, **32**, 273.
Hughes, D. W. (1985b). *Geophys. Astrophys. Fluid Dyn.*, **34**, 99.
Hughes, D. W. & Cattaneo, F. (1987). *Geophys. Astrophys. Fluid Dyn.*, **39**, 65.
Hughes, D. W. & Falle, S. A. E. G. (1998). *Astrophys. J.*, **509**, L57.
Hughes, D. W., Falle, S. A. E. G. & Joarder, P. (1998). *Mon. Not. Roy. Astron. Soc.*, **298**, 433.
Hughes, D. W. & Proctor, M. R. E. (1988). *Ann. Rev. Fluid Mech.*, **20**, 187.
Jensen, E. (1955). *Ann. Astrophys.*, **18**, 127.
Kruskal, M. & Schwarzschild, M. (1954). *Proc. R. Soc. Lond.*, A**223**, 348.
Matthews, P. C., Hughes, D. W. & Proctor, M. R. E. (1995). *Astrophys. J.*, **448**, 938.
Masuda, A. (1978). *J. Ocean. Soc. Japan,* **34**, 8.
Moffatt, H. K. (1978). *Magnetic Field Generation in Electrically Conducting Fluids*. Cambridge: Cambridge University Press.
Moreno-Insertis, F. (1992). In *Sunspots: Theory and Observations*, ed. J. H. Thomas & N. O. Weiss. Dordrecht: Kluwer, p. 385.
Newcomb, W. A. (1961) *Phys. Fluids*, **4**, 391.
Parker, E. N. (1955). *Astrophys. J.*, **121**, 491.
Parker, E. N. (1966). *Astrophys. J.*, **145**, 811.
Parker, E. N. (1975). *Astrophys. J.*, **198**, 205.
Parker, E. N. (1979). *Cosmical Magnetic Fields: Their Origin and Their Activity*. Oxford: Clarendon Press.
Pearlstein, A. J. (1981). *J. Fluid Mech.*, **103**, 389.
Roberts, P. H. & Stewartson, K. (1977). *Astron. Nachr.*, **298**, 311.
Schmitt, J. H. M. M. & Rosner, R. (1983). *Astrophys. J.*, **265**, 901.
Schüssler, M. (1979). *Astron. Astrophys.*, **71**, 79.
Schüssler, M. (2005). *Astron. Nachr.*, **326**, 194.
Soward, A. M. (1979). *J. Fluid Mech.*, **90**, 669.
Spiegel, E. A. & Weiss, N. O. (1980). *Nature*, **287**, 616.
Spiegel, E. A. & Weiss, N. O. (1982). *Geophys. Astrophys. Fluid Dyn.*, **22**, 219.
Spruit, H. C. (1981). *Astron. Astrophys.*, **98**, 155.
Spruit, H. C. & van Ballegooijen, A. A. (1982). *Astron. Astrophys.*, **106**, 58.

Tayler, R. J. (1973). *Mon. Not. Roy. Astron. Soc.*, **161**, 365.
Thelen, J.-C. (2000a). *Mon. Not. Roy. Astron. Soc.*, **315**, 155.
Thelen, J.-C. (2000b). *Mon. Not. Roy. Astron. Soc.*, **315**, 165.
Thomas, J. H. & Nye, A. H. (1975). *Phys. Fluids*, **18**, 490.
Tobias, S. M. & Hughes, D. W. (2004). *Astrophys. J.*, **603**, 785.
Turner, J. S. (1973). *Buoyancy Effects in Fluids*. Cambridge: Cambridge University Press.
Wissink, J. G., Hughes, D. W., Matthews, P. C. & Proctor, M. R. E. (2000). *Mon. Not. Roy. Astron. Soc.*, **318**, 501.

12

Instabilities, angular momentum transport and magnetohydrodynamic turbulence

Gordon I. Ogilvie

The tachocline may be subject to a variety of instabilities leading to turbulent motion and angular momentum transport. This chapter reviews some approaches that have been found useful in the study of astrophysical accretion discs and discusses their possible application to the tachocline.

12.1 Introduction

The solar tachocline is a thin structure characterized by strong differential rotation, presumably in the presence of a magnetic field. It forms the interface between the radiative interior and the convective envelope of the Sun, which differ greatly in their dynamical properties, states of rotation and mechanisms of angular momentum transport. While the tachocline might have the character of a laminar boundary layer between these regions, it is more likely to be turbulent, at least in part, as a result of intrinsic instabilities or possibly because of forcing by the convective motions above.

Instabilities of the tachocline could derive from kinetic, gravitational or magnetic sources of free energy. Shear instabilities depend on the free kinetic energy in differential rotation, and may, as in the case of the magnetorotational instability, require the assistance of a magnetic field. Gravitational energy may be liberated through magnetic buoyancy (Parker) instabilities, while magnetic energy in non-potential configurations may be released in purely magnetic (Tayler) instabilities. To understand the existence and dynamics of the tachocline requires an appreciation of such instabilities and the transport effects, especially angular momentum transport, to which they give rise in a nonlinear regime. The possible role of the tachocline in the operation of a large-scale magnetic dynamo may also depend on the outcome of such instabilities.

Similar issues have been encountered in the study of astrophysical accretion discs (e.g. Pringle 1981; Papaloizou & Lin 1995; Balbus & Hawley 1998), where

The Solar Tachocline, D. W. Hughes, R. Rosner and N. O. Weiss.
Published by Cambridge University Press.

gas rotates in Keplerian orbital motion around a central mass. In these thin structures the differential rotation is a dominant feature and mechanisms have been sought that allow an efficient outward transport of angular momentum. It has been found that, despite the very large Reynolds number of the Keplerian shear flow, hydrodynamic instability is strongly inhibited by the stabilizing angular momentum gradient and may be entirely absent, although this remains a controversial issue. Instead, a magnetohydrodynamic (MHD) instability, the magnetorotational instability (MRI), has been found to be ideally suited to the outward transport of angular momentum. Its nonlinear development leads to MHD turbulence with roughly the desired transport properties (Balbus & Hawley 1998). The possible occurrence of a large-scale dynamo in discs is also an important and unresolved issue.

There are certainly major differences between the tachocline and an accretion disc. While the orbital motion in an accretion disc is highly supersonic and represents a practically inexhaustible source of free energy, the flows in the tachocline are highly subsonic and the effects giving rise to differential rotation are much weaker and more subtle. In addition, the tachocline may have a strong stable stratification that inhibits vertical motions, something that is relatively weak or absent in accretion discs. One could remark that both the tachocline and accretion discs are difficult to resolve observationally at the present time, and are difficult to simulate numerically owing to the ranges of lengthscales and timescales involved. Perhaps for these reasons they present serious challenges to the theoretical astrophysicist.

This chapter focuses on instabilities of differential rotation and magnetic fields and the angular momentum transport to which they give rise. I review some approaches that have been found useful in the study of accretion discs and discuss their possible application to the tachocline. Angular momentum transport is discussed from a general perspective in Section 12.2, while Section 12.3 describes the magnetorotational instability and Section 12.4 reviews simple statistical models of anisotropic MHD turbulence. Section 12.5 describes a useful approach to instabilities of differential rotation and magnetic fields, and conclusions are given in Section 12.6.

12.2 Angular momentum transport in general

The equation of momentum conservation for a fluid can be written in the general form

$$\frac{\partial}{\partial t}(\rho \boldsymbol{u}) + \nabla \cdot (\rho \boldsymbol{u}\boldsymbol{u}) = \nabla \cdot \mathbf{T}, \tag{12.1}$$

where ρ and $\boldsymbol{u}$ are the density and velocity fields, while $\mathbf{T}$ is a symmetric stress tensor field. In ideal MHD the stress tensor is

$$\mathbf{T} = -p\,\mathbf{1} - \frac{1}{4\pi G}\left(\boldsymbol{g}\boldsymbol{g} - \frac{1}{2}g^2\,\mathbf{1}\right) + \frac{1}{\mu_0}\left(\boldsymbol{B}\boldsymbol{B} - \frac{1}{2}B^2\,\mathbf{1}\right), \qquad (12.2)$$

where p is the pressure, $\boldsymbol{g}$ is the gravitational acceleration and $\boldsymbol{B}$ is the magnetic field. The equation for the z-component of angular momentum in cylindrical polar coordinates (s, ϕ, z) is

$$\frac{\partial}{\partial t}(\rho s u_\phi) + \nabla \cdot (\rho s u_\phi \boldsymbol{u} - s\mathbf{T} \cdot \boldsymbol{e}_\phi) = 0, \qquad (12.3)$$

indicating that angular momentum transport in the meridional plane depends on either advection by a meridional flow or transport by the stress components $T_{s\phi}$ and $T_{z\phi}$. According to Equation (12.2), such an anisotropic stress requires either non-axisymmetric gravitational fields or a magnetic field with both meridional and azimuthal components. In the presence of fluctuations associated with waves or turbulence, an anisotropic Reynolds stress can also arise.

All of these possibilities have been considered in the context of accretion discs (Papaloizou & Lin 1995). Gravitational, magnetic and Reynolds stresses can be associated with either large-scale features such as spiral arms, vortices and magnetized outflows, or small-scale features such as waves and turbulence.

12.3 The magnetorotational instability

The shearing sheet (Goldreich & Lynden-Bell 1965) is a very useful local model of a thin, differentially rotating disc. By formally separating the rotation and shear of the disc, straightening the streamlines and removing the horizontal boundaries to infinity, it creates the simplest realistic environment in which to carry out local stability analyses or studies of turbulence in discs.

In a differentially rotating disc with angular velocity $\Omega(s)$, the quantity

$$A(s) = -\frac{s}{2}\frac{\mathrm{d}\Omega}{\mathrm{d}s} \qquad (12.4)$$

measures the shear rate, and is equal to $3\Omega/4$ in a Keplerian disc. Consider a reference point (Figure 12.1), situated in the mid-plane of the disc and orbiting the central mass in a circular orbit of radius s_0 and angular velocity $\Omega_0 = \Omega(s_0)$. It is used as the origin of a local, rotating Cartesian coordinate system (x, y, z), with unit vectors $(\boldsymbol{e}_x, \boldsymbol{e}_y, \boldsymbol{e}_z)$ pointing in the radial, azimuthal and vertical directions, respectively. The flow is represented locally as a uniform rotation $\Omega_0 \boldsymbol{e}_z$ plus a linear shear flow $\boldsymbol{u}_0 = -2A_0 x\, \boldsymbol{e}_y$, where $A_0 = A(s_0)$. One subsequently omits the subscript 0, understanding that Ω now refers to the uniform angular velocity of the frame of reference.

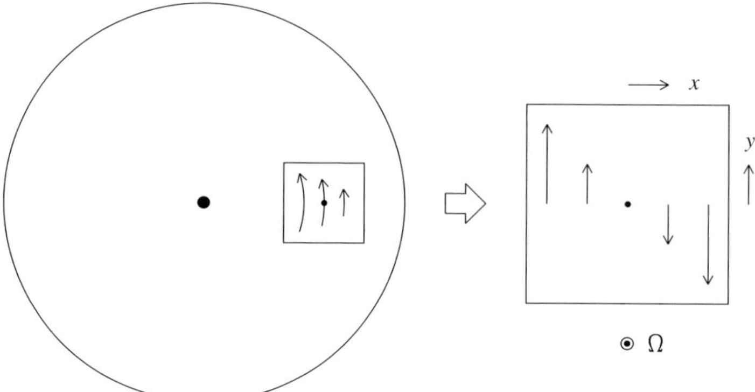

Figure 12.1. The shearing sheet. The differential rotation of the disc is represented locally as a uniform rotation plus a linear shear flow.

Perhaps the simplest way to illustrate the magnetorotational instability (MRI) is by describing the optimal mode in an incompressible shearing sheet, which is an exact nonlinear solution known as the 'channel flow' (Goodman & Xu 1994; Papaloizou & Lin 1995). Consider an incompressible fluid of uniform density ρ, kinematic viscosity ν and magnetic diffusivity η. The sheet has an imposed uniform vertical magnetic field $B_z\,\boldsymbol{e}_z$ corresponding to an Alfvén speed $v_\mathrm{a} = (\mu_0\rho)^{-1/2}B_z$.

The channel flow has a very simple form, consisting of a layerwise motion $\boldsymbol{v}(z,t)$ that is purely horizontal and independent of the horizontal coordinates. The magnetic perturbation $(\mu_0\rho)^{1/2}\boldsymbol{b}(z,t)$ has a similar form, while the total pressure $\Pi = p + B^2/2\mu_0$ is unperturbed. The equation of motion and induction equation then give

$$\frac{\partial \boldsymbol{v}}{\partial t} + \boldsymbol{v}\cdot\nabla(-2Ax\,\boldsymbol{e}_y) + 2\Omega\,\boldsymbol{e}_z\times\boldsymbol{v} = v_\mathrm{a}\frac{\partial \boldsymbol{b}}{\partial z} + \nu\frac{\partial^2 \boldsymbol{v}}{\partial z^2}, \tag{12.5}$$

$$\frac{\partial \boldsymbol{b}}{\partial t} = \boldsymbol{b}\cdot\nabla(-2Ax\,\boldsymbol{e}_y) + v_\mathrm{a}\frac{\partial \boldsymbol{v}}{\partial z} + \eta\frac{\partial^2 \boldsymbol{b}}{\partial z^2}, \tag{12.6}$$

in which the nonlinear terms such as $\boldsymbol{v}\cdot\nabla\boldsymbol{v}$ vanish. The terms proportional to v_a couple the velocity and magnetic perturbations and, by themselves, would lead to vertically propagating Alfvén waves. Also present, however, are Coriolis and shear terms that, by themselves, would lead to epicyclic oscillations of the velocity perturbation and shearing of the magnetic perturbation. It is the coupling between these effects that leads to the magnetorotational dynamics. When $\nu = \eta = 0$ and $2\Omega > A > 0$, a growing solution of these equations (see Figure 12.2) is

$$\boldsymbol{v} = (\boldsymbol{e}_x + \boldsymbol{e}_y)a\,\mathrm{e}^{At}\sin(kz), \qquad \boldsymbol{b} = (\boldsymbol{e}_x - \boldsymbol{e}_y)b\,\mathrm{e}^{At}\cos(kz), \tag{12.7}$$

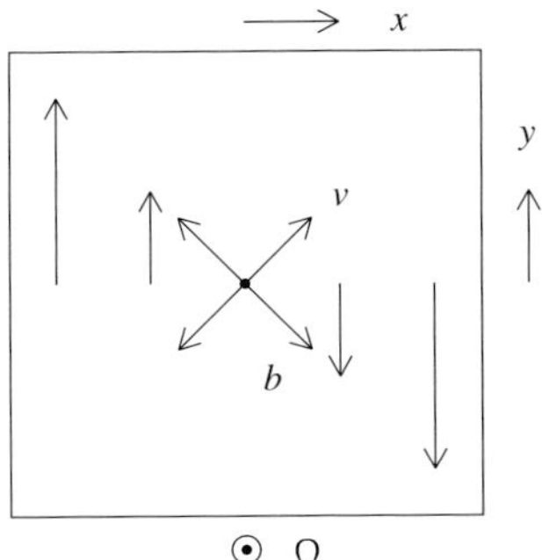

Figure 12.2. The optimal mode of the MRI in the shearing sheet. The velocity and magnetic perturbations are aligned at 45° to the shear flow; they are orthogonal to each other and 90° out of phase in their vertical structure.

with $k^2 v_\mathrm{a}^2 = A(2\Omega - A)$ and

$$\frac{b^2}{a^2} = \frac{2\Omega}{A} - 1 = \frac{5}{3} \quad \text{(Keplerian)}. \tag{12.8}$$

The coupling of inertial and Alfvénic restoring forces, together with the shearing of the magnetic perturbation, allows a runaway horizontal disturbance whose energy grows by a factor of $\exp(3\pi) \approx 12\,392$ per orbit. The instability relies on the fact that, while the angular velocity of the orbital motion decreases outwards, the specific angular momentum increases outwards. When two orbiting entities are connected by a tether (in this case two layers of fluid connected by frozen-in magnetic field lines), the tension in the tether attempts to bring the entities into corotation by transferring angular momentum to the one with the smaller angular velocity. However, the orbital dynamics renders the attempt futile because the addition of angular momentum to an entity actually decreases its angular velocity after an adjustment of its orbital radius.

A necessary condition for instability is $\Omega A > 0$, which is equivalent to $\mathrm{d}(\Omega^2)/\mathrm{d}s < 0$ and is satisfied in realistic astrophysical discs, unlike Rayleigh's criterion $\mathrm{d}(s^4\Omega^2)/\mathrm{d}s < 0$ for instability in the absence of a magnetic field. In the presence of viscosity and resistivity, there is a non-trivial criterion to be met for instability, but it is easily satisfied in the typical situation where the Reynolds and magnetic Reynolds numbers are very large. Apparently the magnetic field allows the fluid to be more inventive in releasing shear energy, but only when the angular velocity decreases outwards.

The nonlinear development of the MRI has been investigated through numerical simulations of the three-dimensional MHD equations in a shearing box, a version of the shearing sheet that is truncated with periodic boundary conditions (Hawley *et al.* 1995; Brandenburg *et al.* 1995). Although the channel flow is an exact nonlinear solution (at least in the case of an incompressible fluid) it undergoes secondary 'parasitic' instabilities, for example those that feed off the growing shear between the layers (Goodman & Xu 1994). The typical outcome appears to be a saturated state of anisotropic MHD turbulence with well defined and stationary statistical properties. Although the turbulent intensity undergoes large fluctuations, the ratios of the various stress components are well defined.

It is of interest to examine the energetics of the MRI. The total energy equation in an incompressible shearing sheet takes the form

$$\frac{\mathrm{d}}{\mathrm{d}t}\left\langle\frac{1}{2}(v^2 + b^2)\right\rangle = 2A\langle v_x v_y - b_x b_y\rangle - \langle \nu|\nabla \times v|^2 + \eta|\nabla \times \boldsymbol{b}|^2\rangle, \tag{12.9}$$

where the angle brackets denote a volume-average. The two terms on the right-hand side of this equation are the production of energy through the action of a shear stress on the imposed background shear flow, and the dissipation of energy through viscosity and resistivity. The shear stress T_{xy} derives from correlations between the velocity components (Reynolds stress) and between the magnetic components (Maxwell stress). In either the phase of growing instability or that of saturated turbulence, the production term must be positive, i.e.

$$-\frac{2A}{\rho}T_{xy} = 2A\langle v_x v_y - b_x b_y\rangle > 0. \tag{12.10}$$

This situation corresponds to an outward transport of angular momentum, which is precisely what is required to allow inward mass transport in an accretion disc. The optimal MRI mode achieves this by having $v_x = v_y$ and $b_x = -b_y$, thereby maximizing the desired correlations. The saturated turbulence also has correlations of the desired sign (T_{xy} is dominated by the magnetic contribution).

Indeed, the energetic balance requires that angular momentum transport occurs down the gradient of angular velocity. In this sense the turbulence acts like a viscosity and an effective viscous coefficient can be defined based on the ratio of the mean stress T_{xy} and the shear $2A$. In contrast, if the turbulence derives from gravitational or magnetic free energy, or is externally forced, it is not necessarily true that the angular momentum transport occurs down the gradient of angular velocity. In such cases 'anti-frictional' behaviour is possible in principle. On the other hand, a negative correlation between b_x and b_y is a very natural outcome of imperfect flux freezing in a shear flow of this kind, and it can be argued on this basis that MHD turbulence is more likely to lead to 'frictional' behaviour.

As discussed in Section 12.5 below, the application of these results to stellar interiors is not entirely straightforward because the shear energy is subdominant in that context. Although helioseismic inversions indicate that $\partial(\Omega^2)/\partial s < 0$ at high latitudes in the solar tachocline, other factors have a bearing on stability. The nonlinear development of the MRI in stars will also be different and is likely to reduce or remove the differential rotation that drives the instability.

12.4 Statistical models of anisotropic MHD turbulence

When instabilities develop into turbulent motion, as is typical in astrophysical situations in which the Reynolds number or similar relevant parameter is very large, the problem arises of how to model the turbulence. The alternatives range from direct numerical simulation (DNS) to analytical statistical modelling, with intermediate options including large-eddy simulation, simulations with unresolved dissipative scales, and simplified stochastic models. It is perhaps worth noting that even DNS cannot predict the actual evolution of any turbulent system, because the dynamics is chaotic and numerical errors are always present at some level even if the dissipative scales are properly resolved.

In spite of their obvious chaotic variability, turbulent flows in laboratory experiments and DNS appear to have well defined statistical properties that vary smoothly in time and space and are reproducible. The difficulties of modelling these properties are well known. While many theories of turbulence focus on the spectral properties of homogeneous and isotropic turbulence, for the present purposes it is best to concentrate on the gross properties of the second-order correlations (stress tensors, entropy flux, etc.) that give rise to transport effects in anisotropic turbulent flows. A basic turbulence model in this context is one that relates the turbulent stress tensor to the mean velocity field (or other large-scale properties of the flow).

The simplest starting point is the eddy-viscosity model (Boussinesq 1877) in which the turbulent stress is proportional to the rate of strain,

$$T_{ij} = \mu(u_{i,j} + u_{j,i}), \tag{12.11}$$

with an effective coefficient of viscosity μ. The mixing-length theory of Prandtl (1925) supplies a formula for μ that provides remarkable agreement with the gross properties of some fully turbulent flows, such as the mean flow rate in turbulent pipe flow. The eddy-viscosity model was applied to accretion discs by Weizsäcker (1948) and Lüst (1952), while Shakura & Sunyaev (1973) introduced a new parametrization of the effective viscosity appropriate for accretion discs.

For certain limited purposes the eddy-viscosity model is probably perfectly adequate. In particular, the evolution of the surface mass density in a thin accretion disc is controlled by the rate of outward transport of angular momentum, and therefore depends on a single quantity, the shear-stress coefficient $T_{s\phi}$ (integrated vertically though the disc, and, if necessary, averaged in azimuth and in time). In a differentially rotating disc, this quantity can always be parametrized in terms of a single effective viscosity coefficient μ; whether μ scales with the local pressure in the disc, as in the model of Shakura & Sunyaev (1973), is a different question, but it is at least a plausible proposition. The eddy-viscosity model certainly does not predict the correct anisotropic shape of the stress tensor in MRI turbulence, nor does it predict the correct dependence on shear rate and angular velocity.

It is sometimes argued that MHD turbulence has nothing to do with viscosity, and that an eddy-viscosity model is entirely inappropriate for accretion discs. However, there is an interesting analogy, both physical and mathematical, between a turbulent magnetized fluid and a dilute solution of long-chain polymer molecules (Ogilvie 2001; Ogilvie & Proctor 2003). Both the magnetic field lines and the polymer molecules are advected and stretched by the fluid flow and respond with an elastic tension force. A polymer solution has a typical relaxation time τ on which the coiling of the chains returns to a statistically isotropic state. Deformations with a characteristic timescale $T \ll \tau$ receive an elastic reaction while those with $T \gg \tau$ experience a viscous response. Similarly, a turbulent MHD fluid is expected to generate elastic or viscous-like stresses in response to rapid deformations and steady shear, respectively. The simplest viscoelastic model is the (incompressible) upper-convected Maxwell fluid, for which

$$T_{ij,t} + u_k T_{ij,k} - T_{ik} u_{j,k} - T_{jk} u_{i,k} = -\frac{1}{\tau}\left(T_{ij} - \frac{\mu}{\tau}\delta_{ij}\right). \tag{12.12}$$

In the limit $T \ll \tau$ this equation becomes identical to that satisfied by the magnetic tension $B_i B_j/\mu_0$ in ideal MHD, while in the limit $T \gg \tau$ the Navier–Stokes viscous stress appears as the deviation from an isotropic pressure.

A more sophisticated approach is based on the exact transport equations for the Reynolds and Maxwell stress tensors (Kato & Yoshizawa 1993; Ogilvie 2003), and relates to the widely studied Reynolds-averaged Navier–Stokes models for hydrodynamic turbulence. The exact transport equations contain some linear terms, which represent the different ways in which velocity and magnetic fluctuations interact with the mean flow and can be retained exactly, together with some nonlinear terms that must be modelled. A simple model of this kind appears to contain enough physics to capture the basic dynamics of the MRI and its saturation in a state of anisotropic MHD turbulence (Ogilvie 2003). It differentiates clearly between the behaviour of hydrodynamic and MHD fluctuations in shearing and rotating systems, and seems to offer a reasonable description of how the saturated stress tensors scale with the various parameters. For application to the tachocline, this work needs to be generalized to stably stratified systems and the models may need to be adapted if the turbulence acquires a quasi-two-dimensional character.

12.5 Analysis of localized perturbations

12.5.1 Introduction

In studies of the tachocline and of accretion discs the question arises of how best to analyse the stability of a differentially rotating fluid in the presence of a magnetic field. Although sometimes successful, the traditional approach of seeking instabilities in the form of exponentially growing normal modes is generally found to be problematic because, where modes exist, they depend essentially on boundary conditions that may have been imposed artificially. For example, in a homogeneous, unbounded shearing sheet there are no radially localized non-axisymmetric normal modes (growing or otherwise), but this does not mean that the system is necessarily stable in practice. It is also well known that instabilities of parallel shear flows with boundaries are not well described by a normal-mode analysis because the linear operator involved is highly ill-conditioned at large Reynolds numbers (Trefethen *et al.* 1993).

An alternative approach is to seek spatially localized disturbances with a non-exponential dependence on time. In parallel shear flows the relevant solutions undergo transient algebraic growth before the onset of viscous damping (Thomson 1887). This has been found to be a more useful way of describing the transition to turbulence (e.g. Grossmann 2000). In dynamically richer systems with inertial, buoyant or magnetic restoring forces the relevant solutions have an exponential (or sinusoidal) time-dependence to a first approximation, with a growth rate (or frequency) deriving from a local dispersion relation.

In the context of accretion discs, Terquem & Papaloizou (1996) presented a linear stability analysis of a differentially rotating fluid with a toroidal magnetic field. Assuming ideal MHD, they demonstrated the existence of an unstable dense or continuous spectrum when certain local criteria are met. Equivalent results were obtained within the Boussinesq approximation by Friedlander & Vishik (1995) using an apparently different approach that placed bounds on the spectrum of the linear operator. These spectral analyses are somewhat technical ways of showing the existence of a linear instability even when normal modes may not exist. Unfortunately the dense or continuous spectrum does not survive the addition of any dissipative effects. In this Section I present this approach with a somewhat broader interpretation and discuss its possible application to the tachocline. It turns out to contain many, although not all, of the most important instabilities.

The theory of ideal MHD has an attractive mathematical structure. Supplementing the equation of motion of the fluid are three further relations: the equation of mass conservation, the adiabatic condition and the induction equation. Each of these relations describes the pure advection of a certain quantity (the mass element $\rho\, \mathrm{d}V$, the specific entropy s or the magnetic flux element $\boldsymbol{B}\cdot\mathrm{d}\boldsymbol{S}$) and can be integrated exactly in Lagrangian variables. This property is useful when formulating problems of stability in ideal MHD.

Consider, as a reference state, an arbitrary solution of the equations of ideal MHD. Perturbations to the reference state may be described using the Lagrangian displacement $\boldsymbol{\xi}$, which is the vector displacement of a fluid element between the unperturbed and perturbed flows. Specifically, $\boldsymbol{x}+\boldsymbol{\xi}(\boldsymbol{x},t)$ is the position vector at time t of the fluid element that is at position $\boldsymbol{x}$ in the unperturbed flow at that time. Owing to the property of integrability described above, the Lagrangian perturbations of density, entropy and magnetic field can be expressed exactly in terms of $\boldsymbol{\xi}$ and $\nabla\boldsymbol{\xi}$. The equation of motion then becomes a closed equation for $\boldsymbol{\xi}$, of second order in space and time variables, which has the character of a nonlinear Hamiltonian field theory constructed on the unperturbed flow. In the linear approximation this equation may be written

$$\rho\frac{\mathrm{D}^2\boldsymbol{\xi}}{\mathrm{D}t^2} = -\nabla\Pi' - (\nabla\cdot\boldsymbol{\xi})\nabla\Pi - \boldsymbol{\xi}\cdot\nabla\nabla\Pi - \rho\boldsymbol{\xi}\cdot\nabla\nabla\Phi - \rho\nabla\Phi' + \frac{1}{\mu_0}\boldsymbol{B}\cdot\nabla[\boldsymbol{B}\cdot\nabla\boldsymbol{\xi} - (\nabla\cdot\boldsymbol{\xi})\boldsymbol{B}], \tag{12.13}$$

where

$$\Pi' = -\left(\gamma p + \frac{B^2}{\mu_0}\right)\nabla\cdot\boldsymbol{\xi} - \boldsymbol{\xi}\cdot\nabla\Pi + \frac{1}{\mu_0}\boldsymbol{B}\cdot(\boldsymbol{B}\cdot\nabla\boldsymbol{\xi}) \tag{12.14}$$

is the (linearized) Eulerian perturbation of the total pressure, Φ is the gravitational potential and Φ' its Eulerian perturbation. The adiabatic exponent is $\gamma = (\partial\ln p/\partial\ln\rho)_s$ (named Γ_1 by Chandrasekhar).

In either the tachocline or an accretion disc one is interested in instabilities of differential rotation and toroidal magnetic fields. Consider now a basic state that is steady and axisymmetric, with velocity field $\boldsymbol{u} = s\Omega(s,z)\,\boldsymbol{e}_\phi$ and magnetic field $\boldsymbol{B} = B(s,z)\,\boldsymbol{e}_\phi$. The equilibrium condition is

$$\nabla\Pi = \rho\left(\boldsymbol{g} - \frac{v_\mathrm{a}^2}{s}\boldsymbol{e}_s\right), \tag{12.15}$$

where $\boldsymbol{g} = -\nabla\Phi + s\Omega^2\,\boldsymbol{e}_s$ is the effective gravitational acceleration including the centrifugal term, and $v_\mathrm{a} = (\mu_0\rho)^{-1/2}B$ is the Alfvén speed.

Solutions of Equation (12.13) may then be sought that have the form of a normal mode,

$$\boldsymbol{\xi} = \mathrm{Re}\left[\tilde{\boldsymbol{\xi}}(s,z)\,\mathrm{e}^{\mathrm{i}m\phi-\mathrm{i}\omega t}\right], \tag{12.16}$$

where $m \in \mathbf{Z}$ is the azimuthal wavenumber and $\omega \in \mathbf{C}$ is the frequency eigenvalue, with $\mathrm{Im}(\omega) > 0$ implying instability. Substitution of this form into Equation (12.13) leads to a vector partial differential equation (PDE) for the spatial structure of the mode in the meridional plane. For frequencies of interest, however, this PDE is typically hyperbolic and may not have smooth solutions satisfying any specified boundary conditions.

12.5.2 Local dispersion relation

Consider the possibility of 'solutions' of this PDE that are asymptotically localized and satisfy the equation in an asymptotic sense. These 'solutions' have an envelope that is localized near a single point (s_0, z_0) in the meridional plane. Within the envelope the displacement has a plane-wave form with many wavefronts. In the asymptotic limit of interest, the scale of localization tends to zero, while the number of wavefronts under the envelope tends to infinity. However, the frequency has a finite limit and the group velocity tends to zero, so that the solution has the nature of a frozen wavepacket. Formally, one may write

$$\tilde{\boldsymbol{\xi}} \sim \hat{\boldsymbol{\xi}}\, \mathrm{e}^{\mathrm{i}k_s s + \mathrm{i}k_z z}\, E\left(\frac{s - s_0}{\ell}, \frac{z - z_0}{\ell}\right), \tag{12.17}$$

where $\hat{\boldsymbol{\xi}}$ is a constant complex vector, $\boldsymbol{k} = k_s \boldsymbol{e}_s + k_z \boldsymbol{e}_z$ is a real wavevector with $k = |\boldsymbol{k}| \to \infty$, E is an envelope function with a scale of order unity, such as $E(x, y) = \exp(-x^2 - y^2)$, and ℓ is the localization scale, such that

$$k^{-1} \ll \ell \ll L, \tag{12.18}$$

where L is the characteristic lengthscale of variation of the basic state. Differentiation of the solution with respect to s or z then corresponds at leading order to multiplication by $\mathrm{i}k_s$ or $\mathrm{i}k_z$.

Let the displacement be normalized such that $\boldsymbol{\xi} = O(1)$. In order for Equation (12.13) to be satisfied asymptotically with $m = O(1)$ and $\omega = O(1)$, one requires that $\nabla \cdot \boldsymbol{\xi} = O(1)$, rather than the natural scaling $O(k)$. This implies that the poloidal (or meridional) part of the displacement $\boldsymbol{\xi}_\mathrm{p} \sim \xi_\mathrm{p} \boldsymbol{e}$, where the unit vector $\boldsymbol{e}$ satisfies $\boldsymbol{e} \cdot \boldsymbol{k} = 0$. In other words, $\boldsymbol{\xi}$ is almost transverse and almost incompressible in order to avoid the potentially large acoustic restoring force. Furthermore, one requires $\Pi' = O(k^{-1})$, rather than $O(1)$ or $O(k)$. This condition, in which the total pressure perturbation is minimized, is typical of the anelastic approximation or its MHD analogue. The Eulerian perturbation of the gravitational potential satisfies $\nabla^2 \Phi' = 4\pi G \rho'$ and is $O(k^{-2})$; it therefore makes a negligible contribution to the equation of motion for localized perturbations.

The leading approximations to Equations (12.13) and (12.14) are then

$$-\rho(\hat{\omega}^2-\omega_{\mathrm{a}}^2)\boldsymbol{\xi}_{\mathrm{p}}+2\mathrm{i}\rho\left(\hat{\omega}\Omega+\frac{v_{\mathrm{a}}\omega_{\mathrm{a}}}{s}\right)\xi_\phi\,\boldsymbol{e}_s=-\mathrm{i}\boldsymbol{k}\Pi'-\rho\left(\boldsymbol{g}-\frac{2v_{\mathrm{a}}^2}{s}\boldsymbol{e}_s\right)\nabla\cdot\boldsymbol{\xi}$$
$$-(\boldsymbol{\xi}_{\mathrm{p}}\cdot\nabla\rho)\boldsymbol{g}-2\rho s\Omega(\boldsymbol{\xi}_{\mathrm{p}}\cdot\nabla\Omega)\boldsymbol{e}_s+\frac{2B}{\mu_0}\left[\boldsymbol{\xi}_{\mathrm{p}}\cdot\nabla\left(\frac{B}{s}\right)\right]\boldsymbol{e}_s, \qquad (12.19)$$

$$-\rho(\hat{\omega}^2-\omega_{\mathrm{a}}^2)\xi_\phi-2\mathrm{i}\rho\left(\hat{\omega}\Omega+\frac{v_{\mathrm{a}}\omega_{\mathrm{a}}}{s}\right)\boldsymbol{\xi}_{\mathrm{p}}\cdot\boldsymbol{e}_s=-\mathrm{i}\rho v_{\mathrm{a}}\omega_{\mathrm{a}}\nabla\cdot\boldsymbol{\xi}, \qquad (12.20)$$

$$(v_{\mathrm{s}}^2+v_{\mathrm{a}}^2)\nabla\cdot\boldsymbol{\xi}=-\boldsymbol{\xi}_{\mathrm{p}}\cdot\left(\boldsymbol{g}-\frac{2v_{\mathrm{a}}^2}{s}\boldsymbol{e}_s\right)+\mathrm{i}v_{\mathrm{a}}\omega_{\mathrm{a}}\xi_\phi, \qquad (12.21)$$

where $\hat{\omega}=\omega-m\Omega$ is the local Doppler-shifted frequency, $\omega_{\mathrm{a}}=mv_{\mathrm{a}}/s$ is the Alfvén frequency and $v_{\mathrm{s}}=(\gamma p/\rho)^{1/2}$ is the sound speed. After Π' is eliminated by projecting Equation (12.19) parallel to $\boldsymbol{e}$, there remain three linear algebraic equations for ξ_{p}, ξ_ϕ and $\nabla\cdot\boldsymbol{\xi}$, which have a non-trivial solution if and only if the local dispersion relation

$$\Bigg[\hat{\omega}^2-\omega_{\mathrm{a}}^2-(\boldsymbol{e}\cdot\boldsymbol{g})\boldsymbol{e}\cdot\nabla\ln\rho-2s\Omega(\boldsymbol{e}\cdot\boldsymbol{e}_s)\boldsymbol{e}\cdot\nabla\Omega+\frac{2B}{\mu_0\rho}(\boldsymbol{e}\cdot\boldsymbol{e}_s)\boldsymbol{e}\cdot\nabla\left(\frac{B}{s}\right)$$
$$+\left(\frac{1}{v_{\mathrm{s}}^2+v_{\mathrm{a}}^2}\right)\left(\boldsymbol{e}\cdot\boldsymbol{g}-\frac{2v_{\mathrm{a}}^2}{s}\boldsymbol{e}\cdot\boldsymbol{e}_s\right)^2\Bigg]\left[\hat{\omega}^2-\left(\frac{v_{\mathrm{s}}^2}{v_{\mathrm{s}}^2+v_{\mathrm{a}}^2}\right)\omega_{\mathrm{a}}^2\right]$$
$$=\left[2\Omega\hat{\omega}\,\boldsymbol{e}\cdot\boldsymbol{e}_s+\left(\frac{v_{\mathrm{a}}\omega_{\mathrm{a}}}{v_{\mathrm{s}}^2+v_{\mathrm{a}}^2}\right)\left(\boldsymbol{e}\cdot\boldsymbol{g}+\frac{2v_{\mathrm{s}}^2}{s}\boldsymbol{e}\cdot\boldsymbol{e}_s\right)\right]^2 \qquad (12.22)$$

is satisfied.

Terquem & Papaloizou (1996) used trial displacements of this form to show that the linear operator has a dense or continuous spectrum including the range of (generally complex) values of ω that satisfy this local dispersion relation at any point. The frozen wavepackets can also be understood in a more physical way, as discussed in Section 12.5.7 below.

This remarkable dispersion relation provides sufficient conditions for instability and merits a detailed analysis. As a quartic equation for $\hat{\omega}$ it is best solved numerically in practice but many of its properties are revealed by considering separately a number of different limits.

12.5.3 Case of zero magnetic field

In the absence of a magnetic field the local dispersion relation (12.22) has solutions $\hat{\omega}^2=0$ and

$$\hat{\omega}^2=\mathbf{M}:\boldsymbol{e}\boldsymbol{e}, \qquad (12.23)$$

where

$$\mathbf{M} = \frac{1}{2}(\boldsymbol{g}\boldsymbol{a} + \boldsymbol{a}\boldsymbol{g} + \boldsymbol{e}_s\boldsymbol{b} + \boldsymbol{b}\boldsymbol{e}_s) \tag{12.24}$$

is a symmetric matrix, and

$$\boldsymbol{a} = \nabla \ln \rho - \frac{1}{\gamma}\nabla \ln p, \qquad \boldsymbol{b} = \frac{1}{s^3}\nabla(s^4\Omega^2) \tag{12.25}$$

are the Schwarzschild and Rayleigh discriminants, proportional to the gradients of entropy and specific angular momentum, which satisfy the 'thermal-wind' equation $\boldsymbol{g} \times \boldsymbol{a} + \boldsymbol{e}_s \times \boldsymbol{b} = \boldsymbol{0}$. The unit vector $\boldsymbol{e}$, corresponding to the direction of the poloidal displacement, can be chosen at will. For stability one requires that $\mathbf{M}$ have non-negative eigenvalues, equivalent to the well known Høiland criteria, which are known to be necessary and sufficient for stability with respect to infinitesimal axisymmetric disturbances (Tassoul 1978). Indeed, the azimuthal wavenumber appears in Equation (12.23) only as a Doppler shift, so the analysis is effectively axisymmetric in this case.

12.5.4 Limit of a weak magnetic field

Now consider the limit of a weak magnetic field ($B \to 0$), but let $m \to \infty$ in such a way that ω_a remains a finite and adjustable parameter. This limit provides the best illustration of the MRI. Equation (12.22) becomes

$$(\hat{\omega}^2 - \omega_\mathrm{a}^2)^2 - (\hat{\omega}^2 - \omega_\mathrm{a}^2)\mathbf{M} : \boldsymbol{e}\boldsymbol{e} - 4\Omega^2\omega_\mathrm{a}^2(\boldsymbol{e} \cdot \boldsymbol{e}_s)^2 = 0; \tag{12.26}$$

regarded as a quadratic equation for $\hat{\omega}^2 - \omega_\mathrm{a}^2$, it has two real roots of opposite sign. The two values of $\hat{\omega}^2$ are therefore real and at least one is positive. Instability occurs if and only if the product of roots of $\hat{\omega}^2$ is negative, i.e.

$$\omega_\mathrm{a}^2\left(\tilde{\mathbf{M}} : \boldsymbol{e}\boldsymbol{e} + \omega_\mathrm{a}^2\right) < 0, \tag{12.27}$$

where

$$\tilde{\mathbf{M}} = \mathbf{M} - 4\Omega^2\boldsymbol{e}_s\boldsymbol{e}_s = \frac{1}{2}(\boldsymbol{g}\boldsymbol{a} + \boldsymbol{a}\boldsymbol{g} + \boldsymbol{e}_s\tilde{\boldsymbol{b}} + \tilde{\boldsymbol{b}}\boldsymbol{e}_s), \tag{12.28}$$

with

$$\tilde{\boldsymbol{b}} = s\nabla(\Omega^2) \tag{12.29}$$

being proportional to the gradient of angular *velocity*. Since ω_a is freely adjustable, this criterion is most easily satisfied in the limit $\omega_\mathrm{a} \to 0$, and instability is found if and only if $\tilde{\mathbf{M}}$ has a negative eigenvalue. The Høiland criteria are recovered but with the crucial difference that the angular momentum gradient is replaced with an angular velocity gradient. This removal of the $4\Omega^2$ stabilizing term through the effects of magnetic tension (which remain in the limit $B \to 0$ because of the increasing parallel wavenumber) is a characteristic property of the MRI (e.g. Papaloizou & Szuszkiewicz 1992) and leads to instability of systems with $\partial(\Omega^2)/\partial s < 0$ under a wide range of conditions.

12.5.5 Case of no rotation

The local dispersion relation in a non-rotating system,

$$\left[\omega^2 - \omega_{\rm a}^2 - (\boldsymbol{e}\cdot\boldsymbol{g})\boldsymbol{e}\cdot\nabla\ln\rho + \frac{2B}{\mu_0\rho}(\boldsymbol{e}\cdot\boldsymbol{e}_s)\boldsymbol{e}\cdot\nabla\left(\frac{B}{s}\right)\right.$$
$$\left. + \left(\frac{1}{v_{\rm s}^2+v_{\rm a}^2}\right)\left(\boldsymbol{e}\cdot\boldsymbol{g} - \frac{2v_{\rm a}^2}{s}\boldsymbol{e}\cdot\boldsymbol{e}_s\right)^2\right]\left[\omega^2 - \left(\frac{v_{\rm s}^2}{v_{\rm s}^2+v_{\rm a}^2}\right)\omega_{\rm a}^2\right]$$
$$= \left(\frac{v_{\rm a}\omega_{\rm a}}{v_{\rm s}^2+v_{\rm a}^2}\right)^2\left(\boldsymbol{e}\cdot\boldsymbol{g} + \frac{2v_{\rm s}^2}{s}\boldsymbol{e}\cdot\boldsymbol{e}_s\right)^2, \qquad (12.30)$$

has real roots for ω^2. The resulting criteria for instability can be shown to be identical to those of Tayler (1973), which were deduced from the MHD energy principle and are known to be necessary and sufficient. (The energy principle does not generally lead to local stability criteria but does so in the case of an axisymmetric system with a purely toroidal magnetic field.) Tayler's analysis shows that only $m = 0$ and $m = 1$ need be considered to determine stability.

12.5.6 Application to the tachocline

The local dispersion relation can readily be applied to any proposed model of the tachocline within the present framework to test its local stability. To make simple analytical deductions requires some idea of the relative magnitudes of the various terms. In the presence of an overwhelming stable stratification (perhaps relevant to the lower tachocline) the more interesting displacements are very nearly horizontal and the dispersion relation simplifies to

$$\left[\hat{\omega}^2 - \omega_{\rm a}^2 - (\boldsymbol{e}\cdot\boldsymbol{e}_r)^2N_r^2 - 2\Omega\cos\theta\sin\theta\frac{\partial\Omega}{\partial\theta} + \frac{2B}{\rho r^2}\cos\theta\frac{\partial}{\partial\theta}\left(\frac{B}{\sin\theta}\right)\right]$$
$$\times\,(\hat{\omega}^2 - \omega_{\rm a}^2) = 4\cos^2\theta\left(\Omega\hat{\omega} + \frac{v_{\rm a}\omega_{\rm a}}{s}\right)^2, \qquad (12.31)$$

where (r,θ,ϕ) are spherical polar coordinates and $N_r = (g_r a_r)^{1/2} \gg |\Omega|, |v_{\rm a}/s|$ is the radial buoyancy frequency. Although the displacement senses only the latitudinal gradients of Ω and B, the results differ from a strictly two-dimensional analysis confined to a spherical shell (Gilman & Fox 1997) because the wavevector

$\boldsymbol{k}$ is almost vertical, rather than horizontal. In the weak-field limit the MRI is present when

$$\cos\theta \frac{\partial(\Omega^2)}{\partial\theta} < 0, \tag{12.32}$$

which is *not* satisfied in the tachocline even though $\partial(\Omega^2)/\partial s < 0$ at high latitudes. In a hypothetical non-rotating situation an $m = 1$ Tayler instability would occur when

$$\frac{\partial}{\partial\theta}(B^2 \cos\theta \sin\theta) > 0 \tag{12.33}$$

(cf. Goossens 1980), which is always satisfied near the poles. Rotation suppresses the instability at the poles if $\Omega^2 > v_{\mathrm{a}}^2/s^2$. This conclusion agrees with that of Cally (2003), although there is no issue of the existence of confined normal modes in the present analysis.

These conclusions change if the stable stratification is not very strong. Radial displacements are then less inhibited and the dispersion relation regains sensitivity to radial gradients of Ω and B, allowing the possibilities of MRI and magnetic buoyancy instabilities, but the general stability criteria are somewhat complicated to write down.

12.5.7 General remarks

The above analysis has the considerable advantages of being purely algebraic and also strictly local and independent of boundary conditions. It can deal with realistic basic states that depend in a non-trivial way on both s and z, which in practice are not amenable to a normal-mode analysis. Of course, ad hoc local stability analyses are commonly made in astrophysical fluid dynamics but some care is required to ensure consistency of the approximations made while eliminating the stable high-frequency modes. One way to interpret the above analysis is in terms of an unstable dense or continuous spectrum of the linear operator (Terquem & Papaloizou 1996). Alternatively, one can proceed to the next order of the asymptotic approximation and obtain an evolutionary equation for the envelope function E, which must, in fact, depend also on a slow time coordinate in a non-exponential way (Ogilvie & Proctor 2003). It is then found that solutions exist that grow according to the local dispersion relation for asymptotically many e-foldings before being destroyed by a superexponential cutoff resulting from dispersion and shear. This is the closest approximation to a classical linear instability in this type of system. Small diffusivities can also be included in the derivation of this evolutionary equation.

It must be emphasized, however, that some shear instabilities that might apply to the tachocline, such as those of the inflection-point kind (Watson 1981;

Garaud 2001) are truly global and are not described by the above analysis. In addition, double and triple-diffusive instabilities involving viscosity, thermal diffusion and possibly resistivity are likely to be important in stellar interiors when no dynamical instabilities exist (Acheson 1978; Spruit 1999; Menou *et al.* 2004). In particular, the effect of stable stratification in suppressing the MRI when $\partial(\Omega^2)/\partial s < 0$ but $\cos\theta(\partial(\Omega^2)/\partial\theta) > 0$ can potentially be overcome through the effects of thermal diffusion.

12.6 Conclusion

The analogies between the tachocline and an accretion disc are imperfect but still of some value. For energetic reasons, instabilities of the differential rotation must lead to angular momentum transport down the gradient of angular velocity, something that is more naturally achieved by MHD instabilities in any case. The magnetorotational instability is optimized for this kind of transport but may be of limited applicability in the tachocline because of strong stable stratification. Useful methods have been found for analysing instabilities of differential rotation and toroidal magnetic fields in situations where normal modes may not be suitable. Statistical models have also been developed to describe the gross dynamical properties of the turbulent stress tensors and other transport properties of the MHD turbulence that may result from such instabilities.

Acknowledgements

I am grateful to the organizers of the Isaac Newton Institute programme on Magnetohydrodynamics of Stellar Interiors, and acknowledge the support of the Royal Society through a University Research Fellowship.

References

Acheson, D. J. (1978). *Phil. Trans. R. Soc. Lond.*, A**289**, 459.
Balbus, S. A. & Hawley, J. F. (1998). *Rev. Mod. Phys.*, **70**, 1.
Brandenburg, A., Nordlund, Å., Stein, R. F. & Torkelsson, U. (1995). *Astrophys. J.*, **446**, 741.
Boussinesq, J. (1877). *Mém. Acad. Sci. Paris*, **23**, 1.
Cally, P. S. (2003). *Mon. Not. Roy. Astron. Soc.*, **339**, 957.
Friedlander, S. J. & Vishik, M. M. (1995). *Chaos*, **5**, 416.
Garaud, P. (2001). *Mon. Not. Roy. Astron. Soc.*, **324**, 68.
Gilman, P. A. & Fox, P. A. (1997). *Astrophys. J.*, **484**, 439.
Goldreich, P. & Lynden-Bell, D. (1965). *Mon. Not. Roy. Astron. Soc.*, **130**, 125.
Goodman, J. & Xu, G. (1994). *Astrophys. J.*, **432**, 213.
Goossens, M. (1980). *Geophys. Astrophys. Fluid Dyn.*, **15**, 123.

Grossmann, S. (2000). *Rev. Mod. Phys.*, **72**, 603.
Hawley, J. F., Gammie, C. F. & Balbus, S. A. (1995). *Astrophys. J.*, **440**, 742.
Kato, S. & Yoshizawa, A. (1993). *Pub. Astron. Soc. Japan*, **45**, 103.
Lüst, R. (1952). *Z. Naturforsch.*, **7a**, 87.
Menou, K., Balbus, S. A. & Spruit, H. C. (2004). *Astrophys. J.*, **607**, 564.
Ogilvie, G. I. (2001). *Mon. Not. Roy. Astron. Soc.*, **325**, 231.
Ogilvie, G. I. (2003). *Mon. Not. Roy. Astron. Soc.*, **340**, 969.
Ogilvie, G. I. & Proctor, M. R. E. (2003). *J. Fluid Mech.*, **476**, 389.
Papaloizou, J. C. B. & Lin, D. N. C. (1995). *Ann. Rev. Astron. Astrophys.*, **33**, 505.
Papaloizou, J. C. B. & Szuszkiewicz, E. (1992). *Geophys. Astrophys. Fluid Dyn.*, **66**, 223.
Prandtl, L. (1925). *Z. Angew. Math. Mech.*, **5**, 136.
Pringle, J. E. (1981). *Ann. Rev. Astron. Astrophys.*, **19**, 137.
Shakura, N. I. & Sunyaev, R.A. (1973). *Astron. Astrophys.*, **24**, 337.
Spruit, H. C. (1999). *Astron. Astrophys.*, **349**, 189.
Tassoul, J.-L. (1978). *Theory of Rotating Stars*. Princeton: Princeton University Press.
Tayler, R. J. (1973). *Mon. Not. Roy. Astron. Soc.*, **161**, 365.
Terquem, C. & Papaloizou, J. C. B. (1996). *Mon. Not. Roy. Astron. Soc.*, **279**, 767.
Thomson, W. (Lord Kelvin) (1887). *Phil. Mag.*, **24** (5), 188.
Trefethen, L. N., Trefethen, A. E., Reddy, S. C. & Driscoll, T. A. (1993). *Science*, **261**, 578.
Watson, M. (1981). *Geophys. Astrophys. Fluid Dyn.*, **16**, 285.
Weizsäcker, C. F. von (1948). *Z. Naturforsch.*, **3a**, 524.

Part VI

Dynamo action

13

The solar dynamo and the tachocline

Steven Tobias & Nigel Weiss

The tachocline is believed to play a crucial role in the dynamo that maintains magnetic activity in the Sun. We first review the observational properties of the 11-year activity cycle and the 22-year magnetic cycle, as well as of the recurrent grand minima, with a characteristic 200-year timescale, that are revealed by proxy records. Then we discuss dynamo mechanisms, including differential rotation (the ω-effect), the net effect of gyrotropic motions (the α-effect) and flux transport by both large-scale motions (e.g. meridional flows) and small-scale processes (e.g. turbulent transport). Next we consider the location of the solar dynamo, comparing models with dynamo action distributed throughout the convection zone, located near the surface or (most likely) concentrated near the interface between the convective and radiative zones. Local pockets of strong field can then escape from the vicinity of the tachocline and emerge through the photosphere as active regions. The nonlinear back-reaction of the magnetic field affects transport coefficients (both α and the turbulent diffusivity β) and also drives the zonal flows that are observed. Furthermore, it provides a mechanism for the modulation associated with grand minima. We conclude with our picture of the relationship between convection, differential rotation and the dynamo in the tachocline.

13.1 Observations

The Sun exhibits cyclic magnetic activity, as do other slowly rotating stars with deep convective envelopes. This activity is manifested in the sunspot cycle, which has an average period of 11 years, as shown in Figure 13.1. Since the field reverses from one activity cycle to the next, the magnetic cycle actually has a 22-year period, which is far less than the timescale, of around 10^{10} years, for ohmic diffusion. It is now generally accepted that this large-scale field is generated by a hydromagnetic

The Solar Tachocline, D. W. Hughes, R. Rosner and N. O. Weiss.
Published by Cambridge University Press.

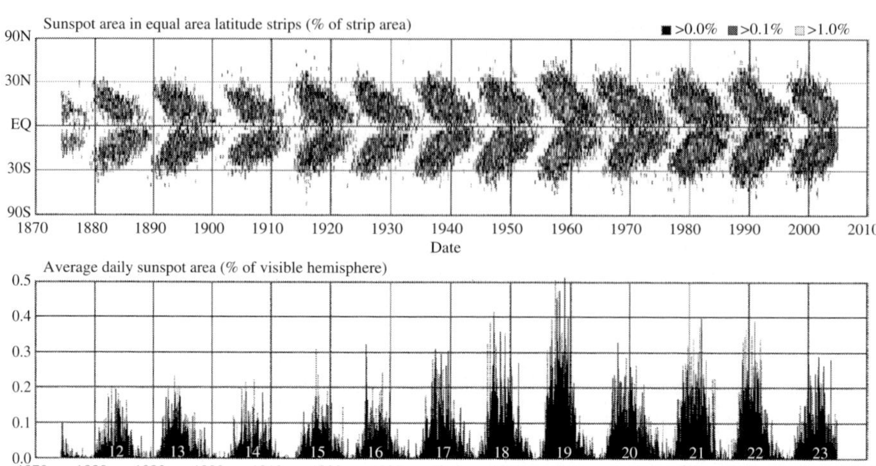

Figure 13.1. Cyclic magnetic activity in the Sun (1874–2005). Upper panel: butterfly diagram, showing incidence of sunspots as a function of latitude and time. Lower panel: variation of area covered by sunspots. (Courtesy of D.H. Hathaway.)

dynamo located in, or just below, the convection zone, and that the tachocline plays a key role in this process (Ossendrijver 2003; Tobias 2005; Charbonneau 2005).

The large-scale magnetic field on the Sun displays a strikingly systematic pattern that is best represented by sunspots, which are the sites of kilogauss magnetic fields that inhibit convection at the solar surface (Schrijver & Zwaan 2000; Stix 2002). The butterfly diagram in Figure 13.1 illustrates the incidence of sunspots as a function of latitude and time: at the beginning of a new cycle, spots appear at latitudes of $\pm 30°$ and the zones of activity then spread towards the equator, where they decay as the next cycle begins at higher latitudes. Although this pattern is not strictly periodic, the mean period is well-defined. Sunspots typically appear as pairs, in active regions that are oriented approximately parallel to the equator. The two spots in a pair have opposite magnetic polarities, and the polarity of preceding spots (in the sense of the Sun's rotation) is the same in each hemisphere but opposite in the north and south, as shown by the magnetogram in Figure 13.2. Moreover, this polarity reverses from one cycle to the next. These properties (Hale's Law) correspond to the emergence through the solar surface of a toroidal magnetic field that is antisymmetric about the equator and reverses at the end of each 11-year cycle. On average, the axes of sunspot groups are slightly inclined to parallels of latitude, with the preceding spots closer to the equator, and this tilt increases with increasing latitude (Joy's Law).

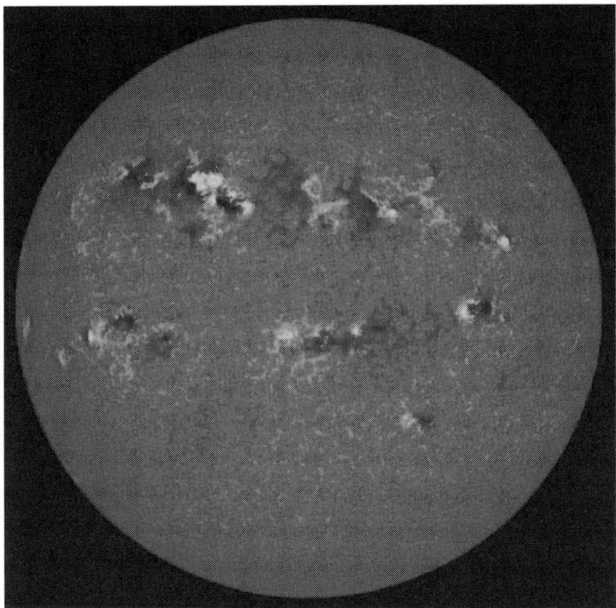

Figure 13.2. Solar magnetogram obtained by the MDI instrument on SOHO. The bright and dark patches denote regions with oppositely directed magnetic fields, and sunspots occur within these regions. The sense of the Sun's rotation is from left to right. The polarities of preceding and following regions are consistent in each hemisphere but antisymmetric about the equator. (Courtesy of Lockheed-Martin Solar and Astrophysics Laboratory.)

In addition to these toroidal fields, there is also a weak poloidal field with dipolar symmetry, which can be detected at high latitudes. These polar fields are strongest at sunspot minimum, when they impart a recognizable large-scale magnetic structure to the corona, and they reverse around the time of sunspot maximum. Their polarity is such that they have the same signs as the fields of following spots of the preceding activity cycle in each hemisphere.

The 11-year activity cycle is modulated on a longer timescale (Tobias 2002; Weiss & Tobias 2007). Sunspots were first observed through telescopes at the beginning of the seventeenth century but there was a prolonged dearth of spots from 1645 to 1715 – the Maunder Minimum – coinciding with the reign of the *Roi Soleil* (Ribes & Nesme-Ribes 1993), and a further decline in activity around 1800. Fortunately, the record of solar activity can be extended much further back by using proxy records. The incidence of galactic cosmic rays, which lead to the formation of the cosmogenic isotopes ^{14}C and ^{10}Be in the Earth's atmosphere, is reduced by magnetic fields in the solar wind. The ^{14}C is absorbed into tree-rings, which can be precisely dated, while the ^{10}Be is deposited in polar icecaps, and the abundances of these radioactive isotopes are anti-correlated with solar activity.

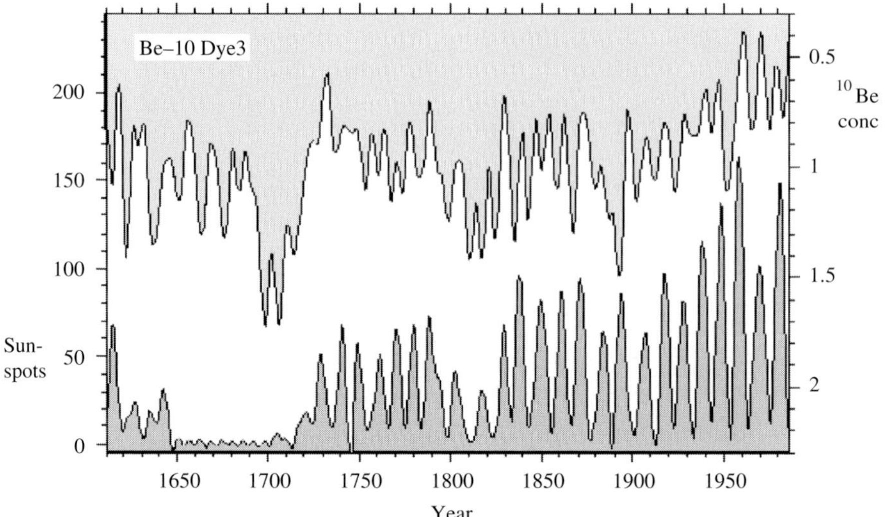

Figure 13.3. Comparison between group sunspot numbers and ^{10}Be concentration in a Greenland ice core, from 1610 to 1985. The Maunder Minimum (1645–1715) is clearly shown, though the 11-year cycle persists in the ^{10}Be record. (Courtesy of J. Beer.)

Figure 13.3 shows a comparison, running from 1610 to 1985, between a direct measure of solar activity, the group sunspot number (Hoyt & Schatten 1998), and ^{10}Be abundance in the Dye 3 ice core from Greenland (Beer *et al.* 1994). Note that the 11-year Schwabe cycle persists throughout the Maunder Minimum in the ^{10}Be record, although there were scarcely any sunspots (Beer *et al.* 1998). Proxy records confirm that such grand minima are a regular feature of solar activity. The ^{10}Be abundances have also been measured in the GRIP ice core over the interval from 50 000 to 18 000 years BP and the corresponding power spectrum in Figure 13.4 shows a significant peak at a period of 205 years (Wagner *et al.* 2001), which is also present in ^{14}C data for the last 10 000 years (Stuiver & Braziunas 1993). In addition to this $\sim$200 years periodicity (the de Vries cycle) there is also evidence of a $\sim$2000 years modulation (the Hallstatt cycle) in the records of both isotopes. Minima are now believed to occur in clusters with a well-defined period of $\sim$200 years between minima, while the clustering has a mean period of $\sim$2000 years.

Stellar magnetic activity is closely related to rotation. When a solar-type star arrives on the main sequence it is spinning rapidly, with a rotation period of a day or two, and is extremely active, with starspots covering a large fraction of its surface. As it evolves it spins down, owing to magnetic braking (Mestel 1999), and grows less active. The Sun, at an age of 4.6×10^9 years, has a (sidereal) rotation period

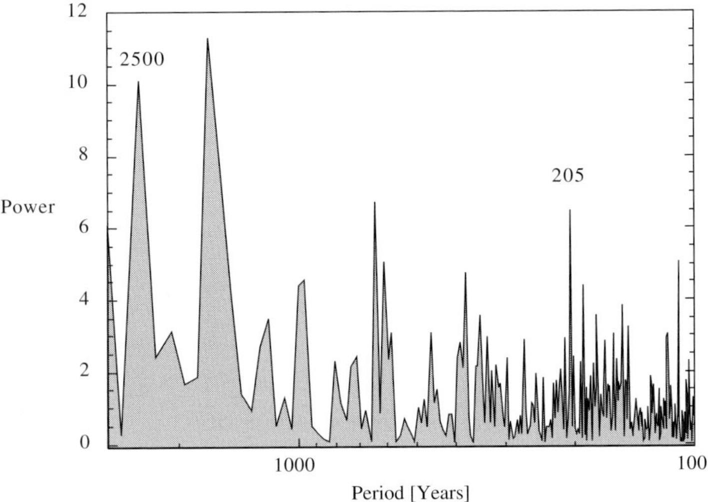

Figure 13.4. Power spectrum (after Lomb–Scargle) for the ^{10}Be record from the GRIP ice core (18 000–50 000 years BP). The significant peak with a period of 205 years corresponds to the de Vries cycle. (Courtesy of J. Beer.)

of 25 days at its equator and a peak spot coverage of about 0.3% (see Figure 13.1). Other stars of similar mass and angular velocity also exhibit cyclic activity with periods of around ten years (Baliunas *et al.* 1995; Saar & Brandenburg 1999).

As has already been explained by Christensen-Dalsgaard & Thompson in Chapter 3 of this book, the Sun's rotation is not uniform and the surface angular velocity decreases by 30% towards the poles (Thompson *et al.* 2003). Moreover, magnetic features rotate about 3% more rapidly than the ambient plasma. In addition, the so-called 'torsional oscillations', which have an 11-year period and follow the sunspot activity zones, are surface manifestations of zonal shear flows that extend throughout the convection zone (Vorontsov *et al.* 2002). Thus it seems natural to suppose that these shear flows are driven by the Lorentz force associated with the activity cycle, since that force is quadratic in the magnetic field and therefore independent of its sign.

In what follows, we first review the various physical mechanisms that are involved in generating and maintaining magnetic fields in the Sun. Next, in Section 13.3, we discuss the location of the solar dynamo, concluding that it is most likely to be seated at the tachocline. Then we describe some models of cyclic activity and present the results of nonlinear mean field dynamo models. Modulation of cyclic activity is considered in Section 13.5, and in the final section we put forward our own speculative views on the structure of the tachocline and its role in large-scale dynamo action.

13.2 Mechanisms

Unfortunately it is not yet feasible to construct a realistic numerical model of the solar cycle, though this has been attempted with a degree of success for the geodynamo (Glatzmaier & Roberts 1995). As of now, the most advanced computational model of the interactions between convection and magnetic fields in the outer portion of the Sun (Brun *et al.* 2004) only functions as a small-scale dynamo, generating a disordered field with no significant large-scale component. Any discussion of large-scale stellar dynamos has therefore to rely on simplifications (such as mean field dynamo theory) or on physical arguments.

13.2.1 Differential rotation

The magnetic field in a highly conducting fluid tends to move with the fluid and field lines are stretched by flow across them. Hence differential rotation is bound to have a powerful effect. The evolution of a magnetic field, $\mathbf{B}$, is governed by the induction equation,

$$\partial\mathbf{B}/\partial t = \nabla \times (\mathbf{u} \times \mathbf{B}) + \eta\nabla^2\mathbf{B}\,, \tag{13.1}$$

where $\mathbf{u}$ is the fluid velocity and the magnetic diffusivity η is assumed to be uniform. Consider now the azimuthally averaged magnetic field $\mathbf{B}$, which can be expressed as the sum of a poloidal field $\mathbf{B}_\mathrm{P} = \nabla \times A\mathbf{e}_\phi$ and a toroidal field $\mathbf{B}_\mathrm{T} = B_\phi\mathbf{e}_\phi$, referred to cylindrical polar coordinates (s, ϕ, z). Then

$$\frac{\partial}{\partial t}\left(\frac{B_\phi}{s}\right) = \mathbf{B}_\mathrm{P}\cdot\nabla\Omega - \nabla\cdot\left(\frac{B_\phi}{s}\mathbf{u}_\mathrm{m}\right) + \left(\frac{\eta}{s}\right)\left(\nabla^2 - \frac{1}{s^2}\right)B_\phi, \tag{13.2}$$

where the velocity is composed of an axisymmetric meridional flow $\mathbf{u}_\mathrm{m}$ and differential rotation with an angular velocity $\Omega(s,z)$ (i.e. $\mathbf{u} = \mathbf{u}_\mathrm{m} + s\Omega(s,z)\mathbf{e}_\phi$). This equation demonstrates how toroidal fields are generated from poloidal fields by differential rotation, and transported by meridional flows. It follows that the radial gradients of angular velocity in the solar tachocline and, to a lesser extent, the latitudinal differential rotation in the convection zone are bound to be major contributors to the generation of the toroidal field in the solar dynamo.

13.2.2 The α-effect

There is no corresponding source term for the poloidal field, since the vector potential A satisfies the equation

$$\frac{\partial}{\partial t}(sA) = -\mathbf{u}_\mathrm{m}\cdot\nabla(sA) + s\eta\left(\nabla^2 - \frac{1}{s^2}\right)A, \tag{13.3}$$

and without such a source the poloidal field must decay (Cowling's Theorem). In order to maintain the field an extra source term has to be inserted into this equation, and dynamo theory aspires to plug this gap. Such a source term is provided by the turbulent α-effect (Parker 1955b; Moffatt 1978; Krause & Rädler 1980; Roberts 1994; Ossendrijver 2003). The azimuthally averaged interaction between the (non-axisymmetric) fluctuating velocity $\mathbf{u}$ and magnetic field $\mathbf{b}$ give rise to an azimuthally averaged electromotive force $\boldsymbol{\mathcal{E}} = \langle \mathbf{u} \times \mathbf{b} \rangle$ in the induction equation. If we assume a separation of scales and ignore small-scale dynamo action then $\mathbf{b}$ is linearly and homogeneously related to $\mathbf{B}$ and we can set

$$\mathcal{E}_i = \alpha_{ij} B_j + \beta_{ijk} \frac{\partial B_j}{\partial x_k}; \tag{13.4}$$

note, however, that these assumptions are unlikely to hold in a stellar convection zone. Separating out the antisymmetric part of α_{ij} and assuming that the remaining turbulence is pseudo-isotropic, we may then write

$$\mathcal{E}_i = \alpha \delta_{ij} B_j + \gamma_j \epsilon_{ijk} B_k + \beta \epsilon_{ijk} \frac{\partial B_j}{\partial x_k} \tag{13.5}$$

or $\boldsymbol{\mathcal{E}} = \alpha \mathbf{B} + \boldsymbol{\gamma} \times \mathbf{B} - \beta \nabla \times \mathbf{B}$. Here $\boldsymbol{\gamma}$ is a turbulent pumping velocity, while β acts as a turbulent diffusivity, and so, averaging the induction equation (13.1) azimuthally and inserting this electromotive force, we have

$$\partial \mathbf{B}/\partial t = \nabla \times (\alpha \mathbf{B}) + \nabla \times [(\mathbf{u} + \boldsymbol{\gamma}) \times \mathbf{B}] + \tilde{\eta} \nabla^2 \mathbf{B}, \tag{13.6}$$

where $\tilde{\eta} = \eta + \beta$. Thus the vector potential for the poloidal field acquires a source term αB_ϕ in Equation (13.3). The corresponding source term for the toroidal field in (13.2) is frequently omitted, since its contribution is usually small compared with that from differential rotation (the ω-effect).

Since α is a pseudo-scalar, the α-effect requires turbulence that lacks mirror-symmetry, typically owing to Coriolis forces. If the turbulent motion has a magnetic Reynolds number, $Rm = v\,l/\eta$ (where v and l are the velocity and length scale, respectively, of the turbulent eddies), that is small, or a correlation time, $\tau = l/v$, that is short, it is possible to adopt first-order smoothing (Roberts 1994) and to derive the relation $\alpha = -(\frac{1}{3})\tau H$, where the kinetic helicity $H = \langle \mathbf{u} \cdot \nabla \times \mathbf{u} \rangle$. Neither of these assumptions is, however, valid in the Sun's convection zone, and it can be shown that such a simple dependence of α on helicity breaks down for flows at high Rm and correlation times of order unity (Courvoisier *et al.* 2006). In practice, therefore, the α-effect is best regarded as a useful parametrization that captures the essential physics of the regeneration process.

There are a number of mechanisms – all involving effects of rotation – that might give rise to a correlation between $\mathbf{u}$ and $\mathbf{b}$ and hence an α-effect in the Sun. Parker (1955b, 1979) originally introduced the idea in the context of cyclonic eddies: buoyant fluid elements would rise and expand, lifting a stitch of the toroidal field, and then spin round owing to the Coriolis force, so creating a meridional component of the field. Thus α should indeed be related to helicity. Models based on this concept typically assume that the scale of the turbulent convective eddies is relatively small and that the α-effect is distributed over a large part of the convection zone. The difficulty with this picture is that the small-scale field is likely to be much stronger than the mean field, and able to suppress the α-effect before it has done much good. Growth of the poloidal field is then limited by nonlinear quenching, so that $\alpha \ll \alpha_0$, its value in the linear (kinematic) regime. The formula conventionally adopted for mean field dynamo models sets

$$\alpha = \alpha_0(1 + B^2/B_0^2)^{-1}, \tag{13.7}$$

where B_0 is the equipartition field, such that $B_0^2 = \mu_0\langle\rho v^2\rangle$ (Jepps 1975). More recent theoretical studies indicate, however, that this expression should be replaced by

$$\alpha = \frac{\alpha_0}{1 + Rm^q B^2/B_0^2}, \tag{13.8}$$

with $0 < q \leq 2$ (Vainshtein & Cattaneo 1992; Diamond *et al.* 2005a; Hughes 2007) – see Section 13.4.2 for more discussion. In the Sun, where $Rm \gg 1$, this would imply that α is quenched when the mean field B is less than 1 G. Numerical experiments on turbulence driven by helical forcing (Cattaneo & Hughes 1996) and on rotating compressible magnetoconvection (Ossendrijver *et al.* 2001) provide support for such catastrophic quenching, with $q = 1$. More recently, Cattaneo & Hughes (2006) have investigated turbulent magnetoconvection in a Boussinesq (i.e. incompressible) rotating layer. This motion is effective as a small-scale dynamo – as indeed it is even without rotation (Cattaneo 1999) – producing a disordered magnetic field with $\langle\mathbf{B}\rangle = 0$. Surprisingly, however, the α-effect is found to be extremely weak and collisional (i.e. not turbulent), scaling as $\alpha \approx \eta/l$. Taken together, these results cast considerable doubt on the viability of any mean field dynamo model that relies on cyclonic turbulence distributed through a large part of the Sun's convection zone.

The alternative is to rely on dynamical processes that are magnetically driven and contribute to an α-effect that is not subject to catastrophic quenching. For instance, a stratified magnetic field is liable to instabilities driven by magnetic buoyancy, as discussed by Hughes in Chapter 11, and these instabilities are influenced by the Coriolis force so as to produce kinetic helicity and an average α-effect (Brandenburg & Schmitt 1998; Thelen 2000a,b). Three-dimensional calculations indicate how isolated flux tubes can be formed from a magnetic layer in the nonlinear regime (Matthews *et al.* 1995; Wissink *et al.* 2000a; Fan 2004) and then released into the convection zone. Adding a velocity shear introduces a further variety of behaviour (Cally 2000; Hughes & Tobias 2001; Cline *et al.* 2003; Tobias & Hughes 2004). Indeed, the combined effects of magnetic fields and a rotational shear in the tachocline can lead to generation of a large-scale poloidal field from a large-scale toroidal field without even invoking magnetic buoyancy, as described by Gilman & Cally in Chapter 10. This effect has been incorporated into some models as a tachocline-based α-effect, though this interpretation is incorrect, as noted by Tobias (2005).

If magnetic flux is confined to isolated toroidal flux tubes that encircle the Sun at the base of the convection zone, then non-axisymmetric instabilities can develop (Ferriz-Mas & Schüssler 1993) and provide a further contribution to the α-effect (Ferriz-Mas *et al.* 1994; Caligari *et al.* 1995, 1998; Ossendrijver 2000). The occurrence of active regions suggests that large-scale fields may indeed be concentrated into flux tubes through much of the convection zone, and their orientation (as described by Joy's Law) is consistent with the effect of Coriolis forces on a toroidal field as it rises to the surface. The corresponding tilts then provide a source of poloidal flux that can be interpreted as an α-effect operating at the solar surface (Leighton 1969; Stix 1974).

13.2.3 Transport of magnetic flux

The mean field is transported bodily by a large-scale meridional flow $\mathbf{u}_{\mathrm{m}}$, as can be seen from Equations (13.2) and (13.3). Observations show that there is in fact a quadrupolar circulation at the surface of the Sun, with a peak velocity of about $20\,\mathrm{m\,s^{-1}}$ directed towards the poles, as already explained in Chapter 3 (see also Thompson *et al.* 2003). Helioseismic measurements indicate that the poleward flow may extend downwards through much of the convection zone (e.g. Braun & Fan 1998; Duvall & Kosovichev 2001) though the dependence of the meridional flow on depth is uncertain and remains a major unsolved problem for helioseismology. Mass conservation nevertheless requires that there should be a return flow near the base of that zone – although there is no observational evidence of the number of layered cells in the meridional flow; after allowing for the increase of density with depth there could still be an equatorward velocity of $1\,\mathrm{m\,s^{-1}}$ (Dikpati *et al.* 2004), enough to traverse 30° within 10 years. This meridional circulation can then act as a conveyor belt within the turbulent convection zone. Superimposed upon this motion there is also a shallow flow that converges towards the activity zones near the surface but apparently reverses direction below about 15 Mm (Beck *et al.* 2004; Zhao & Kosovichev 2004).

Dynamo action requires that magnetic field lines should be stretched by the flow and then allowed to reconnect. Diffusion facilitates reconnection but also permits the field lines to slip through the fluid, and so it plays an ambiguous role. If the motion is turbulent and the field is weak, the total diffusivity $\tilde{\eta}$ in Equation (13.6) is dominated by the turbulent contribution $\beta_0 \approx vl$. As the field grows, the turbulent diffusivity β, like α, is progressively reduced. In two-dimensional geometry, Cattaneo & Vainshtein (1991) found that

$$\beta \approx \frac{\beta_0(1 + B_0^2/B^2)}{Rm^q}, \tag{13.9}$$

with $q = 1$; they conjectured that a similar expression might hold in three dimensions but with $1 < q < 2$, though this has proved difficult to confirm (Hughes 2007).

There is another type of slippage, caused by magnetic buoyancy, for fields that are confined to isolated flux tubes (Parker 1955a, 1979). Consider, for simplicity, a horizontal flux tube in magnetohydrostatic equilibrium with its surroundings. Then the external gas pressure is balanced by the sum of the internal gas pressure and magnetic pressure. Thus the internal gas pressure is less than that outside and, if the tube is in thermal equilibrium with its surroundings, the density will be less too. Hence the tube is buoyant and will rise. (Note that this is a lack of equilibrium, and differs fundamentally from the magnetic buoyancy instabilities discussed above; the effect can only be eliminated if the tube is cooler than its surroundings or, in the case of a toroidal tube, if there is a retrograde axial flow along it.) The motion of rising flux tubes has been extensively investigated (Fan 2004), in both two (Emonet & Moreno-Insertis 1998; Hughes *et al.* 1998; Fan *et al.* 1998; Hughes & Falle 1998) and three dimensions, for atmospheres that are either stably (Wissink *et al.* 2000b) or unstably (Cline 2003; Fan *et al.* 2003; Abbett *et al.* 2004) stratified. As these flux tubes rise they generate trailing vortices (see Figure 13.5) which tear them apart unless the field is strongly twisted. In general, the magnetic field in a turbulent layer is likely to be highly intermittent.

Magnetic buoyancy competes as a transport mechanism with pumping caused by inhomogeneity or anisotropy of the turbulent flow, which is represented by the velocity $\boldsymbol{\gamma}$ in Equation (13.6). If the turbulence is limited to a finite region then magnetic flux is expelled down the gradient of turbulent intensity, and the pumping velocity $\boldsymbol{\gamma}$ can be calculated (Rädler 1968; Zeldovich *et al.* 1983; Moffatt 1983). Tao *et al.* (1998) have demonstrated this diamagnetic effect for forced two-dimensional turbulence. Now the up–down symmetry of Boussinesq convection ensures that magnetic flux is expelled equally towards the top and bottom of a convecting layer but in a stratified layer this symmetry no longer holds. There is then a topological distinction between isolated plumes of gently rising fluid and a network of cooler sinking fluid that splits up into rapidly descending plumes (Spruit *et al.* 1990). In a steady state the horizontally averaged mass flux $\langle \rho w \rangle = 0$, where w is the downward

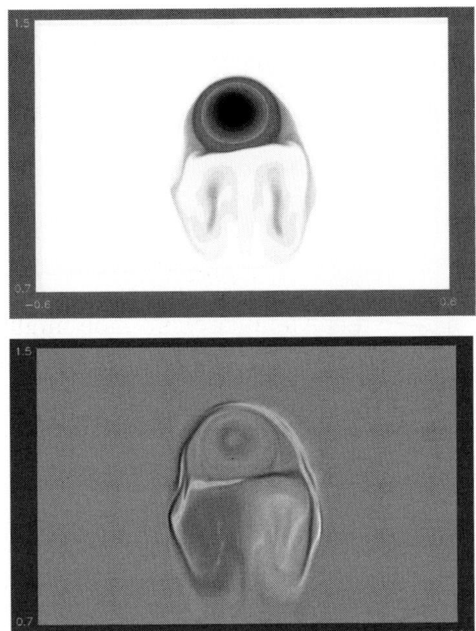

Figure 13.5. Rise of a buoyant flux tube with a twisted magnetic field. Cross-sections showing the axial field strength (upper panel) and the axial vorticity (lower panel) in a two-dimensional model calculation. (After Hughes & Falle 1998.)

vertical velocity, and so, since ρ and w are correlated, $\langle w \rangle < 0$ and there is a net *up*ward velocity; nevertheless, $\langle w^3 \rangle > 0$, for the downflows are moving faster (Weiss *et al.* 2004). Magnetic flux within the rising plumes is carried outwards and entrained into the vigorously sinking plumes, which succeed in pumping the flux preferentially downwards. Figure 13.6 shows the evolution with time of the vertical profile of $\langle B_y \rangle$ in a numerical experiment where the convecting layer is contained between rigid, perfectly conducting boundaries, so that no magnetic flux can escape. Initially a sheet of y-directed field is inserted near the middle of the turbulent layer and eventually there is a statistically steady state with magnetic flux pumped towards the upper and (predominantly) the lower boundary. In this case, with $Rm \approx 75$, pumping has only a moderate effect; note, however, that the ratio $\langle B_y \rangle / \langle \rho \rangle$ doubles between the middle and the bottom of the layer, whereas mixing of a passive scalar, or two-dimensional mixing of a transverse magnetic field, would lead to a concentration that was proportional to $\langle \rho \rangle$.

The flux distribution changes if convection penetrates from the unstable layer into a stably stratified region below, where magnetic flux can be stored (Nordlund *et al.* 1992; Tobias *et al.* 1998, 2001; Dorch & Nordlund 2001; Ossendrijver *et al.* 2002). Figure 13.7 shows the evolution of the profile of $\langle B_y \rangle$ for one such case,

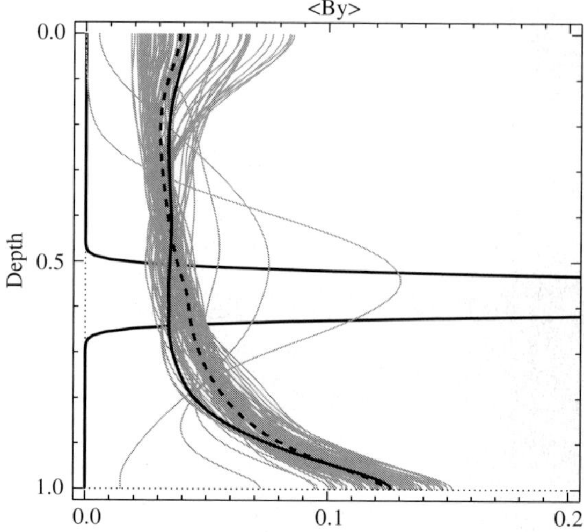

Figure 13.6. Downward pumping of magnetic flux by compressible convection in a closed box. Profiles of the horizontally averaged field $\langle B_y \rangle$ at successive times, starting from a thin sheet. The thick dark line denotes the time-averaged final state, with the strongest mean field at the lower boundary. (Courtesy of N. H. Brummell.)

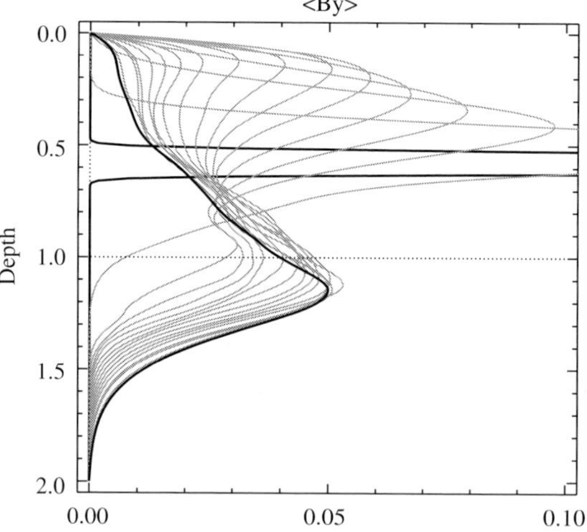

Figure 13.7. Flux pumping with penetration; the upper half of the layer is superadiabatically stratified, while the lower half is strongly subadiabatic. As Figure 13.6 but for the pumping phase only of a rundown calculation. In the last state, denoted by a thick line, the mean magnetic flux is concentrated in the stable region. (After Tobias *et al.* 2001, courtesy of N. H. Brummell.)

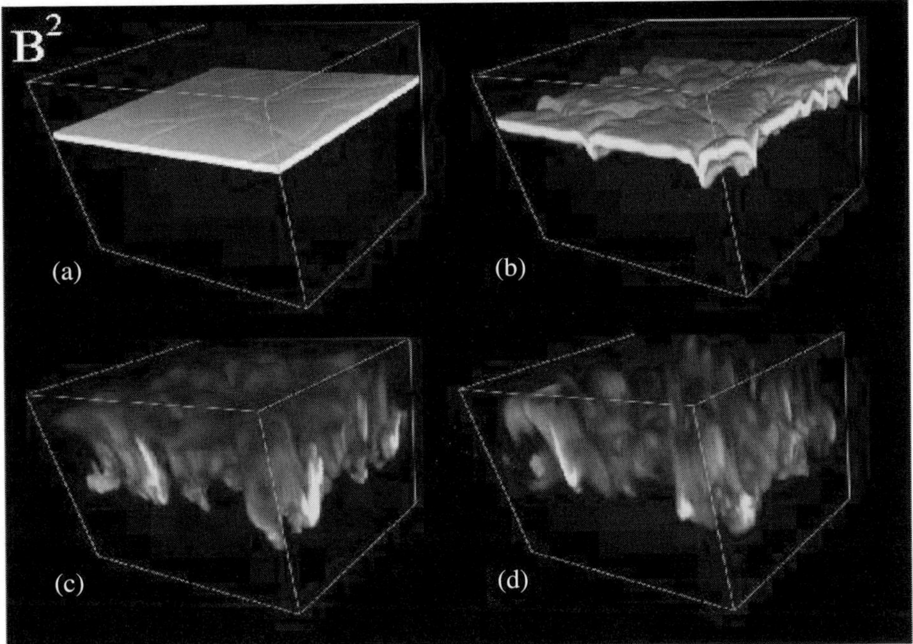

Figure 13.8. Flux pumping with penetration: volume rendering of the magnetic energy density, with high values showing up as opaque and bright. The sequence runs from (a) the initial magnetic field configuration to (d) the field at the end of the pumping phase. The image is vertically exaggerated for clarity. (After Tobias *et al.* 2001, courtesy of N. H. Brummell.)

this time a run-down calculation with $B_y = 0$ at the top and bottom boundaries. It is apparent that most of the flux ends up in the lower, stable half of the box. That certainly does not mean, however, that no magnetic fields are left in the turbulent region above. Figure 13.8 shows the magnetic energy density $|\mathbf{B}|^2$ at four stages of the run. While the strongest fields are in the penetrative region, there are still concentrations of magnetic field in the unstable region above. This process of flux pumping is obviously likely to be important near the interface between the convective and radiative regions in the Sun – which is precisely where the tachocline is located.

13.3 Where is the solar dynamo?

The various mechanisms outlined above can be assembled in different ways to produce models of the solar dynamo with varying degrees of plausibility. The production of strong toroidal fields by differential rotation is a common feature of all these models, though they may rely on the radial or latitudinal gradient of Ω, either within the convection zone or at the tachocline. The nature and position of the process that is represented by an α-effect are much less certain. In this section we outline the various possibilities and put forward arguments in favour of a dynamo located near the tachocline.

13.3.1 Distributed dynamos

The classical $\alpha\omega$-dynamo relies on cyclonic eddies, caused by the effects of the Coriolis force on small-scale convection, combined with local differential rotation. As fluid elements rise (or sink) in a stratified layer they expand (or contract) and develop anticyclonic (or cyclonic) swirling motions, so that their helicity is antisymmetric about the equator. This leads to an estimate

$$\alpha \approx v^2 \tau^2 \Omega \cos\theta / H_\rho \approx l^2 \Omega \cos\theta / H_\rho \,, \tag{13.10}$$

where H_ρ is the density scale-height and θ is the colatitude (Krause & Rädler 1980; Rüdiger & Hollerbach 2004); Zeldovich *et al.* (1983) point out, however, that this formula is valid only in the upper reaches of the convection zone, where the Rossby number is greater than unity. The obvious attraction of this formalism is that such an $\alpha\omega$ (or α^2) dynamo would be effective in any rotating star with a convection zone, even if the star were fully convective and had no tachocline. Given the angular velocity profile within the Sun, an α-effect distributed throughout the convection zone would have to interact with the latitudinal gradient in Ω in order to maintain a cyclic dynamo. An alternative possibility, presented by Brandenburg (2005), is to locate the solar dynamo near the top of the convection zone, in the subsurface shear layer ($0.95 \leq r/R_\odot \leq 1.00$), where $\partial\Omega/\partial r < 0$. This radial velocity shear, combined with a locally distributed α-effect, might be able to generate a large scale field.

This picture has two major drawbacks. First of all, any strong localized concentrations of magnetic flux will be buoyant and float to the surface before they are sufficiently amplified, especially if they are generated near the top of the convection zone. Second, and more importantly, regeneration of the poloidal field will be halted by catastrophic α-quenching before any significant large-scale field can be produced.

13.3.2 Flux transport dynamos

Babcock (1961) proposed a phenomenological model of the solar cycle that explained both Hale's Law and the newly observed reversal of the polar fields; this model has provided a template for many subsequent physical discussions. In it, the poloidal field is wound up by differential rotation within the Sun to produce a toroidal field, which is confined to ropes that erupt through the surface to form active regions. On their way up they are twisted by Coriolis forces in accordance with Joy's Law; the bipolar regions then spread out in latitude (as observed) so that the preceding fields migrate towards the equator, where they can merge and cancel out, while the following parts migrate to higher latitudes and eventually reverse the polar fields. Leighton (1969) ascribed the spreading out to turbulent diffusion caused by supergranular convection, and was able to incorporate these processes into what was effectively a two-dimensional mean field dynamo operating on the solar surface. This model was subsequently extended to include a poleward meridional flow at the surface (e.g. Wang & Sheeley 1991) and, later still, the effects of a much slower counterflow at the base of the convection zone. In the more modern form of these flux transport models, the meridional conveyor belt transports the reversed poloidal field down to the neighbourhood of the tachocline, where the toroidal field for the next activity cycle can be generated (Choudhuri 2003).

Several detailed mean-field models of these processes have been constructed (e.g. Dikpati & Charbonneau 1999). Their attraction is that they rely, to a great extent, on observable behaviour at the surface of the Sun and that they can explain the reversal of the polar fields at sunspot maximum as a consequence of meridional transport and surface diffusion (Dikpati *et al.* 2004; Durrant *et al.* 2004). This requires a supergranular diffusivity of around 600 $\mathrm{km}^2\,\mathrm{s}^{-1}$ (Schrijver & Zwaan 2000), which is consistent with estimates derived from kinematic modelling (Simon *et al.* 1995). Nevertheless, there are grave difficulties: it is not at all clear how a weak poloidal field can remain coherent as it passes slowly through the turbulent convection zone on the meridional conveyor belt; nor is the surface α-effect likely to be adequate (Dikpati *et al.* 2002).

13.3.3 Interface dynamos

The scale of sunspot groups and active regions (see Figure 13.2) implies that their accompanying fields must be deep-seated rather than near-surface features. If such a strong azimuthal field is confined to a large flux tube within the convection zone then the tube will float upwards owing to magnetic buoyancy, and escape within about a month, unless it is held down at its ends (Parker 1975). Weaker fields, on the other hand, will be expelled from the turbulent region and pumped preferentially downwards, to accumulate around the interface between the radiative and convective zones (Spiegel & Weiss 1980; Golub *et al.* 1981; van Ballegooijen 1982). That is, of course, just where the strong gradient in Ω can stretch poloidal field lines to form a strong toroidal field. Taken together, these considerations provide strong arguments for locating the solar dynamo in, or just above, the tachocline. Then toroidal fields can be kept submerged until they become unstable and liberate

Ω-shaped flux ropes that rise through the convection zone and break through the surface to form active regions at the photosphere.

Parker (1993) introduced an idealized two-layer model of an interface dynamo, in Cartesian geometry, with the α-effect provided by turbulent cyclonic eddies in the upper layer and the ω-effect produced by a velocity shear confined to the lower layer. Linear solutions then take the form of surface waves at the interface between these layers. Turbulent diffusion now plays an essential role, for it allows the transfer of poloidal and toroidal fields between the layers, and is itself reduced in the thin region of weak convective overshoot, which is where the fields are strong. Parker observed that the interface dynamo scenario would remain consistent in the nonlinear regime providing that the turbulent diffusivity is suppressed in inverse proportion to the mean magnetic energy (Tobias 1996b). This is a delicate balance – see the discussion by Diamond *et al.* in Chapter 9.

While the ω-effect can naturally be ascribed to shear in the tachocline, the precise origin of the α-effect is less immediately obvious. The two-layer configuration allows a turbulent α to persist in a region where the large-scale field is weak, so avoiding catastrophic quenching. The poloidal field can then be generated within the upper layer and pumped downwards. Alternatively, it could be produced by the nonlinear development of instabilities driven by magnetic buoyancy at or near the interface, or even by magneto-rotational instabilities within the tachocline itself.

Parker's linear model has since been extended to describe both nonlinear waves in spatially periodic systems (Tobias 1997a) and cyclic behaviour in boxes with lateral boundaries that represent the poles or the equator (Tobias 1996a, 1997b; Phillips *et al.* 2002) or spherical domains (Zhang *et al.* 2003; Chan *et al.* 2004). Indeed, mean field ($\alpha\omega$) dynamo models readily yield butterfly diagrams with a passing resemblance to that in Figure 13.1. Adding a significant meridional flow at the base of the convection zone produces a family of advection-dominated dynamos, in which the equatorward flow controls the migration of dynamo waves and also sets the period of the activity cycle.

13.3.4 The role of the tachocline

The tachocline itself plays a key role in both flux transport and interface dynamos, as the most obvious site of the ω-effect. As we have seen, several possible locations have been proposed for the α-effect. It seems likely, however, that any α that is distributed throughout the convection zone will be quenched before it is able to generate significant poloidal flux. Furthermore, comparisons between dynamo models with α localized either near the surface (as in flux transport models) or at the base of the convection zone show that the latter choice is much more effective (Mason *et al.* 2002); indeed, Dikpati *et al.* (2004) found that their flux transport model would not function effectively unless it was able to rely on an additional strong α-effect located near the interface.

Such arguments have led to a general (though far from universal) consensus that the seat of the solar dynamo is at the tachocline. Strong toroidal fields can then be created locally by differential rotation and held down by a combination of turbulent pumping (the γ-effect) and transport by overshooting convection. Thus they can be safely stored, in a stably stratified region, for times that are comparable with the cycle period. Samples of these strong fields can then be released through instabilities, and reach the surface, as envisaged by Parker (1979). This picture can therefore explain both the horizontal scale of active regions, as observed at the solar photosphere, and their systematic time-dependent behaviour during the solar cycle.

13.4 Models of cyclic activity

13.4.1 The form of the magnetic field

In order to construct models of the large-scale solar magnetic field, it is necessary to make some assumptions about the spatial form of the field in the regions of generation. The form of the field will then determine its dynamics once it has been generated (including the important issue of the nonlinear back-reaction of the magnetic field on the flow) and place restrictions on dynamo models. In particular, the spatial intermittency of the field in the solar convection zone and tachocline is a key issue for dynamo modelling. In order to facilitate progress in representing the solar magnetic field, two extreme models of its structure have been proposed. In the first of these scenarios (see, for example, Schüssler 2005), the field exists in the form of isolated flux tubes. These tubes contain all the magnetic flux and each tube is bounded by a flux surface. The dynamics of the magnetic field can then be understood as the dynamics of an ensemble of such tubes, without any contribution from the completely field-free region between them. This is an appealing paradigm from a modelling perspective – if the dynamics of an isolated tube (and, more ambitiously, the interactions between tubes) can be understood then the behaviour of the field as a whole can be predicted. Moreover, the concept of isolated tubes of magnetic flux is immediately comprehensible. This approach has, however, been subjected to some criticism – see Chapter 11. The problem is that of the existence and stability of the flux surfaces that form the boundaries of flux tubes in turbulent flow at high magnetic Reynolds number (Cattaneo *et al.* 2006). Moreover, it seems unlikely that field-free regions exist between the magnetic flux tubes. In the other extreme scenario the field varies smoothly with position and, at least locally, the field gradients are small. In this ansatz the large-scale field exists in a smooth layer and it is the dynamics of this layer that is of interest. A deeper discussion of the nature of these two modelling approaches, with particular emphasis on the consequences for magnetic buoyancy instabilities, is to be found in Chapter 11.

In reality the magnetic field in the solar plasma at high magnetic Reynolds number is likely to exhibit a highly complicated topology and chaotic field line structure. Although a large-scale component for the magnetic field will certainly exist, there will also be extreme fluctuations on many scales down to the diffusive length-scale for the magnetic field. Of crucial importance to the dynamics of the magnetic field is then its filling factor, and also the ratio of the large-scale field to the fluctuating field (see, e.g. Ruzmaikin (1998, 2000) for a discussion). It is likely that these two measures of the structure of the field will depend on the local level of turbulence, with turbulent small-scale flows in the convection zone leading to local amplification of small-scale magnetic fields. In the tachocline, where the shear flow is predominantly large-scale, one might expect the large-scale field to be more significant owing to the enhancement of magnetic diffusion. Here we envisage the field in the tachocline to have a large-scale component with an average field strength of around 10^4 G. We expect that the field strength will peak locally with field strengths perhaps reaching 10^5 G. These pockets of strong field may be formed as the result either of local amplification by small-scale flows, or of an instability such as magnetic buoyancy or collapse due to the inhibition of turbulent transport (Kleeorin *et al.* 2001). It is presumably this strongest field that makes it to the solar surface to form active regions, without being distorted by convection or Coriolis forces as it travels upwards. The rest of the magnetic flux is reprocessed by the convection zone. In this way the convection zone acts as a filter for the dynamo field (Tobias *et al.* 2001). Indeed, if the toroidal field were to drop below a certain threshold then active regions might no longer be formed, giving the impression that the dynamo had switched off entirely. We shall return to this theme in Section 13.5.

13.4.2 Dynamics of the magnetic field

Kinematic modelling, described earlier, establishes that the region spanning the base of the convection zone and tachocline is the most likely seat for the generation of the large-scale toroidal field that leads to the formation of active regions. However, this theory, in which the velocity is prescribed, takes no account of the back-reaction of the magnetic field on the motion. The form of the dynamical interaction between flow and field is a subtle and contentious issue and a full account of the intricacies of the suppression of the mean field transport coefficients in the nonlinear regime is beyond the scope of this review – see Diamond *et al.* (2005a) or Brandenburg & Subramanian (2005) for in-depth discussions. Here we briefly discuss the various mechanisms by which the back-reaction of the Lorentz force may saturate the linear dynamo instability and speculate as to which of these mechanisms may be most appropriate for the solar dynamo.

The most contentious aspect of the nonlinear dynamics of the magnetic field concerns its role in modifying the transport coefficients that are so vital to the mean-field theory. This is particularly important in the turbulent convection zone, but may also be of significance in the tachocline where turbulence driven by overshooting convection or shear-driven instabilities will interact with magnetic fields in a complicated manner. As discussed earlier, calculation of the transport coefficients of mean-field theory relies on a parametrization of the small-scale correlations between the turbulent velocity field and the small-scale magnetic field. These correlations are very sensitive to the precise form and level of the turbulence. It is therefore not surprising that the transport coefficients are also sensitive to the strength of the local magnetic field as this will modify the form of the turbulence. The traditional argument is that the transport coefficients α and β will be quenched when the energy in the *mean* field reaches equipartition with the energy in the turbulence, as expressed in Equation (13.7). As we have already pointed out, this formula for the quenching of α has been the subject of intense debate. Here we give the physical argument for amending the formula and again refer the interested reader to the article by Diamond *et al.* (2005a), which includes a discussion of such intricacies as the role of magnetic helicity conservation in determining the form of the transport coefficients in the nonlinear regime. Physically, one expects the magnetic field to have a significant back-reaction on the turbulent motion once its energy reaches equipartition with the turbulence. However, as noted by Vainshtein & Cattaneo (1992), in a turbulent environment one expects the small-scale magnetic energy to be significantly larger than that contained in the large scales. Numerical simulations indicate that the energy in the small-scale field will be up to Rm^q (with $0 < q \leq 2$) larger than that in the mean field. If this is the case then one would expect the transport coefficients to be significantly quenched when the energy in the mean field is extremely weak, and the formula for α should be that shown in Equation (13.8). It should be noted that these formulae are postulated by using order of magnitude estimates for the level of saturation and preserving the correct (quadratic) dependence on strength of the mean field. In these expressions, the magnetic field acts back instantaneously and locally on the turbulence to suppress the transport coefficients; the same dependence can be obtained from closure models for the turbulence on the assumption that it is quasi-steady (see for example, Gruzinov & Diamond 1994, 1995). Others have argued that the static dependence of α_{ij} and β_{ijk} can only be verified in the limit of small Rm and that the transport coefficients themselves should be modelled as time-dependent quantities (see Kleeorin & Ruzmaikin 1982; Kleeorin *et al.* 2000; Blackman & Brandenburg 2002) leading to dynamic evolution equations for the tensors α_{ij} and β_{ijk}.

In addition to modifying (possibly significantly and catastrophically) the form of the transport coefficients of mean-field theory, the magnetic field will also play a dynamic role in altering the distribution of large-scale solar differential rotation. The solar differential rotation profile is maintained in the convection zone through the interaction between rotation and turbulent convection; see Diamond *et al.* (2005b) for a review of the generation of zonal flows, and Brummell *et al.* (1998) or Miesch (2005, and Chapter 5 of this book) for a discussion of the generation of mean flows and differential rotation by convection. As noted by these authors, the generation of mean flows by small-scale Reynolds stresses is sensitive to the precise nature of the convective flows and cannot easily be determined in terms of large-scale quantities. Within the mean-field framework this interaction has, however, been parametrized via the Λ-effect – for a full discussion see the account by Rüdiger & Kitchatinov in Chapter 6 or Rüdiger & Hollerbach (2004) – where the velocity correlation tensor $Q_{ij} = \langle u_i u_j \rangle$ is directly related to the local rotation rate and its spatial derivatives via 'mean-field hydrodynamic' transport coefficients Λ_{ijk} and ν_{ijkl}.

The back-reaction of the magnetic field on the turbulent transport of angular momentum by convection can therefore be understood in terms of the modification of turbulent Reynolds stresses owing to the presence of the magnetic Maxwell stresses. Within the mean-field framework, this back-reaction has been parametrized using a formula for the quenching of the transport coefficient Λ_{ijk}. It will be clear from the discussion above that the nature of 'Λ-quenching', and the concomitant modification of angular momentum transport, will be open to the same levels of uncertainty as the α and β-effects. Of crucial importance is whether the hydrodynamic angular momentum transport can be maintained in plasmas with strong small-scale magnetic fields at high *Rm*.

Whilst the turbulence is clearly of significance in transporting angular momentum in the convection zone, other processes may also lead to the generation of zonal shear flows (i.e. of perturbations in the azimuthal angular velocity profile), and these may be of greater importance in the stably stratified tachocline. Once the large-scale magnetic field has been generated by the dynamo, it will drive a flow via the large-scale Lorentz force ($\langle \mathbf{j} \rangle \times \langle \mathbf{B} \rangle$). This macrodynamic process has been termed the Malkus–Proctor effect after the pioneering work of Malkus & Proctor (1975), who were the first to use this saturation mechanism in a mean-field model. This nonlinearity is likely to be of importance in relatively quiescent regions of strong mean magnetic field such as the tachocline – in that region the ratio of small-scale to large-scale magnetic field should be significantly smaller than in the convection zone above, as demonstrated by Tobias *et al.* (2001).

Both Λ-quenching and the Malkus–Proctor effect allow for the dynamic back-reaction of the magnetic field on the angular momentum transport. In each case a separate partial differential equation for the evolution of the differential rotation must be coupled to those for the evolution of the large-scale poloidal and toroidal magnetic fields (e.g. Belvedere *et al.* 1990; Kitchatinov *et al.* 1994). This equation governs the dynamics of the zonal shear flows generated by the presence of the magnetic field. These zonal flows driven by the Lorentz force may be directly associated with the 'torsional oscillations' seen in helioseismic inversions for the differential rotation. In the discussion that follows we focus our attention on models that include the macroscopic Malkus–Proctor mechanism, as we believe that this nonlinearity plays a key role in driving zonal flows in the tachocline. (The turbulent Maxwell stresses may be of greater importance in the convection zone; however, the dynamics of models that use microscopic Λ-quenching as the dominant nonlinearity is qualitatively similar.) In dynamos that rely on the Malkus–Proctor effect, the generation process saturates by driving a zonal flow that diminishes the differential rotation that causes the field to grow in the first place. As these torsional oscillations are driven by the Lorentz force, which is quadratic in the magnetic field, they take the form of oscillations with half the period of the magnetic field. An example of such a nonlinear solution for a spherical mean-field interface dynamo based in the tachocline (Bushby 2005) is shown in Figure 13.9a, which shows the latitudinal distribution of magnetic field as a function of time for a periodic dipolar solution. The zonal flows driven by the magnetic field are shown in Figure 13.9b. The latitudinal spatial dependence of these zonal flows for these parameters is manifestly related to that of the strong toroidal magnetic field of the butterfly diagram, although there is clearly a phase lag between the generation of the magnetic field and the zonal shear flow. Note the presence of a weaker polar branch, both for the toroidal field and (as measured) for the zonal shear. The radial dependence of the zonal shear flow depends on the level of stratification (Kleeorin & Ruzmaikin 1982; Covas *et al.* 2004; Bushby 2005). It is possible for strong magnetic fields generated in the tachocline to drive zonal shear flows at much larger radii, so that, although the Lorentz force is strongest at the base of the convection zone, the response of the velocity perturbations to the magnetic forces is largest closer to the solar surface. This result is simply understood by an angular momentum argument – a weak force at larger radii can have a locally large effect owing to the decrease in density there. Models that include the dynamic nonlinearities discussed above are therefore capable of saturating the dynamo growth by driving zonal shear flows. The dynamo will begin to saturate when the energy of the zonal flows, which is clearly comparable with that of the large-scale magnetic energy, is of the same order of magnitude as the flows (turbulent convection and differential rotation) that are causing the dynamo to grow. The detection of such flows via helioseismic inversions (see Chapter 3) yields valuable constraints on the amplitude of dynamo-generated zonal flows.

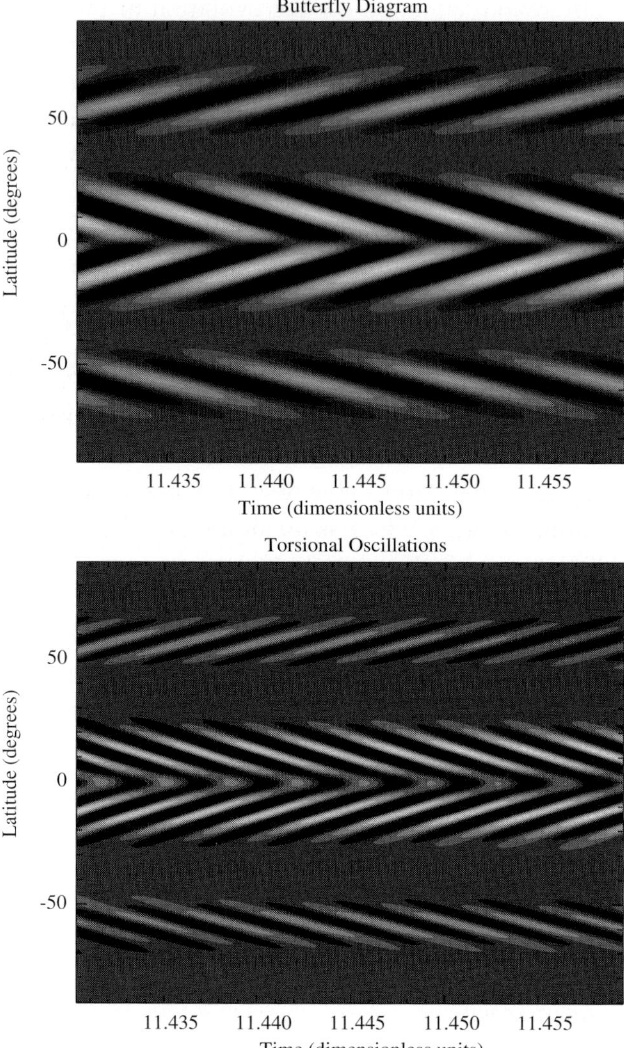

Figure 13.9. Nonlinear cyclic behaviour for a spherical model of an interface dynamo. Upper panel: butterfly diagram showing toroidal fields of opposite signs (with dipole symmetry) at the base of the convection zone; note the presence of a polar branch. Lower panel: the corresponding zonal shear flows (torsional oscillations) with twice the frequency of the magnetic cycle. (Courtesy of P. J. Bushby.)

13.5 Modulation

We noted in Section 13.1 that the basic 11-year solar activity cycle is modulated on timescales of approximately 200 and 2000 years, with grand minima of activity appearing not only in the sunspot numbers but also recurrently in the proxy ice-core and tree-ring data. Although this modulation is very striking in the sunspot record – the interval of the Maunder minimum in the seventeenth century is characterized by the almost complete disappearance of sunspots – the level of modulation of the dynamo generated field in the tachocline is uncertain. As mentioned earlier, only the peak fields at the base of the convection zone are tough enough to resist the effects of the convective flows and Coriolis force, so that they can reach the solar surface and form active regions. Weaker fields are reprocessed and pumped back down into the tachocline by sinking plumes. It is therefore possible for the dynamo generated field to be only mildly modulated, but yet to produce significant modulation of the sunspot cycle – weak modulation may push the peak dynamo field below the threshold for penetrating the convection zone and stop the formation of active regions. In that case the dynamo cycle will still continue in the tachocline and the Sun's global magnetic field will continue to vary. This filter-effect explains why the eleven year cycle is still visible in ^{10}Be records during the Maunder minimum, as can be seen from Figure 13.3, even though there were scarcely any sunspots (Beer *et al.* 1998).

Although the level of modulation of the solar dynamo is uncertain, it is certainly the case that the basic cycle *is* modulated on a longer timescale and it is natural to ask what is the origin of this modulation that leads to the occurrence of grand minima (see, e.g. Tobias (2002) for a fuller discussion of modulation). Two separate mechanisms have been postulated as natural candidates for modulating the basic cycle. The first of these is stochastic modulation (see, e.g. Hoyng 1988). In this paradigm, stochastic fluctuations in the transport coefficients of mean-field theory arise because of small-scale interactions. These stochastic perturbations cause the dynamo to switch on and off randomly. Here, the period between minima is a random variable with a distribution that depends on the form of the stochastic perturbation to the equations. The second paradigm is that of deterministic modulation owing to the presence of nonlinearities in the dynamo equations (Zeldovich *et al.* 1983; Tobias 2002; Weiss & Tobias 2007). In this scenario the modulation of the basic cycle arises as a natural consequence of the quadratic Lorentz force and occurs with a well-defined mean period. Distinguishing between these paradigms is not possible using solely the short sunspot record. However the presence of a well-defined 200 year period for recurrent grand minima in the proxy ^{10}Be and ^{14}C data is highly suggestive that the modulation is indeed deterministic in origin. In the rest of this section we explain how deterministic modulation of solar magnetic activity is a natural consequence of including a dynamic nonlinearity in solar dynamo models.

It has become apparent, through a large number of numerical investigations of nonlinear solar dynamo models, that there is a robust mechanism that leads to modulation of basic cyclic activity. This mechanism is found in a wide range of models based on different assumptions, and with varying degrees of complexity, as outlined below. Here we first describe the sequence of bifurcations that underlies the transitions in mean-field dynamo models as the dynamo number D (a non-dimensional measure of the rotation rate of the star) is increased, eventually resulting in a transition to chaotically modulated solutions. As already explained, kinematic solar dynamo models may be constructed where a transition to large-scale dynamo action occurs at an oscillatory (Hopf) bifurcation as D is increased past a critical value. It can be shown (see, for example, Knobloch 1994) that in a rotating system the initial bifurcation to large-scale dynamo action is expected to be oscillatory. As noted above, for standard mean-field solar dynamo models the period of the oscillatory mode is set by the values of the turbulent transport coefficients (α and β) whilst for advection dominated dynamos it is set by the strength of the meridional flow. This oscillatory dynamo solution remains the preferred solution until, as D is further increased, it loses stability in a secondary Hopf bifurcation to a doubly-periodic solution with trajectories that lie on a two-torus in phase space. As we shall see, this transition is generic and arises owing to nonlinear interactions, either between magnetic dynamo modes of different symmetry or between dynamo modes and zonal shear flows. As the rotation rate is increased further the doubly periodic solution can then disappear as the torus breaks down and there is a transition to chaos. This transition is accompanied by frequency-locking and the appearance of resonant solutions, and by subsequent period doubling bifurcations.

This bifurcation structure was first put forward as generic for nonlinear solar and stellar dynamos by Tobias *et al.* (1995; see also Wilmot-Smith *et al.* 2005), who constructed a third-order dynamo model based on normal form theory. This model demonstrated that modulation and a transition to chaos arise naturally through the interaction between the magnetic field and the flows driven by the quadratic Lorentz force. Here the modulation appears owing to the continual exchange of energy between the dynamo generated field and the resulting zonal flows. A related approach based on constructing normal form equations using the underlying symmetries of the dynamo equations about the solar equator also led to the construction of a low-order dynamo model that exhibits a similar sequence of bifurcations (Knobloch & Landsberg 1996). In that case the energy exchange between modes with dipole and quadrupole symmetry provides an effective modulational mechanism. The two approaches can be combined to yield a model capable of undergoing modulation using either (or both) of these mechanisms. The competition between these two mechanisms has been investigated in detail (Knobloch *et al.* 1998; see also Ashwin *et al.* 2004). Indeed this competition yields an interesting new effect: the magnetic field may enter a minimum of activity with one symmetry and emerge after flipping to another. The minimum can therefore act as a potential trigger for a change in parity of the magnetic field (Beer *et al.* 1998). These models, although instructive, nevertheless leave many questions unanswered: what, for instance, determines the period of the modulation, or the dependence of the depth of the minimum on parameters?

The possible modulational processes are best understood by analysing the results of numerical models of the dynamo process. These range in complexity from low-order models based on truncations of the appropriate partial differential equations (Weiss *et al.* 1983; Jones *et al.* 1984) to nonlinear mean-field models in Cartesian and spherical domains (e.g. Brandenburg *et al.* 1989; Tobias 1996a, 1997b; Küker *et al.* 1999; Pipin 1999; Markiel 1999; Brooke *et al.* 2002; Bushby 2005). Such models demonstrate that the transition to chaos and the presence of grand minima are always mediated by the presence of a secondary Hopf bifurcation. As in the low-order models described above, modulation can arise owing either to the interaction between dynamo modes of different symmetry or to the interaction between a dynamo mode and a zonal flow driven by the Lorentz force. In the former case the modulation takes the form of changes in the symmetry of the dynamo solutions about the equator as the two interacting dynamo modes (dipolar and quadrupolar) move in and out of phase, on a timescale longer than that for the basic dynamo cycle. For this type of modulation the timescale for the modulation is associated with a 'beat frequency' between the two kinematic frequencies for the dipole and quadrupole modes (Zeldovich *et al.* 1983; Brandenburg *et al.* 1989). That is not, however, the form of modulation that is observed in solar magnetic activity, where the amplitude of the dynamo-generated field is modulated with little significant change in the symmetry of the solutions (except when the Sun was emerging from the Maunder minimum, as mentioned earlier). This type of modulation naturally arises when the dynamo modes interact with the zonal flows, as shown by the example in Figure 13.10 for the spherical interface dynamo model of Bushby (2005). Here the nonlinear dynamics takes the form of a relaxation oscillation. The combination of differential rotation and α-effect generates a magnetic field, which in turn drives a zonal flow via the Lorentz force. This zonal flow then acts so as to turn off the generating mechanism through modifying the differential rotation and results in a modulation of the amplitude of the dynamo cycle. The timescale for the modulation is set by the response time of the zonal shear flow to the driving by the Lorentz force. In a mean-field model this response time is set by the level of the angular momentum transport in the model. In particular, for simple models where the only transport is due to turbulent diffusion (of angular momentum and magnetic field) the ratio of the period of the modulation to that of the cycle is controlled by the ratio of the turbulent diffusivities (ν_T/η_T). If the ratio of these two diffusivities is small then the modulational period can be significantly longer than the cycle period. More generally, modulation of the basic cycle requires a phase lag

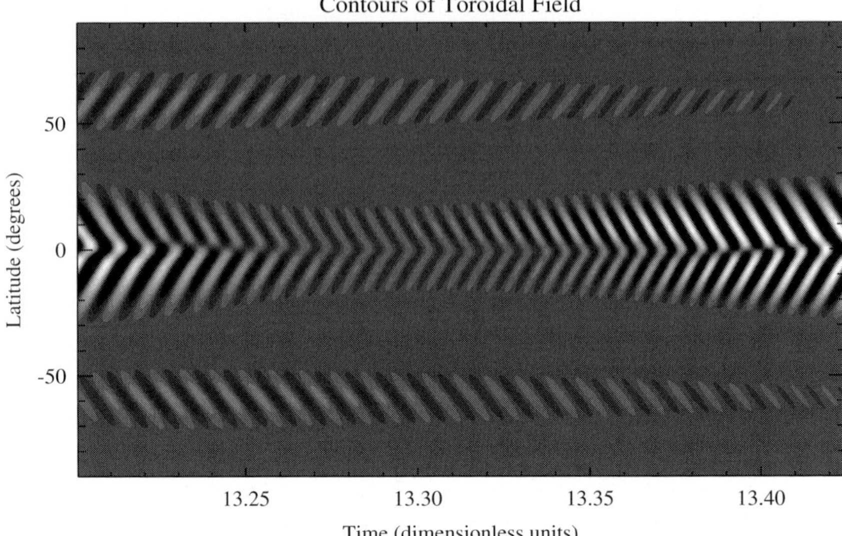

Figure 13.10. Butterfly diagram showing mildly modulated cyclic activity in a spherical model of an interface dynamo. (Courtesy of P. J. Bushby.)

between the generation and back-reaction mechanisms (Yoshimura 1978), which is a natural consequence of models that include dynamic nonlinearities. We claim that this phase lag is related to the structure of the tachocline.

13.6 Structure of the tachocline

In this section we present our personal opinion as to the nature of the tachocline and its relation to the solar dynamo. We recognize that our views are speculative and that there are other possible scenarios (see, for instance, Gilman 2005), some of which are discussed elsewhere in this book. Our picture is encapsulated in Figure 13.11, which shows a schematic cross-section of the tachocline, which we define as the region of strong radial shear in angular velocity. This region, indicated in the figure, extends from above the base of the adiabatically stratified convection zone down into the radiative zone with a total thickness of not more than 30 Mm, as determined by helioseismic inversions (see Chapter 3).

The uppermost layers of the tachocline are characterized by the presence of turbulent convective motions (as indicated in the diagram) not only in the superadiabatically stratified region above $R = 0.713R_\odot$ but also in the underlying layer of convective overshoot and penetration that is believed to extend downwards for a fraction of a pressure scale height. Owing to the presence of convective motions, the dynamics occurs on a fast timescale. Below the tachocline lies the radiative

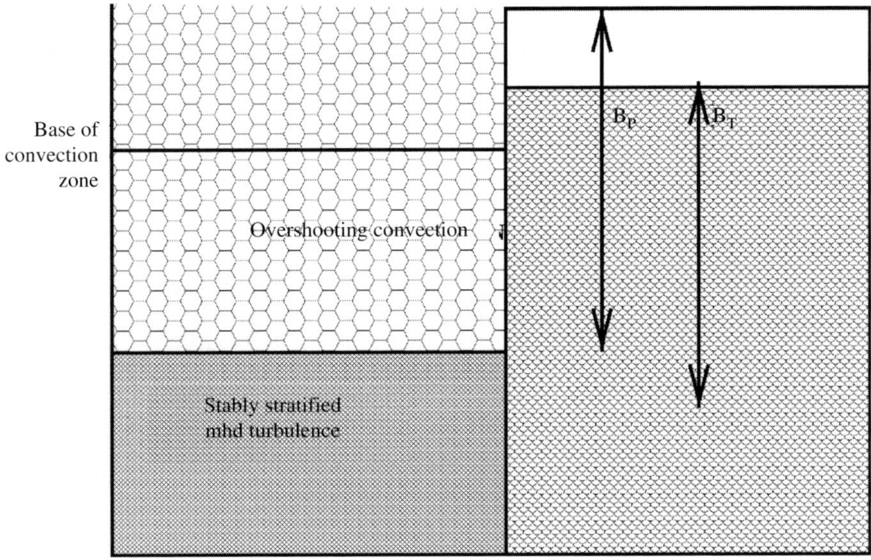

Figure 13.11. Sketch indicating the relationship between the dynamo and the tachocline. The shaded region on the right is the tachocline itself (the layer of rotational shear). The hexagons on the left denote both the superadiabatically stratified convection zone and the region of convective overshoot or penetration. Beneath them is a stably stratified region with weak turbulence driven by MHD instabilities. Below that is the quasistatic, uniformly rotating, radiative interior, with a primordial magnetic field. Poloidal fields ($\mathbf{B}_\mathrm{P}$) are generated near the top of the tachocline and diffuse downwards, while the alternating toroidal field ($\mathbf{B}_\mathrm{T}$) is produced by differential rotation in the upper convecting part of the tachocline and leaks into the stably stratified region below.

interior with a primordial magnetic field that is strong enough to enforce approximately solid body rotation on an evolutionary timescale. Sandwiched between these two regions is a layer that is stably stratified, but liable to MHD instabilities that develop into non-convective turbulence. In this layer there may also be meridional flows, as envisaged by Gough & McIntyre (1998). We believe that the tachocline is magnetically confined, and that its thickness is controlled not by the oscillatory field generated by the dynamo but by the relatively steady field down below – see the discussion by Garaud in Chapter 7.

The solar dynamo relies on motion within the tachocline and the convection zone. Whilst the dynamo itself may be influenced by a meridional circulation in the convection zone this need not modify the properties of the tachocline. The poloidal field $\mathbf{B}_\mathrm{P}$ is pumped down towards the base of the region of convective penetration, as indicated in the figure. The strong radial shear will then generate a toroidal field $\mathbf{B}_\mathrm{T}$ in the lower convection zone and the penetrative region. This field will also leak into the stably stratified layer below. As this field is oscillatory, we expect the interfaces between these sub-layers to wobble up and down during the solar cycle (Spiegel & Weiss 1980). In addition the Lorentz force associated with the magnetic fields in the region where $\mathbf{B}_\mathrm{P}$ and $\mathbf{B}_\mathrm{T}$ coexist will drive zonal shear flows with half the period of the cycle.

In Section 13.5 above, we ascribed the modulation of the dynamo to a time-lag between the generation of the magnetic field and the response of the differential rotation in the tachocline. Our picture provides a natural mechanism for this time-lag. In addition to the oscillatory component of the shear flow driven by the Lorentz force, there is a steady component. This magnetically driven zonal flow will leak into the stably stratified lower layer of the tachocline on a timescale longer than that of the solar cycle, allowing slow variations in differential rotation. In this context it appears significant that the Sun's equatorial rotation rate decreased by 2% during the Maunder Minimum, while its variation with latitude was enhanced (Ribes & Nesme-Ribes 1993).

Finally, we should comment on how these considerations relate to magnetic fields in other late-type stars. Slowly rotating stars with deep convective envelopes may be expected to have tachoclines and to behave like the Sun, though it does not follow that their magnetic fields will have dipolar symmetry. For rapidly rotating stars we should expect a different pattern of non-uniform rotation, with Ω constrained to be uniform on cylindrical surfaces in accordance with the Taylor–Proudman theorem; if there is a tachocline it is likely to be contained within the tangent cylinder that encloses the radiative core (Bushby 2003). In fully convective stars the dynamo process must, however, be completely different.

References

Abbett, W. P., Fisher, G. H., Fan, Y. & Bercik, D. J. (2004). *Astrophys. J.*, **612**, 557.

Ashwin, P., Rucklidge, A. M. & Sturman, R. (2004). *Physica*, **612D**, 557.

Babcock, H. W. (1961). *Astrophys. J.*, **133**, 572.

Baliunas, S. L., Donahue, R. A., Soon, W. H. *et al.* (1995). *Astrophys. J.*, **438**, 269.

Beck, J. G., Gizon, L. & Duvall, T. L. (2004). *Astrophys. J.*, **575**, L47.

Beer, J., Joos, F., Lukasczyk, C. *et al.* (1994). In *The Solar Engine and its Influence on Terrestrial Atmosphere and Climate*, ed. E. Nesme-Ribes. Berlin: Springer, p. 221.

Beer, J., Tobias, S. M. & Weiss, N. O. (1998). *Sol. Phys.*, **181**, 237.

Belvedere, G., Pidatella, R. M. & Proctor, M. R. E. (1990). *Geophys. Astrophys. Fluid Dyn.*, **51**, 263.

Blackman, E. G. & Brandenburg, A. (2002). *Astrophys. J.*, **579**, 359.

Brandenburg, A. (2005). *Astrophys. J.*, **625**, 539.

Brandenburg, A., Krause, F., Meinel, R., Moss, D. & Tuominen, I. (1989). *Astron. Astrophys.*, **213**, 411.

Brandenburg, A. & Schmitt, D. (1998). *Astron. Astrophys.*, **338**, L55.

Brandenburg, A. & Subramanian, K. (2005). *Phys. Rep.*, **417**, 1.

Braun, D. C. & Fan, Y. (1998). *Astrophys. J.*, **508**, L105.
Brooke, J., Moss, D. & Phillips, A. (2002). *Astron. Astrophys.*, **395**, 2002.
Brummell, N. H., Hurlburt, N. E. & Toomre, J. (1998). *Astrophys. J.*, **493**, 955.
Brun, A. S., Miesch, M. S. & Toomre, J. (2004). *Astrophys. J.*, **614**, 1073.
Bushby, P. J. (2003). *Mon. Not. Roy. Astron. Soc.*, **342**, L15.
Bushby, P. J. (2005). *Astron. Nachr.*, **326**, 218.
Caligari, P., Moreno-Insertis, F. & Schüssler, M. (1995). *Astrophys. J.*, **441**, 886.
Caligari, P., Schüssler, M. & Moreno-Insertis, F. (1998). *Astrophys. J.*, **502**, 481.
Cally, P. S. (2000). *Sol. Phys.*, **194**, 189.
Cattaneo, F. (1999). *Astrophys. J.*, **515**, L39.
Cattaneo, F., Brummell, N. H. & Cline, K. S. (2006). *Mon. Not. Roy. Astron. Soc.*, **365**, 727.
Cattaneo, F. & Hughes, D. W. (1996). *Phys. Rev. E*, **54**, R4532.
Cattaneo, F. & Hughes, D. W. (2006). *J. Fluid Mech.*, **553**, 401.
Cattaneo, F. & Vainshtein, S. (1991). *Astrophys. J.*, **376**, L21.
Chan, K. H., Liao, X., Zhang, K. & Jones, C. A. (2004). *Astron. Astrophys.*, **423**, L37.
Charbonneau, P.(2005). *Living Rev. Solar Phys.*, **2**,2 (www.livingreviews.org/lrsp-2005-2).
Choudhuri, A. R. (2003). In *Dynamic Sun*, ed. B. N. Dwivedi. Cambridge: Cambridge University Press, p. 103.
Cline, K. S. (2003). *The Formation and Evolution of Magnetic Structures in the Solar Interior*. Ph.D. Thesis, University of Colorado.
Cline, K. S., Brummell, N. H. & Cattaneo, F. (2003). *Astrophys. J.*, **588**, 630.
Courvoisier, A., Hughes, D. W. & Tobias, S. M. (2006). *Phys. Rev. Lett.*, , **96**, 034503.
Covas, E., Moss, D. & Tavakol, R. (2004). *Astron. Astrophys.*, **416**, 775.
Diamond, P. H., Hughes, D. W. & Kim, E. (2005a). In *Fluid Dynamics and Dynamos in Astrophysics and Geophysics*, ed. A. M. Soward, C. A. Jones, D. W. Hughes & N. O. Weiss. London: CRC Press, p. 145.
Diamond, P. H., Itoh, S.-I., Itoh, K. & Hahm, T. S. (2005b). *Plasma Phys. Cont. Fusion* **47**(5), R35.
Dikpati, M. & Charbonneau, P. (1999). *Astrophys. J.*, **518**, 508.
Dikpati, M., Corbard, T., Thompson, M. J. & Gilman, P. A. (2002). *Astrophys. J.*, **575**, L41.
Dikpati, M., de Toma, G., Gilman, P. A., Arge, C. N. & White, O. R. (2004). *Astrophys. J.*, **601**, 1136.
Dorch, S. B. F. & Nordlund, Å. (2001). *Astron. Astrophys.*, **365**, 562.
Durrant, C. J., Turner, J. P. R . & Wilson, P. R. (2004). *Sol. Phys.*, **222**, 345.
Duvall, T. L. & Kosovichev, A. G. (2001). In *Recent Insights into the Physics of the Sun and Heliosphere (IAU Symp. No. 303)*, ed. P. Brekke, B. Fleck & J. B. Gurman (Astron. Soc. Pac., San Francisco), p. 159.
Emonet, T. & Moreno-Insertis, F. (1998). *Astrophys. J.*, **492**, 804.
Fan, Y. (2004). *Living Rev. Solar Phys.*, **1**, 1 (www.livingreviews.org/lrsp-2004-1).
Fan, Y., Zweibel, E. G. & Lantz, S. R. (1998). *Astrophys. J.*, **493**, 480.
Fan, Y., Abbett, W. P. & Fisher, G. H. (2003). *Astrophys. J.*, **582**, 1206.
Ferriz-Mas, A. & Schüssler, M. (1993). *Geophys. Astrophys. Fluid Dyn.*, **72**, 209.
Ferriz-Mas, A., Schmitt, D. & Schüssler, M. (1994). *Astron. Astrophys.*, **289**, 949.
Gilman, P. A. (2005). *Astron. Nachr.*, **326**, 208.
Glatzmaier, G. A. & Roberts, P. H. (1995). *Nature*, **377**, 203.
Golub, L., Rosner, R., Vaiana, G. S. & Weiss, N. O. (1981). *Astrophys. J.*, **243**, 309.
Gough, D. O. & McIntyre, M. E. (1998). *Nature*, **394**, 755.
Gruzinov, A. V. & Diamond, P. H. (1994). *Phys. Rev. Lett.*, **72**, 1651.

Gruzinov, A. V. & Diamond, P. H. (1995). *Phys. Plasmas*, **2**, 1941.
Hoyng, P. (1988). *Astrophys. J.*, **332**, 857.
Hoyt, D. V. & Schatten, K. H. (1998). *Sol. Phys.*, **179**, 189; **181**, 491.
Hughes, D. W. (2007). In *Mathematical Aspects of Natural Dynamos*, ed. E. Dormy & A. M. Soward, to be published.
Hughes, D. W. & Falle, S. A. E. G. (1998). *Astrophys. J.*, **509**, L75.
Hughes, D. W. & Tobias, S. M. (2001). *Proc. R. Soc. Lond.*, A**457**, 1365.
Hughes, D. W., Falle, S. A. E. G & Joarder, P. (1998). *Mon. Not. Roy. Astron. Soc.*, **298**, 433.
Jepps, S. A. (1975). *J. Fluid Mech.*, **67**, 625.
Jones, C. A., Weiss, N. O. & Cattaneo, F. (1984). *Physica*, **14D**, 161.
Kitchatinov, L. L., Rüdiger, G. & Küker, M. (1994). *Astron. Astrophys.*, **292**, 125.
Kleeorin, N. & Ruzmaikin, A. A. (1982). *Magnetohydrodynamics*, **18**, 116.
Kleeorin, N., Moss, D., Rogachevskii, I. & Sokoloff, D. (2000). *Astron. Astrophys.*, **361**, L5.
Kleeorin, N., Rogachevskii, I. & Sokoloff, D. (2001). *Phys. Rev. E*, **65**, 36303.
Knobloch, E. (1994). In *Lectures on Solar and Planetary Dynamos*, ed. M. R. E. Proctor & A. D. Gilbert. Cambridge: Cambridge University Press, p. 331.
Knobloch, E. & Landsberg, A. S. (1996). *Mon. Not. Roy. Astron. Soc.*, **278**, 294.
Knobloch, E., Tobias, S. M. & Weiss, N. O. (1998). *Mon. Not. Roy. Astron. Soc.*, **297**, 1150.
Krause, F. & Rädler, K.-H. (1980). *Mean-field Magnetohydrodynamics and Dynamo Theory*. Berlin: Akademie-Verlag.
Küker, M., Arlt, R. & Rüdiger, G. (1999). *Astron. Astrophys.*, **343**, 977.
Leighton, R. B. (1969). *Astrophys. J.*, **156**, 1.
Markiel, J. A. (1999). *Astrophys. J.*, **523**, 827.
Mason, J., Hughes, D. W. & Tobias, S. M. (2002). *Astrophys. J.*, **580**, L89.
Malkus, W. V. R. & Proctor, M. R. E. (1975). *J. Fluid Mech.*, **67**, 417.
Matthews, P. C., Hughes, D. W. & Proctor, M. R. E. (1995). *Astrophys. J.*, **448**, 938.
Mestel, L. (1999). *Stellar Magnetism*. Oxford: Clarendon Press.
Miesch, M. S. (2005). *Living Rev. Solar Phys.*, **2**, 1 (www.livingreviews.org/lrsp-2005-1).
Moffatt, H. K. (1978). *Magnetic Field Generation in Electrically Conducting Fluids*. Cambridge: Cambridge University Press.
Moffatt, H. K. (1983). *Rep. Prog. Phys.*, **46**, 621.
Nordlund, Å., Brandenburg, A., Jennings, R. L. *et al.* (1992). *Astrophys. J.*, **392**, 647.
Ossendrijver, M. A. J. H. (2000). *Astron. Astrophys.*, **359**, 364.
Ossendrijver, M. A. J. H. (2003). *Astron. Astrophys. Rev.*, **11**, 287.
Ossendrijver, M. A. J. H., Stix, M. & Brandenburg, A. (2001). *Astron. Astrophys.*, **376**, 713.
Ossendrijver, M. A. J. H., Stix, M., Brandenburg, A. & Rüdiger, G. (2002). *Astron. Astrophys.*, **394**, 735.
Parker, E. N. (1955a). *Astrophys. J.*, **121**, 491.
Parker, E. N. (1955b). *Astrophys. J.*, **122**, 293.
Parker, E. N. (1975). *Astrophys. J.*, **198**, 205.
Parker, E. N. (1979). *Cosmical Magnetic Fields*. Oxford: Clarendon Press.
Parker, E. N. (1993). *Astrophys. J.*, **408**, 707.
Phillips, A., Brooke, J. & Moss, D. (2002). *Astron. Astrophys.*, **392**, 713.
Pipin, V. V. (1999). *Astron. Astrophys.*, **346**, 295.
Rädler, K.-H. (1968). *Zeits. Naturforsch.*, **23**, 1851.
Ribes, J. C. & Nesme-Ribes, E. (1993). *Astron. Astrophys.*, **276**, 549.

Roberts, P. H. (1994). In *Lectures on Solar and Planetary Dynamos*, ed. M. R. E. Proctor & A. D. Gilbert. Cambridge: Cambridge University Press, p. 1.
Rüdiger, G. & Hollerbach, R. (2004). *The Magnetic Universe*. Weinheim: Wiley-VCH.
Ruzmaikin, A. A. (1998). *Sol. Phys.*, **181**, 1.
Ruzmaikin, A. A. (2000). *Sol. Phys.*, **192**, 49.
Saar, S. H. & Brandenburg, A. (1999). *Astrophys. J.*, **524**, 295.
Schrijver, C. J. & Zwaan, C. (2000). *Solar and Stellar Magnetic Activity*. Cambridge: Cambridge University Press.
Schüssler, M. (2005). *Astron. Nachr.*, **326**, 194.
Simon, G. W., Title, A. M. & Weiss, N. O. (1995). *Astrophys. J.*, **442**, 886.
Spiegel, E. A. & Weiss, N. O. (1980). *Nature*, **287**, 616.
Spruit, H. C., Nordlund, Å. & Title, A. M. (1990). *Ann. Rev. Astron. Astrophys.*, **28**, 263.
Stix, M. (1974). *Astron. Astrophys.*, **37**, 121.
Stix, M. (2002). *The Sun*, 2nd edition. Berlin: Springer.
Stuiver, M. & Braziunas, T. F. (1993). *Holocene,* **3**, 289.
Tao, L., Proctor, M. R. E. & Weiss, N. O. (1998). *Mon. Not. Roy. Astron. Soc.*, **300**, 907.
Thelen, J.-C. (2000a). *Mon. Not. Roy. Astron. Soc.*, **315**, 155.
Thelen, J.-C. (2000b). *Mon. Not. Roy. Astron. Soc.*, **315**, 165.
Thompson, M. J., Christensen-Dalsgaard, J., Miesch, M. S. & Toomre, J. (2003). *Ann. Rev. Astron. Astrophys.*, **41**, 599.
Tobias, S. M. (1996a). *Astron. Astrophys.*, **307**, L21.
Tobias, S. M. (1996b). *Astrophys. J.*, **467**, 870.
Tobias, S. M. (1997a). *Geophys. Astrophys. Fluid Dyn.*, **86**, 287.
Tobias, S. M. (1997b). *Astron. Astrophys.*, **322**, 1007.
Tobias, S. M. (2002). *Astron. Nachr.*, **323**, 417.
Tobias, S. M. (2005). In *Fluid Dynamics and Dynamos in Astrophysics and Geophysics*, ed. A. M. Soward, C. A. Jones, D. W. Hughes & N. O. Weiss. London: CRC Press, p. 193.
Tobias, S. M. & Hughes, D. W. (2004). *Astrophys. J.*, **603**, 785.
Tobias, S. M., Weiss, N. O. & Kirk, V. (1995). *Mon. Not. Roy. Astron. Soc.*, **273**, 1150.
Tobias, S. M., Brummell, N. H., Clune, T. L. & Toomre, J. (1998). *Astrophys. J.*, **502**, L177.
Tobias, S. M., Brummell, N. H., Clune, T. L. & Toomre, J. (2001). *Astrophys. J.*, **549**, 1183.
Vainshtein, S. I. & Cattaneo, F. (1992). *Astrophys. J.*, **393**, 165.
van Ballegooijen, A. A. (1982). *Astron. Astrophys.*, **113**, 99.
Vorontsov, S. V., Christensen-Dalsgaard, J., Schou, J., Strakhov, G. N. & Thompson, M. J. (2002), *Science*, **296**, 101.
Wagner, G., Beer, J., Masarik, J. *et al.* (2001). *Geophys. Res. Lett.*, **28**, 303.
Wang, Y.-M. & Sheeley, N. R. (1991). *Astrophys. J.*, **375**, 761.
Weiss, N. O., Cattaneo, F. & Jones, C. A. (1983). *Geophys. Astrophys. Fluid Dyn.*, **30**, 305.
Weiss, N. O., Thomas, J. H., Brummell, N. H. & Tobias, S. M. (2004). *Astrophys. J.*, **600**, 1073.
Weiss, N. O. & Tobias, S. M. (2007). In *Mathematical Aspects of Natural Dynamos*, ed. E. Dormy & A. M. Soward, to be published.
Wilmot-Smith, A. L., Martens, P. C. M., Nandy, D., Priest, E. R. & Tobias, S. M. (2005). *Mon. Not. Roy. Astron. Soc.*, **363**, 1167.
Wissink, J. G., Hughes, D. W., Matthews, P. C. & Proctor, M. R. E. (2000a). *Mon. Not. Roy. Astron. Soc.*, **318**, 501.

Wissink, J. G., Matthews, P. C., Hughes, D. W. & Proctor, M. R. E. (2000b). *Astrophys. J.*, **536**, 982.
Yoshimura, H. (1978). *Astrophys. J.*, **226**, 706.
Zeldovich, Ya. B., Ruzmaikin, A. A. & Sokoloff, D. D. (1983). *Magnetic Fields in Astrophysics*. New York: Gordon & Breach.
Zhang, K., Chan, K. H., Zou, J., Liao, X. & Schubert, G. (2003). *Astrophys. J.*, **596**, 663.
Zhao, J. & Kosovichev, A. G. (2004). *Astrophys. J.*, **603**, 776.

Part VII

Overview

14

On studying the rotating solar interior

Robert Rosner

14.1 Afterthoughts

Even the most casual of readers of this book will have noticed that the subject of the solar tachocline is highly controversial, in the best traditions of our science: we are all well aware that the tachocline constitutes an important physical structure in the solar interior, but we are not at all in agreement about any of the details. While this makes for a good deal of excitement – much in evidence both at the workshop and in this book – I did early on recognize that a straightforward summary of the workshop was therefore an impossibility; and my strong belief is that it is very premature for me to act as a 'referee' judging the merits of the various points of view expressed by my co-authors of this volume. This does not mean of course that I will not venture an opinion when appropriate – but it does mean that, in many cases, *ex cathedra* declarations of what is correct, and what is incorrect, are entirely premature.

For these reasons, I thought it would be more appropriate for me to step back from the fray, and to discuss some of the larger issues related to the tachocline, most especially those that I believe will play a key role in further developments of this subject; and to explain, whenever appropriate, why exactly it is that a definitive result remains to be obtained. As an aside, I should note the peculiar nature of the tachocline: as it is a boundary layer, it occupies (in terms of volume) an insignificant portion of the solar interior; but, precisely because it is a boundary layer, it appears to be the key to understanding a remarkably broad set of solar physics problems, from the differential rotation of the solar convection zone proper, and spin-down of the Sun on stellar evolution timescales, and compositional mixing at the bottom of the convection zone, to the functioning of the solar magnetic dynamo.

To set the stage, I would like to note that much of what one hears and reads on this subject reminds me greatly of a wonderful quote attributed to John Kenneth Galbraith, to wit, 'Faced with the choice between changing one's mind and proving that there is no need to do so, almost everybody gets busy on the proof'.

The Solar Tachocline, D. W. Hughes, R. Rosner and N. O. Weiss.
Published by Cambridge University Press.

14.2 The key issues

To begin with, I will attempt to summarize briefly what I regard as the main issues that were uncovered during this workshop; and, as part of this discussion, I will also focus on the question of how far observations can (and do) seriously constrain these issues. This latter discussion is not simply a matter of technical detail, but is rather a problem that is pervasive in astrophysics, namely the challenge flowing from the fact that in many instances, observational constraints may be insufficient to distinguish between proposed theoretical alternatives. I make this point not to denigrate the subject, but in order to point out that there may be instances in which disputes regarding the physics of the solar interior may not be capable of being settled – even in principle – by means of solar observations; and that hence resolution may demand alternative venues, possibly including laboratory experimentation or large-scale numerical simulations, or (hopefully, in only a very small subset of instances) may not be capable of being resolved by any means that we are aware of today. These kinds of uncertainties are among the key distinguishing aspects of astrophysics (as compared to physics), and they contribute to making the pursuit of astrophysics enormously exciting and stimulating. My hope is that appreciation of this fundamental limitation will contribute significantly to avoiding needless (and pointless) arguments in those cases where no appeal to observations can possibly resolve a dispute.

14.2.1 How is the tachocline defined, and where is it?

By my count, there are at least six variants of the definition for the tachocline to be found in this book, not all entirely consistent with one another. A significant source of this inconsistency is the question of what precisely is the relationship between the tachocline, the region of convective penetration, and the overshoot region, a topic to which I will return shortly. Thus, consider the following variations on the theme of 'tachocline'.

- The tachocline is 'a thin boundary layer separating the quiescent radiative region from the overlying turbulent convective region' (Chapter 1).
- The tachocline is 'a transition between convection zone and radiative interior' (Chapter 3).
- The tachocline is 'a shallow layer connecting the regimes of differential rotation above and quasi-uniform rotation below' (Chapter 4).
- The tachocline is 'a strong shear layer beneath the convective region' (Chapter 8).
- The tachocline is 'a region which operates the dynamical transition between the convection zone and the radiative zone' (Chapter 7).
- The tachocline is 'a region of strong radial shear in angular velocity' (Chapter 13).

All of these definitions are clearly closely related and, in addition, sound (gratifyingly) similar – why am I then drawing attention to the slight differences in wording? The reason is that while some focus on the tachocline as a transition layer between convection zone and radiative interior, others emphasize its role as a shear boundary layer (presumably the original intent of Ed Spiegel and his co-conspirators); and it is this difference in emphasis that has led to some apparent confusion about the relative position of the tachocline (viewed as a shear boundary layer) and the thermodynamic radial structuring of the solar interior (e.g. radial positions of the convective zone proper, the convective penetrative region, the region of convective overshoot, and the radiative interior proper).

If we could ignore magnetic fields, then the issue is easily settled: the tachocline should then be viewed simply as a velocity shear boundary layer, separating the differentially rotating convective zone and underlying thermal boundary layer (the subadiabatic convective overshoot region) from the fundamentally uniformly rotating radiative interior. It is important to be clear about the basic reason for this simplicity: under the posited conditions, there is no specific physical mechanism that provides feedback from the shear flows in the tachocline to the processes that govern the thermal structure of the convection zone/radiative interior interface. It is the physical processes that govern energy transport in the convection zone/radiative interior interface that drive the physics of the shear layer (and not the other way around).

Unfortunately, we cannot ignore magnetic fields, and this simplicity melts away as a result. Let's first ask why we cannot ignore magnetic fields. I see at least two distinct reasons: first, and most obviously, we know that the solar convection zone is 'magnetized', and it would be a remarkable feat if magnetic fields in this highly turbulent region managed to be excluded from the tachocline, which presumably is at least adjacent to, if not overlapping with, the bottom of the convection zone. Indeed, the work of Nigel Weiss and his collaborators on turbulent pumping of magnetic fields suggests strongly that relatively quiescent regions adjacent to turbulent magnetized flows will inevitably experience intrusion of magnetic fields expelled from the turbulent regions. Second, virtually all of the discussions in this book of the radial confinement of the tachocline conclude that suppression of the spread of this shear layer into the radiative interior on solar evolutionary time scales can only be understood if the radiative interior has a non-trivial magnetic field. What difference does the magnetic field make? The key difference follows from the fact that the Lorentz force can modify the extant velocity fields, thus providing the missing feedback mechanism connecting the shear boundary layer to the processes governing energy transport in this region. In other words, the very fields that are called upon to confine the tachocline may also play a role in modifying energy transport in the convective overshoot (and possibly the penetrative convection) layer.

In order to continue the discussion, it is at this point useful to mention a slight complication of the picture sketched by Jean-Paul Zahn of the convection zone/radiative interior boundary region (Chapter 4). In one-dimensional stellar evolution models of the solar interior, the position of the base of the convection zone (z_i in Zahn's notation) is fixed by the criterion that the radiative temperature gradient equals the adiabatic temperature gradient (the Schwarzschild criterion). Now, if one takes the resulting one-dimensional model star (i.e. the run of temperature, pressure and composition with radius) as input to a three-dimensional hydrodynamic code that models the same physics as the one-dimensional version, one finds that the radial position of the point at which the radiative flux equals the total flux moves slightly radially inward; this movement is the result of convective erosion, and means that the resulting value of z_i in the three-dimensional case is not exactly equal to that in the one-dimensional case. Thus, region A in Figure 4.1 corresponds to a slightly enlarged convectively unstable region (i.e. where $\mathrm{d}\ln T/\mathrm{d}\ln P$ is slightly superadiabatic).

Now, let us compare the descriptions of the convection zone/radiative interior interface provided by Zahn and by Pascale Garaud (Chapter 7), the latter fundamentally based on the description first offered by Gough & McIntyre (1998): Zahn's region A is clearly the same as Garaud's region 1; and region B appears to be identical to region 2 (this is the stably stratified, but almost adiabatic, convective penetration region); however, Garaud's region 3 (her 'tachopause', or magnetic field boundary layer) appears to be quite distinct from Zahn's very thin region C (this is the transition region in which buoyancy braking of down-streaming plumes occurs and the temperature gradient changes from almost adiabatic, or slightly subadiabatic, to radiative)[1]; and, finally, Zahn's region D is clearly the same as Garaud's region 4. Now, from the helioseismic perspective, one cannot easily (if at all) distinguish Zahn's regions A and B (or Garaud's regions 1 and 2): hence, as far as observations are concerned, the base of the convection zone is not at z_i, but rather at the location of the interface between regions B and C.

So, where is the tachocline located? It seems that the answer is very much tied to the physics that a particular worker is focusing on. Thus, Zahn's discussion seems to have the tachocline largely confined to the uppermost region of the radiative interior and possibly in the overshoot layer, i.e. in the stably stratified regions; there is very little said about the extent of the tachocline into the overlying almost adiabatic regions. I suspect the reason is that Zahn's focus is on anisotropic (largely two-dimensional) turbulence, which may have the important property of suppressing latitudinal shear. It seems, however, that Zahn's model is really a 'straw-man' and 'best effort' case for a purely hydrodynamic (i.e. non-magnetic) model for

[1] This region is what Zahn regards as the overshoot region, a view I fully concur with.

the tachocline, and thus does not represent his own realistic assessment of the tachocline structure – indeed, he points out in Chapter 4 that magnetic fields are clearly likely to have significant effects on tachocline radial confinement.

In contrast, Garaud's physics focuses on the possibility that magnetic fields in the radiative interior are responsible for the radial confinement of the tachocline. This focus has two consequences. First, one needs to explain why the fields in the radiative interior have not themselves emerged into the convection zone; the solution to this problem was provided by Gough & McIntyre in the form of large-scale meridional flows driven by the convection zone, which confine the magnetic field to the radiative interior. Region 3, the tachopause or magnetic diffusion layer, is a consequence of this large-scale flow. So where is the tachocline in this model? The answer appears to be in regions 2 and 3, which are the regions identified with the tachocline ventilation depth. (I note as an aside that I am uncertain whether Garaud's use of the term 'overshoot region' is really the same as Zahn's: the discussion in Section 7.2.2 suggests that what is meant in that context is what Zahn refers to as convective penetration.)

Finally, it is instructive to compare all this with the physical picture sketched by Steven Tobias and Nigel Weiss (Chapter 13), which focuses (unlike the previous two cases) on the effects of magnetic fields in the overlying convection zone (and thus makes the critical connection to the solar dynamo). Here again there is a slight clash of nomenclature – I presume that the region marked 'overshooting convection' in Figure 13.11 is identical with the upper portion of Zahn's region B, i.e. the convective penetration zone, while Zahn's overshoot layer is presumably entirely absent here.[2] What about the region marked 'stably stratified MHD turbulence'? This region appears to correspond to the lower part of Zahn's weakly stably stratified region B, a point that is made explicitly by Tobias & Weiss. Most interesting is the location of the tachocline in this picture – in this case, the tachocline is mostly contained within the well-mixed boundary layer at the base of the convection zone, does not overlap to any significant extent with the strongly stably stratified radiative interior, and slightly overlaps with the lower part of the convection zone (i.e. its upper boundary lies slightly above z_i in Zahn's sketch).

Can these various alternatives be distinguished observationally? It would seem that the answer depends on whether one can measure the relative positions of (a) the transition from almost adiabatic to strongly subadiabatic (radiative) stratification, and (b) the lower boundary of the shear layer defining the tachocline.

[2] This inference is based on the observation that Zahn's overshoot layer – region C – is extremely thin, and thus cannot possibly play the role indicated by the 'overshooting convection' layer shown in Figure 13.11; indeed, it is hard to see how that layer would lead to any significant dynamical consequences within the context of the dynamo model sketched out by Tobias & Weiss, other than to form a rather stiff lower boundary that acts as a barrier to vertical motions attempting penetration into the radiative interior.

Based on the error estimates for these two measurements (see immediately below), I believe that the answer currently is 'no'; but I confess that I do not have a firm understanding of how much improvement can yet be made in the helioseismic radial position estimates for the tachocline. (There are estimates in the published literature that suggest that such a comparison is currently feasible, but given the error estimates discussed in Chapter 3, I am somewhat dubious about these claims.)

14.2.2 How thin is the tachocline, and why is it thin?

It is clear that the observational upper bound on the radial extent of the tachocline has been an important constraint on theoretical models of this layer, starting from the early work of Spiegel and his co-workers. The fundamental physics issue is of course that a variety of hydrodynamic mechanisms can be put forward that all lead to what would be today a thickening of this transition layer that should be easily detected by modern helioseismic measurements – but the measured upper bounds on the thickness clearly indicate that, whether or not any of these mechanisms are in fact operating, there is some other process not yet accounted for that limits the spread of this shear layer.

Now, the alert reader may have noticed a closely related, and intriguing, *apparent* contradiction in Jørgen Christensen-Dalsgaard and Michael Thompson's masterful article on helioseismic measurements of the interior rotational structure of the Sun (Chapter 3): at one point, they rightly point out that 'Since discovering the tachocline, helioseismology has pinned down with reasonable precision its location and thickness in the radial direction...'; but shortly following, also state that the '... tachocline is thinner than the intrinsic resolution of present-day inversions...'. How can this be? The answer of course resides in the method(s) used to constrain the tachocline thickness. That is, the inversion methods used by helioseismology are designed so as to minimize the number of assumptions required to obtain sensible solutions – for example, these assumptions may be in the nature of 'regularizations' of the inversions applied to the formal Fredholm integral equation of the second kind that lies at the heart of the helioseismic problem. Such assumptions are usually not based on specific expectations regarding the physics, but rather derive from expectations of 'natural' properties of the solar interior rotational profile, such as smoothness of the solution. The practitioners of this art have wisely used a variety of methods (and hence a variety of regularization techniques) for these inversions; and the consensus upper bound on the tachocline thickness derives from comparisons of the results of these various inversion methods – and it is fortunately gratifying that, by and large, there are no large discrepancies between these various upper bounds.

Now, it is useful to recall the key point made by Christensen-Dalsgaard & Thompson that the maximum spatial resolution obtainable by helioseismic inversions is fundamentally limited by the functional properties of the acoustic eigenfunctions, a point that is best illustrated in their discussion of optimally localized averages (OLA). One way of thinking about this process of extracting the spatial structure of the solar interior rotation rate is from the information-theoretic perspective, for the helioseismic inversion problem is subject to the Nyquist–Shannon sampling theorem. Thus, one can view the inversion process as a functional mapping from the space of acoustic eigenfunctions to the space of real-valued functions $\Omega(r, \theta, \phi)$ on $\mathbb{R}^3$ (i.e. the solar interior rotation rate as a function of position); and this theorem defines the conditions under which this transformation preserves all of the information content of the original (helioseismic) data, so that the reverse transformation (reconstructing the helioseismic signal) precisely recovers the original input. As a consequence, this theorem prescribes the functional properties of Ω, given the information content of the original signal. Thus, 'optimal' inversions should be regarded as those inversions that extract the maximum amount of information available in the data, and therefore any attempts to extract more information – for example, to obtain stronger bounds on the radial extent of the tachocline – must perforce involve the insertion of additional information not contained in the original helioseismic data to be inverted. More specifically, this additional information is inserted in the form of specific choices for the functional radial dependence of Ω (cf. Equations (3.9) and (3.10), as well as Figure 3.9). Since these choices have no physical basis, one can ask to what extent the 'super-resolution' radial information that is then extracted has any veracity, that is, bears any connection with the likely actual behaviour of the shear layer; and it is my view that the credibility of these enhanced inversions depends critically on the extent to which differing functional forms for Ω yield similar super-resolution results. (As noted by Christensen-Dalsgaard & Thompson, such comparisons need to take very careful account of the differences in functional parametrization of the assumed radial profile for Ω, so that one compares 'oranges with oranges'.)

Given the preceding discussion, I believe the following basic properties of the tachocline have a good chance of surviving further scrutiny.

- Radial location: a variety of inversions and 'forward' analyses using *ad hoc* trial functions give a consistent position for the centre of the tachocline at $r_c \sim 0.7R_\odot$. I would conservatively estimate the error on this position to be of the order of $0.05R_\odot$, although it is to be noted that some authors claim errors that are roughly an order of magnitude smaller than that given here. These latter analyses tend to be based on 'super-resolution' methods, and it is very likely that systematic errors in these analyses are not well-accounted for; I would therefore tend to be sceptical of any claims for great certainty in tachocline positioning.

- Radial width: estimates for the tachocline width w are far more uncertain, with estimates ranging from $w \sim (0.05 - 0.09)R_{\odot}$; error estimates are similarly wide-ranging, from $\sim$20% to 50%. Again, these results are largely obtained via 'super-resolution' analyses, and hence the apparent differences in conclusions very likely reflect systematic uncertainties, especially ones related to the *ad hoc* assumed radial variation in Ω; but there is no doubt about the fact that the tachocline is 'thin', so that the ratio w/r_{c} is of order 0.1 or smaller.
- Location vis-à-vis the convection zone and radiative interior. Here it is useful to be extremely cautious, primarily because there may be, or there may not be, a distinct latitudinal variation in both the convection zone depth and the tachocline location – the analyses I have seen to date do not convince me that such variations in convection zone or tachocline depth are well in hand. For this reason, I am most comfortable with the approach taken by Tobias & Weiss in their Figure 13.11, i.e. there is good evidence that the tachocline overlaps significantly with the bottom of the convection zone (the superadiabiatic region of convective penetration), the almost-adiabatic overshoot region, and some small fraction of the subadiabatic and weakly (magneto)turbulent top of the radiative interior.
- Mechanism(s) leading to a thin tachocline: I am totally unconvinced by any of the arguments made in this volume that we understand the physics underlying the thickness of the tachocline. The reason is as follows: all of the studies to date, including all of the numerical simulations presented here, are critically constrained in one way or another; for example, there is typically no back-reaction on the 'forcing' of convective motions in the simulations. Furthermore, the effects of magnetic fields are only beginning to be understood, and certainly none of the global simulations can currently claim to have included magnetic field effects in a fully consistent manner. Furthermore, as pointed out by Zahn (Chapter 4), current simulations of the boundary layer are very diffusive, and are principally limited by the relatively small dynamic range of spatial scales that can be simulated in three dimensions. (I will return to this last point in the next section.) Thus, the problem of accounting for the small radial extent of the tachocline is a very good example of a research area in which raging arguments are currently beside the point – we simply do not have the tools available as yet to solve this problem.

14.3 Comments on modelling and simulations

Whereas the practitioners of helioseismic analyses seem to have made enormous strides in developing a commonly-accepted framework for analysing the data and interpreting the results – so that at this point there is a very good, and widely accepted, understanding of what the limits are to our knowledge of tachocline properties derived from helioseismic measurements – this is decidedly not the case in the computational arena: successive readings of Chapters 5, 6, 7, 11 and 13 make plain the huge differences in perception of what numerical studies of the solar interior in general, and of the tachocline more specifically, are all about. Because numerical studies are likely to be the dominant tool for understanding tachocline dynamics in the future, I thought it useful to make an excursion into the realm of computational fluid dynamics and astrophysics, and to discuss in some detail the key issues we face.

The first distinction to be made is between 'modelling' and 'simulating'. By and large, there is a general consensus regarding the physics needed to deal with the solar interior. That is to say, with the principal exception of the opacities (which unfortunately remain somewhat uncertain because of remaining uncertainties in elemental composition of the solar interior as a function of radius, especially immediately below the convection zone – see Chapter 3), there are no disputes regarding the equations to be solved, or the constitutive relations and the microscopic ('molecular') transport coefficients to be used. However, it is also generally well-understood that the numerical solution of these equations for spatial domains that span a significant portion of the solar interior is well beyond our current capabilities, and is likely to remain so for the lifetime of any of the participants in this workshop. By 'solution' I mean the direct integration of the full set of governing equations, with no appeal to any type of parametrization – such a calculation is what I will refer to as a 'direct numerical simulation', or DNS. The reason DNS for any significant portion of the Sun is currently impossible is closely related to the fact that the dynamic range of spatial scales in a fluid (meaning the ratio of the largest [$L_{\rm max}$] to smallest [$L_{\rm min}$] physically important scales in the problem) is related to the Reynolds number *Re* of the velocity field by

$$Re \sim 2(L_{\rm max}/L_{\rm min})^{4/3};$$

since the Reynolds number for the convection zone is extremely large, the dynamic range of spatial scales, as measured by the ratio $L_{\rm max}/L_{\rm min}$, typically will exceed 10^9, whereas the largest extant three-dimensional simulations of turbulent flows struggle to reach a dynamic range of $\sim 10^{3.5}$. Since the economies provided by adaptive gridding cannot be obtained for turbulent flows (because small scale features in the velocity field are generated everywhere in the turbulent domain), this means that while numerical solutions of the equations of motion can currently be computed on a grid composed of $\sim 10^{10.5}$ grid points, a faithful (DNS) solution of these same equations for the solar interior would require more than 10^{27} grid points. This means that if one insists on solving the exact equations of motion, the size of the domain in which one can obtain a sensible solution will be a tiny fraction of the domain of interest; if, for example, the viscous cutoff scale is of order a millimetre, then the outer scale of a corresponding DNS will have dimensions of order 3 m, i.e. an outer scale that is totally impossible to observe.

Nevertheless, it is evident that numerical computations for the solar interior are being conducted – how then should we interpret these calculations? Since the outer scales of these calculations are typically a significant fraction of the solar radius, this means that the inner scale of these calculations must be far larger than the actual governing dissipative scale (the viscous cutoff scale), and therefore there must be either an implicit or an explicit 'sub-grid' model for the physics that occurs on the unresolved scales. In this sense, such calculations resemble computational fluid dynamics (CFD) in the engineering realm, with the critical difference that whereas the sub-grid models for engineering calculations are typically extremely well verified and validated, validation is usually not possible in the case of astrophysical CFD.[3] As a consequence, these calculations are really 'models' for the solar interior, in which the adopted sub-grid dynamics must be regarded as an *ad hoc* assumption as opposed to being firmly grounded in fundamental physics. In other words, any conclusions derived from such calculations are subject to the caveat that they rest on the correctness of the unvalidated astrophysical sub-grid model.

Now, some would argue that laboratory fluid dynamics (and the related CFD) provide a hierarchy of sub-grid models with established pedigrees (the Reynolds-averaged stress models, etc.), so that the actual situation is not as bad as I have made it sound. I frankly do not agree. The reason is that virtually all extant explicit sub-grid models have been developed either in the context of incompressible fluids, or in the context of compressible fluids in which stratification is unimportant. (Simulations of the terrestrial atmosphere are at the frontier of what can be done in this context, but unfortunately most global circulation models focus attention on the CFD in the horizontal, and typically parametrize transport in the vertical – precisely what we would like to avoid in studying the tachocline.) This does not mean that all such calculations are rubbish; but it does mean that conclusions drawn from a specific calculation gain in credibility only to the extent that they can be shown to survive if alternative sub-grid model assumptions are made, i.e. if one can show explicitly that the conclusions are not sensitive to the specifics of the sub-grid model used. As a consequence, there is considerable utility in testing conclusions against drastically different sub-grid models; and careful practitioners of such calculations tend as a result to be very conservative regarding the extraction of broad conclusions from their simulations (see, for example, Chapter 5).

Even more challenging is the realm of magnetohydrodynamics (MHD). Here the challenge is that there is no good reason to believe that a 'universal' sub-grid model for MHD can be constructed; instead, there is considerable evidence that any such model will be very sensitive to the details of the problem, such as the specific boundary conditions, the nature of the forcing, etc. (see, for example, Chapter 9).

[3] The reason for the lack of validation is not a lack of effort, but rather the fact that we do not have sufficient experimental or observational access to velocity fields at the Reynolds numbers characteristic of astrophysical objects such as the Sun.

As a consequence, the conservatism I've urged on the interpretation of conclusions drawn from hydrodynamic simulations relying on sub-grid modelling becomes even more relevant in this case. This is especially so because the typical magnetic Prandtl number (*Pm*) of astrophysical numerical calculations is $O(1)$, whereas *Pm* for the solar interior is very small ($\sim 10^{-3}$ or less). As a result, global MHD calculations of the solar interior tend to be highly diffusive, and essentially never in a fully developed MHD turbulent regime.

Some critiques of astrophysical simulations go far beyond my conservative views and essentially regard all attempts to carry out DNS for turbulent systems as suspect, whether they are applied to an astrophysical object or not. For example, to quote one such critic, '... even DNS cannot predict the actual evolution of any turbulent system, because the dynamics is chaotic and numerical errors are always present at some level even if the dissipative scales are properly resolved' (see Chapter 12). I do not concur with this view. First, DNS should have the explicit property that the dissipative scales are resolved, so that is not an issue. Second, while it is most certainly true that integration of partial differential equations for systems whose parameters are such that the system dynamics is chaotic cannot (even in principle) precisely follow the trajectory of every given chaotic fluid parcel, it is generally believed that integral quantities (such as mean or rms velocities, turbulent transport coefficients, Nusselt numbers in the case of convecting systems, etc. – the very quantities that are in fact of interest in the case of the tachocline) are well-defined. These comforting beliefs are not pulled out of the air, but are in fact buttressed by detailed convergence studies, as well as by detailed comparisons with experimental data for turbulent flows (see, for example, DeLuca *et al.* 1990).

Given all the apparent bad news, where does this leave us? The first point is that some practitioners of solar CFD do carry out DNS – the question is then, how should we interpret these? Clearly, such calculations are not meant to be directly applied to the solar interior since DNS has the property that the dissipative scales are fully resolved. Instead, these types of calculations are meant to instruct us and to inform our physical intuition for problems in which highly nonlinear physical processes operate. As a concrete illustration, this is precisely the sense in which numerical MHD calculations are approached in Chapter 11 of this volume – the question at hand there is, can we understand how magnetic buoyancy affects the physics of magnetic flux transport out from the stably-stratified tachocline. This is a well-defined question, applied to a well-defined physical system; and while that system's physical characteristics (e.g. the non-dimensional parameters that define it, such as the Reynolds number) bear no resemblance to those of the solar interior, there is the hope that the 'workings' of the nonlinear physical processes in these two cases are not entirely dissimilar. To be sure, this hoped-for resemblance is not guaranteed (and may indeed be absent), but past experience indicates that we always gain by learning more about the remarkably complex behaviour of fully nonlinear (MHD) fluid systems.

Second, there is a long and distinguished history of using one's physical intuition to build fluid-dynamical models that may or may not be explicitly derived from the full (exact) set of hydrodynamic (or MHD) equations; and to the extent that these models faithfully represent (i.e. 'model') the most important physics, such calculations have been an extremely powerful tool for studying astrophysical fluid dynamics. There are many ways of constructing such models: some rest on powerful physical intuition, coupled to a terrific sense of how to translate this intuition into mathematics – the work on mean field hydro- and electrodynamics and the solar dynamo is characteristic of this approach (cf. Chapter 6); others rest on the notion of truncated models, for instance considering a Fourier decomposition of all fluid variables, but then studying only the nonlinear coupling of the lowest-order modes (see Chapter 13).

14.4 The (possible) role of experiments

As I have already repeatedly mentioned, we as astrophysicists labour under the not inconsiderable handicap that the physical systems we would like to understand are characterized by extreme physical conditions that are difficult, if not impossible, to replicate, whether experimentally or within DNS. My immediately preceding discussion of numerical computations focused on the possible benefits of using DNS as a means of learning about the workings of highly nonlinear processes, without the imposition of *ad hoc* model assumptions or 'priors'; and one key point I wish to make here is that the same obtains for laboratory experiments. Perhaps the best known example of this sort directly relevant to the question of astrophysical convection is the work of John Hart, Juri Toomre and collaborators (Hart *et al.* 1986a,b), who studied electroconvection experiments carried on board the Spacelab missions, and compared experimental results with numerical simulations of the experiments. These experiments were notable in that they allowed observations of convective flows in a geometry in which the effective gravitational force was radial (difficult, if not impossible, to arrange in terrestrial settings) and in which the consequences of rotation (and variation of the rotation rate) could be examined. These studies had an enormous impact on work in planetary atmospheric and stellar convection; and the lesson to be drawn here is that the effort to design (ingenious) laboratory proxy experiments turns out to be extremely worthwhile in helping us understand the complex dynamics of systems such as the solar convection zone.

A second, and equally critical, role played by experimentation is in the validation of numerical simulation codes. Traditionally, verification and validation (V&V) of numerical codes has been a largely informal activity in computational astrophysics, but in computational fluid dynamics, especially as applied in engineering, V&V has been extensively formalized and rigorously applied. An example of a formal guidance document in the engineering literature is the AIAA *Guide for the Verification and Validation of Computational Fluid Dynamics Simulations*; and a more informal discussion can be obtained at the NPARC Alliance web site (Slater 2005a), together with a fairly straightforward and informative on-line tutorial (Slater 2005b). Typically, this difference in rigour and consistency largely reflects the significant differences in (financial) investments in computational tools in various disciplines; thus, given the resources, astrophysics codes can be subjected to extensive V&V (see, for example, Calder *et al.* 2002).

What exactly is V&V, and how does 'verification' differ from 'validation'? The formal definition of 'verification' (taken from the AIAA definition document) is 'the process of determining that a model implementation accurately represents the developer's conceptual description of the model and the solution to the model', which is a fancy way of saying, 'are we solving the equations correctly?'. In contrast, 'validation' is defined (in the same document) as 'the process of determining the degree to which a model is an accurate representation of the real world from the perspective of the intended uses of the model', which is in turn a fancy way of saying, 'are we solving the correct equations?'. Thus, the process of verification is largely a matter of applied mathematics, whereas the process of validation is inherently an experimental process, involving the design of appropriate laboratory experiments that test whether the numerical code in question is in fact capable of correctly describing the intended target physical system. Thus, while I earlier advertised numerical simulations as a guide to understanding complex nonlinear (fluid) dynamics, I am now cautioning that we do need to ask (and answer!) the question, 'to what extent can simulations be viewed as a reliable guide to complex fluid dynamical behaviour?'.

Lest the reader think that the exercise of answering this last question is a relatively benign activity, it is useful to point out that recent experience in simulating fluid mixing problems – which are substantially simpler than the mixing thought to occur within the tachocline – has shown the great difficulty of actually getting correct results. One illustrative case in point is the question of interfacial mixing driven by the Rayleigh–Taylor instability (which arises from placing a heavy fluid on top of a relatively light fluid, with the two fluids in pressure equilibrium). As shown by the extensive study by Dimonte *et al.* (2004), the common ploy of intercode comparisons (often done in the absence of readily available experimental data) can lead to totally incorrect conclusions. In the Rayleigh–Taylor problem, it turns out that most of the modern compressible hydrodynamics codes tested by Dimonte *et al.* produced results that were in very good agreement with one another, but disagreed entirely with the applicable experimental results. This immediately suggested that the code authors had indeed done their homework as far as code verification was concerned (all of these codes correctly solved the same set of hydrodynamic equations) but, by the same token, there was clearly something amiss in the physical description of the problem, as defined by the codes and initial and boundary conditions used to define the calculations. This result was an object lesson to many of us contemplating the use of DNS as a guide to solving astrophysical fluid dynamics problems, and it has led some of us to establish much closer ties to experimental groups focusing on laboratory experiments that can validate our codes. Indeed, under the most favourable circumstances, such experiments – if appropriately designed – can serve both to validate codes and to directly inform the astrophysical problem under study (i.e. the first role for simulations I discussed above); an illustration of such experiments, and of the extensive discussions that have led up to such experiments, can be seen in the proceedings of the recent conference on magneto-Couette flows (Rosner *et al.* 2004).

14.5 Comments and conclusions

It is evident that the subject of the Sun's interior remains one of the most fascinating areas of modern astrophysics; and the remarkable, intense interplay between helioseismic observations and data analysis on the one hand, and theoretical modelling on the other – which was quite in evidence at the workshop – bodes well for the future of the subject. To repeat a point made at the beginning: there is no doubt that part of the fascination derives from the fact that the tachocline – deceptively modest in volume and mass – is the key to a stunning variety of important solar (and stellar) physics problems, from rotational spindown to compositional evolution of the interior and to the mystery of the solar dynamo and solar magnetic activity. My personal crystal ball predicts that with the advent of computers capable of sustained petaflop performance levels over the next decade (a target already in sight in the technology plans for IBM's Blue Gene technology), there will be an enormous opportunity to test theoretical ideas by means of targeted laboratory experiments and closely matched DNS; such opportunities are already in hand in the case of the closely related magneto-Couette problem (see Chapter 12) – it's a great time to enter this field!

References

AIAA. (1998). *Guide for the Verification and Validation of Computational Fluid Dynamics Simulations* (AIAA G-077-1998).

Calder, A. C., Fryxell, B., Plewa, T. *et al.* (2002). *Astrophys. J. Suppl.*, **143**, 201.

DeLuca, E. E., Werne, J., Rosner, R. & Cattaneo, F. (1990). *Phys. Rev. Lett.*, **64**, 2370.

Dimonte, G., Youngs, D. L., Dimits, A. *et al.* (2004). *Phys. Fluids*, **16**(5), 1668.
Gough, D. O. & McIntyre, M. E. (1998). *Nature*, **394**, 755.
Hart, J. E., Glatzmaier, G. A. & Toomre, J. (1986a). *J. Fluid Mech.*, **173**, 519.
Hart, J. E., Toomre, J., Deane, A. E., *et al.* (1986b). *Science*, **234**, 61.
Rosner, R., Rüdiger, G. & Bonanno, A., editors (2004). *MHD Couette Flows: Experiments and Models* (AIP Conference Proceedings 733).
Slater, J. W. (2005a). Computational Fluid Dynamics (CFD) Verification and Validation Web Site of the NPARC Alliance, http://www.grc.nasa.gov/WWW/wind/valid/
Slater, J. W. (2005b). Tutorial on CFD Verification and Validation, http://www.grc.nasa.gov/WWW/wind/valid/tutorial/tutorial.html

Index

The Solar Tachocline, D. W. Hughes, R. Rosner and N. O. Weiss.
Published by Cambridge University Press.